Ideas and Actions in the Green Movement

The western green movement has grown rapidly in the last three decades. Green ministers are in government in several European countries, Greenpeace has millions of paying supporters, and green direct action against roads, GM crops, the WTO and neo-liberalism has become ubiquitous.

Since green action is now divided between those inside and those outside political institutions, can we still speak of a single green movement? In answering this question, this book analyses four types of green action:

- green parties
- environmental organisations, such as Greenpeace
- direct action groups
- local environmental campaigns.

Brian Doherty argues that 'greens' share a common ideological framework but are divided over strategy. Using social movement theory and drawing on research from many countries, he shows how the green movement has become increasingly differentiated over time, as groups have had to face the task of deciding what kind of action is appropriate.

In the breadth of its coverage and its novel focus on the relationship between green ideas and action, this book makes an important contribution to the understanding of green politics.

Brian Doherty is a Lecturer in Politics at Keele University, and also a member of the UK Green Party. He has published widely on green ideas, parties and movements. Previous publications include *Democracy and Green Political Thought* (co-edited with Marius de Geus) and *Direct Action in British Environmentalism* (co-edited with Ben Seel and Matthew Patterson).

Routledge Research in Environmental Politics
Edited by Stephen C. Young

Ideas and Actions in the Green Movement

Brian Doherty

London and New York

First published 2002
by Routledge
11 New Fetter Lane, London EC4P 4EE

Simultaneously published in the USA and Canada
by Routledge
29 West 35th Street, New York, NY 10001

Routledge is an imprint of the Taylor & Francis Group

© 2002 Brian Doherty

Typeset in Baskerville by Taylor & Francis Books Ltd
Printed and bound in Great Britain by
Biddles Ltd, Guildford and King's Lynn

British Library Cataloguing in Publication Data
A catalogue record for this book is available from the British Library

Library of Congress Cataloging in Publication Data
A catalogue record for this book has been requested

ISBN 0–415–17401–5

Contents

Preface

The main aim of this book is to present a comprehensive description and analysis of the green movement using social movement theory as the main framework. However, to then avoid some of the normative issues that arise in relation to these movements would be artificial. On occasion, therefore, I have made normative arguments, which also reflect my own political commitments as a green. In part this book is an attempt by a green to understand my own movement. It is not primarily a contribution to debates among greens about what should be done, because that would require less detailed description and more explicitly normative debate. However, by presenting a definition of what the green movement is, I hope that this book might help other greens to understand why our movement is as it is and therefore have a more realistic appreciation of what has been achieved and what the constraints are affecting what can be done.

No book is ever written wholly alone and so the shadow authors of this work should also be brought into the light, although claiming them as authors does not mean that they bear any responsibility for the book's weaknesses.

Rosemary O'Kane, John Barry and Andrew Dobson commented on drafts of this book, a favour for which any author should be grateful. Tina Devey did great work on the bibliography. Others at Keele who helped without necessarily knowing it were Mat Paterson, Ben Seel, Kara Shaw, and Pauline Weston. At Routledge Annabel Watson was very patient about my failure to meet deadlines. Conversations and debate with many people in the green academic network have shaped the thinking in this book but a special mention should be made of Chris Rootes, Tom Cahill, Neil Carter, Mario Diani, Florence Faucher, Mike Kenny, Hilary Wainwright and Ian Welsh.

Most recently, working with Alexandra Plows and Derek Wall on a study of local direct action communities has convinced me that teamwork is better than going it alone. It has also deepened my respect for those heroic individuals, among whom Alex and Derek are two, who go further than the rest of us to put right the wrongs of this world and have fun while doing it.

The research in this book draws on over a decade of empirical work on green parties and green movements during which I have imposed on the time and goodwill of many green activists. Among green party representatives I am particularly grateful to Bodo Zeuner, Frieder Otto Wolf, Sergio Andreis, Jean Lambert, Sara Parkin, Andrée Buchmann, Solange Fernex, Spencer Fitzgibbon and the late Alexander Langer.

In the UK grassroots green networks, John Stewart of ALARM UK, activists at camps and offices at Newbury, the M66 and particularly, Manchester Earth First! have been especially helpful when they had much more important things to be doing.

It is impossible to do justice to the sophistication and diversity of green activist culture and I doubt that any activists will agree with everything I say here, but I hope that they will find it useful in providing a perspective on the range of green activity and ideas.

The book is dedicated to Julie, Caitlin and Eleanor, who had to put up with my failure to finish for too long.

Abbreviations

ACF	Australian Conservation Foundation
AdT	Friends of the Earth, Italy
AGALEV	Flemish Green Party, Belgium
ALARM UK	All London Against the Roads Menace, United Kingdom
ALF	Animal Liberation Front
APÖ	Ausserparliamentarische-Öpposition (Extra-Parliamentary Opposition), Germany
B90/DG	Bündnis 90/Die Grünen German Green Party after 1993
BBU	Federation of Environmental Citizens' Initiatives, Germany
BUND	Friends of the Earth, Germany
BIs	Citizens' Initiatives, Germany
CCHW	Citizens' Clearinghouse on Hazardous Wastes, USA
CDU	Christian Democrat Union, Germany
CND	Campaign for Nuclear Disarmament
CPRE	Council for the Protection of Rural England
CRS	French paramilitary police
CSU	Christian Social Union, Germany
DN	Danish Nature Conservation Association
DNR	German Nature Protection Ring
Ecolo	Walloon Green Party, Belgium
EDF	French Electricity Monopoly.
EDA	Ecological Direct Action
EF!	Earth First!
ELF	Earth Liberation Front
EMOs	environmental movement organisations
FDP	Free Democratic Party, Germany
FFSPN	French Federation of Societies for the Protection of Nature
FLN	Algerian Independence Movement
FN	National Front, France
FOE	Friends of the Earth
G8	Group of Eight Leading Industrialised Countries

GÉ	(Génération Écologie), French Environmental Group led by Brice Lalonde
GM	genetic modification
IMF	International Monetary Fund
INF	Intermediate Nuclear Forces
IUCN	International Union for Nature Conservation
LADT	Friends of the Earth, France
LAV	Anti-Vivisection League, Italy.
LETS	Local Employment and Trade System
LULUs	locally unwanted land uses
MDA	militant direct action
MdEP	Political Ecology Movement, France.
MRG	Radical Left Movement, France
MEI	Movement of Independent Ecologists, France.
MDC	Citizens' Movement, France
NABU	German Nature Conservation Association.
NOAH	Danish branch of Friends of the Earth
NRDC	Natural Resources Defence Council, USA
NSM	New Social Movement
NT	National Trust of England and Wales
NVDA	non-violent direct action
ÖDP	Ecological Democratic Party (German right-wing ecological party)
OOA	Danish Anti-Nuclear Energy Organisation
PCF	French Communist Party
PDS	Party of Democratic Socialism, Germany
PdS	Democratic Left, Italy
PGA	People's Global Action
PPR	Radical Party, Dutch New Left party
PS	Socialist Party, France
PSP	Pacifist Socialists, Dutch New Left party
PSU	Unified Socialist Party, French New Left party
PvdA	Labour Party, Netherlands
RAF	Red Army Faction, Germany
RMT	resource mobilisation theory
RSPB	Royal Society for the Protection of Birds, UK
RTS	Reclaim the Streets, UK
SCRAM	Scottish Campaign to Resist the Atomic Menace
SDS	Students for a Democratic Society, USA
SdS	Marxist Student Organisation, Germany
SMOs	social movement organisations
SNF	Swedish Association for the Protection of Nature
SNCC	Student Non-Violent Coordinating Committee, USA
SPD	Social Democratic Party, Germany

SPV	'Other Political Association', Germany: precursor of Die Grünen
TWS	Tasmanian Wilderness Society
UNCED	United Nations Conference on Environment and Development
UNESCO	United Nations Educational, Scientific and Cultural Organization
WTO	World Trade Organization
WWF	Worldwide Fund for Nature (World Wildlife Fund in the USA)

Introduction

This book is concerned not with environmental movements in general, so much as with green movements. In popular usage, the environmental movement is a broad and loose term, which can refer to climates of opinion, to formally organised groups or to loose networks of protesters (Rootes 1997a: 319). So, while the environmental movement and the green movement are often used as synonyms, in this book the green movement is defined more narrowly as those western environmentalists who believe that radical political and social changes are necessary to deal with the ecological crisis. These changes would result in a new kind of society, based on a new relationship with the natural world, a more radical democracy and much greater social equality. Many environmentalists do not believe that major changes of this kind are necessary and are confident that environmental problems can be dealt with by science and appropriate public policies within the existing social and economic structures. As well as a shared radicalism greens are also defined by the fact that they take protest action together and belong to a network of groups linked by a shared collective identity. Thus in this book the green movement is seen as part of the broader environmental movement, and as a social movement within it.

If the green movement is defined as a social movement it allows us to be more precise about how to distinguish greens from the wider environmental movement. This book will therefore examine those western environmental groups that meet the criteria for a social movement. These are, counter-cultural green protest groups, such as Earth First!; the more radical of the established environmental movement organisations, such as Friends of the Earth and Greenpeace; grassroots environmental protest groups in western countries; and green parties. The meaning of the concept of social movement will be discussed in Chapter One. For now it will suffice to say that a social movement is defined by having the following four characteristics: a collective identity; action outside political institutions, principally protest; a network of interaction between members that is not reducible to a formal organisation; and a challenge to dominant forms of power.

Treating the western green movement as a single social movement is in conflict with many commentators who prefer to speak of green movements rather than a single green movement. They do so because of the diversity of types of environmental movement across national boundaries and the major differences that exist within the environmental movement within countries. While there is too much diversity within the range of groups that can be called environmental to identify any common identity or challenge to the dominant order, by defining green movement as a social movement within the more diffuse environmental movement, it is possible to identify groups with common features. The social movement wing of the environmental movement is its more radical and explicitly ideological wing. Yet, there is still much heterogeneity even within this green social movement. One task of this book, therefore, is to examine the above four types of green movement in relation to an ideal type definition of the green movement outlined in the first part of this book. As long as we remember that the concepts of social movement and green movement are theoretical constructs, they can be used to analyse different groups within the green movement without imposing a false universality on empirical cases.

This book focuses on greens in the western countries (western Europe, Australasia and North America), rather than radical environmentalists in other parts of the world. The main reason for this is lack of space. The second reason is that western industrialised countries share certain structural features, in their economies, their political systems and in their political cultures – and all that is meant by the familiar shorthand term 'the west'. This similar context is reflected in similarities in their environmental movements. While one should not take such categories for granted, they provide a means of limiting the analysis and a basis for comparing like with like. But, as the final chapter suggests, the differences between the western green movement and radical environmentalism in other parts of the world is not clear-cut and is likely to be reduced as mutual interaction increases.

As yet there has been no book-length attempt to analyse the western green movement, although there have been many studies of national environmental movements in different countries. Even by limiting the cases to western environmental groups the range of potential groups and issues is very broad. While the aim has been to be comparative where possible, the limits of length and the need to keep space for the discourse of activists means that the emphasis is more on depth than range. Where the evidence allows, as it does more with the formally organised groups such as environmental movement organisations (EMOs) and green parties, there is more general analysis. For direct action groups and local environmental groups the evidence is much less systematic, and ethnographic evidence is much more useful, but some cross-national comparisons are also possible. Much of the book is structured around national cases, charting forms of evolution within national boundaries, but there is also much evidence of transnational diffusion of ideas, repertoires of action and forms of organisation. There is also evidence of common patterns of evolution and collective learning which includes drawing on experiences of greens in other countries. Implicit in the idea of a singular western green movement is the assumption that common features exist despite different local and national contexts.

The strongest claim that I make about how we should define the green movement is based on green ideology. I argue that to be green, radical environmentalists must be committed to a view of nature which does not value it only for its use to humans; they must be in favour of extending participation in democracy through the decentralisation of social and political power, and must be egalitarians in that they are committed to overcoming material and culturally based inequalities. Their ecological commitments are what make the greens most distinctive, but they do not necessarily provide the foundation for their democratic or egalitarian commitments. Unlike most of those who have analysed green ideology I argue that green activists do not always base other aspects of their ideology on ecological principles. Some do, but others see questions of equality or social justice, or democracy as not always reducible to ecological principles. The latter are just as much greens as the former. The two types of green are often found in the same activist network, but such networks are much more likely to be divided over strategy and praxis than ecocentrism versus anthropocentrism. A definition of green ideology that fits a large number of radical environmentalists needs to be broader than one based on ecologism or ecocentrism. According to some definitions the three principles of ecological rationality, grassroots democracy and egalitarianism would not be an ideology. Political theorists would point out that the key questions about this three-fold combination are: How are they to be combined? And in cases of conflict, which will take priority?

What is most interesting is that these issues do not usually cause problems in green movements. Divisions within green movements over the priority to be given to ecological philosophy are rare. Much more important as a source of divisions is the traditional problem of radical political praxis: namely, how to achieve radical political changes while working within the constraints of a hostile system, which is structurally antithetical to green concerns, not only on ecology but also on democracy and equality. And divisions over these questions do not always correspond to differences between ecocentric and non-ecocentric greens.

Thus this book deals with green ideology mainly in relation to green praxis. The focus is on the relationship between ideology and action, examined through case studies of different types of green movement. The types of green movement are defined as much by empirical sociological and political characteristics, such as their types of organisation, favoured forms of action and their principal sources of support, as by their ideas. This means that certain types of greens that would deserve separate treatment in a book that focused exclusively on green ideas do not get this here. These include ecofeminists and deep ecologists. Advocates of ideas associated with both these positions can be found within the kinds of green groups analysed here, but because they seem to work with others based on shared commitments to a more general green 'framework' ideology, they are not analysed as a separate type of green movement.[1] There is no space either for separate treatment of the animal rights movement, because despite the evidence of network ties linking this movement to greens, the relationship between the two movements varies between different countries and it is less clear that most animal rights activists are committed to a broader green ideology.

Outline of the book

Chapter One focuses on the definition of the green movement as a social move-ment. For the green movement to be a social movement there must be evidence of shared identity, protest action, loose networks of interaction and a challenge to power. This can be found among radical environmentalists in the west, but the degree to which different groups approximate to an ideal type of green social movement varies case by case. Social movement theory is useful too in explaining the relationship between green ideas and major changes in society. More compar-ative analysis of protest and social movements also explains much of the variety within the green movement.

Chapter Two presents a general explanation for the emergence of the green movement. Although the heritage of environmental groups and radical environmental thought is a long one, it is argued that the green groups that emerged gradually in the 1970s were distinct, particularly in their ideology, from previous forms of environmentalism. A comparison of the early develop-ment of the green movements in France and Germany shows the importance of relationships within the left and differences in national institutions in shaping the new movement. The chapter also focuses on explaining the sources of green activism socially and over time. The macro-structural ana-lyses developed by social theorists to account for the emergence and character of new social movements are explored, but these need to be supplemented by an analysis that can answer questions that arise more specifically about the green movement. These include why green activists are disproportionately drawn from particular social groups, such as the new middle class, the young and those with higher education. In avoiding an explanation that assumes green activism can be explained by material interests, the alternative explana-tion is based on the relationship between green ideas and experiences common to those in these social groups. These experiences include new evidence about ecological problems, the new critical thinking and anti-authori-tarian ethos associated with the New Left and alternative counter-culture of the 1960s and 1970s, encouragement of rational critique in university educa-tion, and the political character of work in welfare professions. Two concepts, the idea of a political generation, linked by shared formative conditions and that of collective learning, in which this generation confronted problems of political praxis and evolved in response, are used as frameworks for this analysis.

Chapter Three analyses green ideology. Rather than ecocentrism, it is the combination of a pluralist and complex egalitarianism, grassroots democracy and ecological rationality that best defines the ideology expressed by most green groups. This three-fold combination is important because it places greens more clearly on the left than analyses based only on their ecological politics. In prac-tice, green movements have integrated commitments to ecology with social justice and a radical form of participatory democracy. And when activists have tried to downgrade issues of social justice they have been excluded from green movements.

Green ideology is defined as a product of collective action and evolves in response to action. However, it is also a combination of existing traditions of thought. Green ideology is an adaptation of the traditions of the left by a particular political generation facing problems which for them seemed not to be dealt with adequately by the predominant currents of thought on the left.

Chapter Four analyses green parties, focusing on Germany and France. The first part of this chapter analyses the influence of a range of social movements on the formation of these two green parties. The institutions and social networks of an alternative milieu, which to some extent cuts across the boundaries of specific movements, shaped the green parties, but did so in different ways in both countries. Also important was the nature of political opportunities, embracing both institutions and the nature of the Left in each country. The third factor was the effect of experience, defined as a process of collective learning.

In the second part of the chapter the dilemmas of fundamentalism and realism are addressed. Once established as political parties the greens have had to resolve the strategic dilemma of how to be critical of the existing system while working within it. Continuing the comparison of Germany and France, it is clear that there has been a process of adaptation and reduction of expectations. Nevertheless, tension between the scope of the green project and the limits of their political power remain. However, the nature of the dilemma is now more acute for the Germans and French as participants in government. Fundamentalist strategies were found wanting but although realist strategies do not inevitably lead to deradicalisation, the experience of government at national level has been difficult, particularly for the German greens. Ideological changes in the German green party, which began even before entering government, raise questions about whether they can still be considered part of the green social movement.

Chapter Five analyses the development of environmental movement organisations (EMOs), focusing on the more radical groups such as Friends of the Earth, and Greenpeace. These have become increasingly institutionalised and the nature of this process is examined. It is argued, however, that organisational institutionalisation and deradicalisation are not equivalent. There is less evidence of deradicalisation in the strategy and identities of EMOs. Nor is there any clear link between the level of protest in particular countries and patterns of institutionalisation. While organisational evolution has imposed new constraints on EMOs, and in some countries, such as Italy, they have largely lost their challenging movement character, in others this trend is much less obvious. Many EMOs still have some choices about strategy and can turn back to a more radical strategy. Thus, although these groups are less radical in ideology and actions than other green groups, many still fit within the definition of the green movement.

Chapter Six examines groups such as Earth First! that are primarily based upon the use of direct action. Groups of this kind from the USA, Britain and Australia are compared. They share a repertoire of actions based upon 'manufactured vulnerability' through which activists create devices that prolong site occupations while exposing themselves to risks. In recent years national differences between these groups have become blurred as global communications, particularly through the internet, have facilitated transnational diffusion of ideas and tactics. The terms of globalisation have themselves become the stake of more recent militant action. This in turn has given added impetus to long-standing debates about whether violent action is justified. It is argued that any turn to violent strategies would lead to a more elitist underground organisation. Non-violent strategies are linked to a non-reductionist form of green ideology.

Chapter Seven focuses upon local environmental groups. Local environmental groups appear to have weak claims to be part of the green movement because of their limited political aims, lack of resources and, often, short duration. But to ignore them would mean ignoring a major part of environmental activism. Moreover, as the ,analysis shows, using social movement concepts emphasises dimensions of these groups that would be missed if they were viewed only as local interest groups. For instance, the experience of campaigning often leads to changes in identity towards a more radical perspective. Local environmental groups are forced to develop scientific expertise and often come to question the role of science and expertise in decision-making. They also have to define their own identity in relation to the wider community and this involves a social construction of the nature of their community. While the outcome of local environmental conflicts is often decided by forces beyond the local level, local campaigns can influence outcomes and their political impact can be broader than the success or failure of a particular campaign.

The final chapter compares the different types of group in relation to the general model of a green movement and assesses the factors shaping the future of the movement. While local environmental groups are not usually part of the green movement, green parties and direct action groups and some of the EMOs share a common ideological project and are sufficiently inter-linked to constitute a social movement. This raises the question of whether there is a global green movement linking western greens and greens in other parts of the world. Non-western radical environmentalisms are argued to be still too distinct from that in the west for a global green movement, although this may change as contacts increase. For western greens the strategic questions of praxis still pose the central dilemmas which seem likely to shape their future. These are analysed through an examination of two central questions. First, on what terms, if any, should the green movement get involved with reforms? Second, how should the green movement respond to the challenge posed by the diversity of forms of green praxis?

1 The green movement as a social movement

How do we tell who is a green? Is it justifiable to regard those who visit nature reserves and those who trash branches of McDonalds as being part of the same social movement and sharing the same politics? This chapter will address these questions by using social movement theory in three ways. First, by defining what the green movement is, which will involve defining the term social movement. Second, by considering whether the green movement is a new kind of social movement. Third, by explaining the kind of actions that greens take, their relationship to the political system and how activists become involved.

The green movement as a social movement

Although there are no universally accepted definitions of the concept of a social movement the following four characteristics are found in many definitions.[1] A social movement:

- Must have a consciously shared collective identity;
- Must act at least partly outside political institutions, using protest as one of its forms of action;
- Is characterised by uninstitutionalised networks of interaction; and
- Must reject, or challenge, dominant forms of power.

Collective identity

Before people can act together they must have a certain degree of solidarity. In particular they should be able to assess the world in terms of 'them' and 'us'. Activists will have stronger solidarity when they share values that provide meaning and justification for their actions. Yet, a shared identity is based on more than simply these shared 'frames': shared culture, practices and traditions also shape a movement's identity and these evolve and change over time. In the case of the green movement its culture includes a commitment to non-hierarchical styles of organisation and an acceptance of the value of changing one's lifestyle to be consistent with political principles. These broad principles are not unique to green culture and have been part of the culture of other movements at least as far back as the nineteenth century (Calhoun, 1995). Feminism has challenged the divide between public and private life by showing how this division, in which women are seen as having primary responsibility for the private sphere, serves to exclude women from effective power. Like feminists, greens are expected to be able to try to organise their private lives in ways that are morally responsible and the practice of these ideas is not restricted to those who are publicly active in social movement politics. For instance, greens make a link between individual actions such as cutting car use, and global goals such as averting climate change.

Participation in the green movement involves changes in lifestyle, but it is not always clear what the most green action is. For example, in buying food greens will favour local and organic produce that has been produced in working conditions that conform to ethical standards and respect the welfare of animals. Since relatively little of the available food meets these criteria greens are often caught in moral dilemmas about which criterion is most important. In fact most questions of consumption produce similar dilemmas for greens. In what circumstances, if any, is it reasonable to use a car? Or, in circumstances of global inequality, what constitutes a reasonable standard of living for someone living in the affluent west? One UK Green Party activist I interviewed on this question said that he tried to limit his income to a level consistent with levels of global equality, but most find it hard to live to such standards and some argue that this level of consistency is not necessary since it places too much emphasis on individual responsibility. But for most, being a good green means trying to live as greens argue other people should. When it is impractical to do all that might be desirable in this respect, greens are likely to express regret, guilt or point to excuses for their failures. To do otherwise would be hypocritical. Thus for greens, practising what you preach requires political judgements to be made on issues that have traditionally been seen as personal choices.

This green culture is more variable than green ideology. Ideas about the nature of problems and broad principles about the desirable alternative society that constitute green ideology (see Chapter Three) are more easily translated than ways of speaking, lifestyle, and expectations about behaviour, which are sustained more by routine interaction in everyday situations. For instance, Faucher (1999) has noted important differences between the extent of lifestyle politics among activists in the green parties in Britain and France. In Britain, the question of what to eat was a political one. To be vegetarian was superior to being a meat eater. In France, vegetarianism was not necessarily accepted as superior and was much less widespread than in Britain. Many other differences separated activists in the two parties. The French were mostly astonished to be asked whether green politics was spiritual, the British mostly took this aspect of green politics for granted. Sentiments based on deep ecology were more or less non-existent in France, and in many other green parties in Europe, whereas they were relatively common in Britain (Faucher 1999: 54; Bennie *et al.* 1995). While both parties had a similar ideology and their policy programmes were also similar, the meaning given to concepts such as ecology was often different (more scientific in Britain, more political in France) and the question of what was reasonable in making personal life consistent with political principles also varied between the parties. The process of reconciling the personal and the political is socially constructed. It is through interaction with others in their movement or party that activists develop different practices and a sense of what counts as 'enough'. Although the activists studied by Faucher viewed those from the other party as ideological comrades with a shared political project, and the policies of both parties were very similar, their party cultures reflected national differences. Thus, while those from both green parties shared an identity, there were also differences between the cultural practices associated with the identity.

Alberto Melucci (1989, 1996) emphasises the need to be careful to avoid reifying the concept of movement identities. The identities shared by social movement actors are always heterogeneous and include significant internal conflicts. There are no clear boundaries to social movements because the collective identity of a movement is not fixed. What holds a group of actors together has to be continually reaffirmed and negotiated. For this reason, Melucci argues first, that the achievement of a collective identity is in itself a major achievement of a social movement and second, that we should not think of movements as necessarily unified actors. To do so would mean failing to do justice to the necessarily provisional and fluid process of collective action. An example of this is the way that network ties of different groups within the green movement link them with other groups. Ecological direct action groups such as Earth First! often work with anarchist groups that are not necessarily committed to ecological goals, and formally organised environmental movement organisations (EMOs) have ties with businesses that they seek to influence, despite the conflicts between much of the activity of business and the values of greens. Moreover, the nature of the common values linking the most radical and the most reformist wings of the movement, shifts as each is affected by its contacts with groups outside the movement as well as debates with other greens. Groups can have more than one identity and be 'in' more than one movement. When radical environmentalists join other anarchists and radicals in direct action protests against neo-liberalism, they are both greens and part of a newly emerging anti-capitalist movement, which is not wholly green.

While remedying injustices and a shared belief that the world can be made a better place are part of what motivates greens, not all motivations are cognitive. Emotion is perhaps the last taboo in the field of social movement studies. In the 1970s there was a reaction against the assumptions of mass society and collective behaviour theorists who treated movements as emotionally driven, and often therefore irrational (Smelser 1962). Social movement theory has recognised the role of expressive action (Rucht 1990) and the importance of emotion. The tendency, however, was to focus on the cognitive dimensions of action. Even actors pursuing expressive and symbolic actions that challenged dominant codes were seen as pursuing rational action. Also, as Goodwin *et al.* say, since 'science, not feeling is the dominant language of legitimation and persuasion in today's liberal societies' (2000: 81) it is hardly surprising that protesters themselves tend to be suspicious of emotionality. Protesters were incorrectly seen as more emotional and less rational than the authorities, despite the frequent evidence of emotional sentiments by authorities and other opponents of social movements.

It is because of this pejorative legacy that importance of emotion tended to be downplayed in recent social movement theory. More recently Kevin Hetherington (1998) has used Raymond Williams's (1989) concept of 'a structure of feeling' to try to capture the emotional bond of solidarity often found in new kinds of elective communities. Hetherington focuses on the rituals and 'ethic of aesthetics' through which participants in movements communicate their allegiances towards each other. Often this 'libidinal economy' (Goodwin *et al.* 2000) of friendship, solidarity or love shapes the dynamics of a group. The jealousies and tensions between activists affect turnover and levels of commitment. Shared emotional feelings towards outsiders are also a major factor in explaining motivation. Moral outrage towards opponents, pride in collective identities, a sense of shame at the indignity of a tarnished identity, are all part of the emotional fabric of movements.

Action outside political institutions

The second feature of the definition of social movements is that they operate at least partly outside political institutions. They act outside the political institutions because they perceive them to be closed to their ideas or because political institutions are inappropriate avenues for change. Protest is an important part of this extra-institutional activity. According to one tradition in social movement theory the protest activity of social movements is essentially a means of making claims on the public authorities. This may be for rights, such as the right to vote, or to strike, or it may be in opposition to specific projects such as preventing the construction of a nuclear power station. Social movements are therefore conceived of as 'outsiders' using protest to try to force those with power to listen. According to McAdam, Tarrow and Tilly:

> A social movement is a sustained interaction between mighty people and others lacking might: a continuing challenge to existing power-holders in the name of a population whose interlocuters declare it to be unjustly suffering harm or threatened with such harm.
>
> (1996: 21)

This definition excludes collective bargaining by well-established and powerful social groups but also collective withdrawal or 'opting-out' by marginal groups. As Seel notes (1999: 53), it therefore excludes collective challenges to cultural norms – and thus actions that many greens see as political. Some commentators therefore prefer to exclude protest from the definition of social movements. However, protest is really what we think of as characteristic of social movements, and too much is lost if protest is not part of the definition of a social movement. One solution is to broaden the category of protest beyond claims-making directed at power-holders. If a protest *by a social movement* is defined as collective action outside political institutions which includes the expression of a dissenting political argument that challenges the dominant forms of power, and not restricted only to action directed at the state or a specific opponent, this may yet include some of the 'cultural' actions that should be part of such a definition.

One of the justifiable criticisms of the over-emphasis on protest is that it presents a picture of what social movements do which suggests that all other aspects of the movement are a form of preparation for action directed against the state. It also lays too much emphasis on the negative and reactive dimension of social movements. The development of a common identity and a culture can seem as if it is merely a stage that must be passed before the full maturity of 'political' (state-oriented) action is reached. But, if the concept of political action is defined more broadly, action directed at the state becomes one of a range of protests. For instance, when green activists in Manchester dressed as aliens and acted as if they were tourists visiting Earth to see the curious sight of Christmas shoppers buying more and more goods in pursuit of happiness (Purkis 2000), they were making a political argument. To buy more and more goods which people don't really need and may not want, mainly because it is expected, helps to sustain the over-production and exploitation which creates ecological crisis and social injustice. Their action was not about a particular change in policy, nor was it illegal, but it was certainly political. It was also confrontational. The activists challenged the legitimacy of what the shoppers and shop staff were doing through parody (Melucci 1996).

Protest can be defined then, to include all collective action that is directed at an audience outside the group, involves confrontation, and is intended to express or advance the group's politics. On this basis an illegal action that involves arrests and confrontation with the police is not necessarily more 'political' because it involves conflict with the state. This broader conception of protest goes some way towards reconciling the traditional 'political' conception of protest (as a form of claims-making oriented towards the state) with those, such as Melucci (1996: 37) who have argued that newer forms of collective action tend to be increasingly symbolic and based upon non-negotiable demands.

Not all green protest is intended as a form of claim directed at the state. An example is Reclaim the Streets parties. In these the focus of protest is the specific place where the party is held. The aim is to directly reclaim that place as a social space (rather than simply as a through route for cars) and thus also to transform the people who use that place. More ambiguous are the cases of radical communes and alternative 'intentional' communities. Because the attempt to realise an alternative lifestyle is a fundamental part of the political practice of many greens, although few live in communes, they are important and valued symbols of green utopian politics. (Sargisson 2000). As Sargisson argues, in transgressing contemporary norms such communities are inherently political. Where they share an identity with other greens, take part in protests and are linked with other greens by networked ties and challenge dominant forms of power, they are part of the green movement. However, unlike protest camps, not all alternative communities are established to confront existing forms of power directly. It seems better then to see such alternative communities as part of the network of the green counter-culture which is political, but does not in itself constitute an act of protest.

It is also important to point out that not all protest is carried out by social movements. Protest by homeowners worried about the effect of new developments on their property prices, or by leaders of major political parties using protest as a political tactic against a rival party, are not social movement protests because they are not rooted in a movement challenging dominant forms of power. As protest has become an increasingly normal form of action used by many institutionalised groups and by many different sectors of society (Meyer and Tarrow 1998), and often does not involve the creation of uncertainty through disruption, it is increasingly the radical ideological dimension of protest by social movements that makes it distinctive.

Networks of interaction

The third characteristic of social movements is that they are characterised by loose networks of interaction because they usually contain multiple groups that may not be linked by formal ties and may not be formally organised at all.[2] The membership of social movements is defined by the fact that its adherents share an identity and take action together, not by whether they carry a membership card. However, this does not mean that formal organisations cannot be part of a social movement. In the green movement there are two types of organisation that might be considered part of the movement, but which are also formally organised. The first is the green parties. Since green parties are formally organised political parties working within political institutions they might appear to be outside the green movement.[3] Yet, most theorists do include them as part of the movement because of the strong bonds between most green parties and other parts of the green movement.

The second group is the established environmental movement organisations. EMOs cover a broad range from conservationist groups, which usually focus on specific policy issues and do not argue for broader political changes to more openly radical groups such as the various national affiliates of Friends of the Earth. Those EMOs that do not challenge the existing political or social system are not part of the green movement. Examples include groups[4] focused on conservation activity such as Wildlife Trusts or those who lobby for policy changes on technical and scientific grounds but do not argue that government policy is unjust, undemocratic or unsustainable.

We can say then that social movements may contain organisations that are hierarchical, formal and bureaucratic but a social movement is not equivalent to a single organisation. Even where one formal organisation is dominant, as in the case of the Campaign for Nuclear Disarmament (CND) in the British peace movement, there are also always more informal groups acting independently on the basis of a shared identity as peace activists, irrespective of the formal policy of CND such as, for instance, those involved in peace camps. The boundaries of social movements are difficult to establish. This is why social movements are better conceived of as networks that alter as new groups form, as activists from other groups become part of the movement, and others become less active or move beyond the point where the criteria for a social movement can be said to apply. This may be either by dropping out from political action, or by becoming so deradicalised that they no longer share the movement's 'challenging' political identity.

As well as inter-organisational networks, social movements are characterised by inter-personal networks, and it is through these that the solidarity necessary to develop a collective identity and sustain activism is forged. At local level the network ties between individuals with different organisational allegiances are likely to be particularly important. Comparison of green activism in Manchester, Oxford and North Wales showed that there was a community of green activism which cut across the boundaries between different groups (Doherty *et al.* 2001) and that these ties had built up through joint campaigns. Studies also show that social ties play an important role in the formation of new movements and in recruitment of new groups of activists. It is invariably pre-existing social groups that form and join movements. These may be groups based upon friendship, as McAdam (1988) showed in his study of student recruits to the Freedom Summer civil rights campaign. In this case potential recruits all shared a similar political analysis, but it was those in friendship groups who became activists, as distinct from more isolated individuals. Or they may emerge from within existing political groups, as many of those who formed green parties did, combining new environmental groups and New Left groups.

People join movements for a variety of reasons, based on values, loyalty to groups, pressure from friends or hope for personal advantage. As Tarrow notes (1998: 15), this variety is one reason why it is difficult to hold movements together. But what is most likely to sustain movement activity is loyalty to others based on shared experience of action.

Social movements challenge power

The fourth characteristic of social movements is that they must offer a challenge to dominant forms of power. This is perhaps the most difficult aspect of social movements to interpret. If a social movement must offer a challenge to existing forms of power, identifying whether this is the case requires some definition of what power is and where it lies. This is why Melucci (1996) argues that we cannot speak about social movements without at the same time offering an analysis of the wider social and political system. Others have argued in a more neutral way that social movements take contentious action based on common interests. But, even the definition of social movements by McAdam, Tarrow and Tilly (1996) quoted above, from the more empirical tradition of social movement analysis, assumes that social movements 'lack might'. It is possible to show this in a relatively uncontroversial way, for instance by showing that most social movements lack the resources of their opponents. Most environmental groups cannot match the money available to major corporations to spend on advertising, and none can match the force available to the state when it wants to prevent a demonstration. However, social movements also challenge power in a more ideological sense

The challenge to power offered by a social movement is part of its political identity. Movements develop an identity that allows participants to re-interpret their shared experiences. This occurs collectively in debates within the movement, which may include reading the work of influential theorists, but is also sustained by activists having to decide collectively what is wrong with the world and what is to be done about setting it right. When this debate produces new perspectives that challenge the existing organisation of society, the potential to challenge power exists. Action by the movement brings participants into potential conflict with those who hold most power, political, social or economic.

Not all conflicts involving protest are social movement conflicts. An example might be the protests for and against the right to carry on hunting. Protest actions are often carried out by both sides: hunt saboteurs try to prevent the death of foxes by obstructing hunters, while hunters in Britain and France organised mass demonstrations in 1998 when the right to hunt was threatened. In France hunters have also ransacked the offices of their opponents and presented candidates in elections. However, these protest movements are not offering a challenge to power. With the exception of the more radical parts of the animal rights movement (Veillet 2001; Jasper and Nelkin 1992), neither side in the conflict over hunting challenges the basic principles by which society is organised; neither sees the need for radical social and political change. Their conflict is a single-issue one between animal rights and the traditions and economic interests underlying hunting. Social movements are not single-issue movements. They are engaged in a struggle for control of the direction of society.

Some environmental groups are single-issue pressure groups and cannot be considered part of a green movement. The Royal Society for the Protection of Birds (RSPB) is a good example. It is the largest pressure group in the UK with over a million members (Grant 2000: 1). Although its remit extends beyond policy on birds to more general environmental reforms necessary to protect bird habitats (such as policy on energy and agriculture) it does not engage in ideological conflict with its opponents. For instance, unlike greens the RSPB does not argue that the wealth of the northern countries is based on exploitation of the south and of nature and that a new global redistribution of wealth is required to correct this injustice. Nor does it try to mobilise its supporters to challenge government and capitalist institutions for their responsibility in sustaining this injustice. Instead it seeks to influence government policy through expert research and quiet lobbying, while maintaining bird reserves and offering a range of services to its members, such as renewable electricity supplies.

Melucci (1996: 35) defines social movements as 'antagonistic'. By this he means that they take conflictual collective action that challenges the goals and direction of society. By doing so, social movements create new identities, make power visible (by politicising something previously seen as natural or unalterable) and break the limits of the existing system. Antagonistic action by greens exposes the dominant forms of power by confrontation. This can be physical, as in the disruption and damage caused to McDonalds by direct action, but it can also be symbolic, using parody, as in the anti-consumerist action targeted at Christmas shoppers, described earlier. It is often symbolic actions that reveal the vulnerability of opponents most. Thus, one role played by social movement actions is to challenge the taken-for-granted character of dominant cultural codes. Greens make such a broad range of arguments that it is difficult to define them adequately, but among them we can include: the view that we need to develop a new relationship with nature in which the natural world is more than simply a resource for material growth; that the richest could live more fulfilling lives with less material possessions; and that the gross inequalities and lack of democracy within societies and between them are unjust and incompatible with developing a sustainable society.

Social movements are therefore repositories of critical ideas and practices which are used both to change the lives of participants and to change society. Participants in social movements try to develop a degree of autonomy from those that they see as holding dominant power in order to develop alternatives. It is this role of social movements which leads Giddens to suggest that they could 'constrain the juggernaut of modernity' (1990: 158). The breadth of this challenge means that movements cannot hope to achieve all their most far-reaching aims. This means too that focusing only on the effectiveness of movements in achieving changes in the political system or through new policy will only tell part of the story. Even movements that might appear failures in this respect can still have achieved much. The effects of movement actions extend beyond particular issues. As Welsh says,

> irrespective of issue foci, NSMs [new social movements] also engage in a process of relatively autonomous capacity building. Capacity building involves techniques and technologies of the self which combine to create collective forms of experience that I consider central to the practice of critical reflection within modernity.
>
> (2000: 151)

Protests against new dams, or incinerators or nuclear power stations, or animal experiments can also change the ideas of those involved. Beginning out of opposition to a personal threat, activism often leads to a broader analysis of power and how it might be transformed. Many studies of activist life histories (Jasper 1997; Newman 2001; Epstein 1991; McAdam 1988) show that individuals were transformed politically as a result of participation in social movements (an issue covered in more detail in Chapter Seven). When Welsh mentions capacity building part of what this refers to is the way that one movement provides the ideas and experience that later movements pick up and develop. Since individual activists often progress between what appear to

observers as separate movements this is an important consequence of movement action.

Movements also often provoke wider cultural debate even if they fail to
achieve their ostensible aim. For instance, the use of direct action by radical
environmentalists to challenge the World Bank, International Monetary Fund
(IMF), World Trade Organization (WTO) and G8 has not transformed global
capitalism in any obvious way, but it has achieved a discursive shift. It showed
that global capitalist institutions do have an opposition, and required that the
institutions justify and defend their actions. For instance, James Wolfensohn,
Chairman of the World Bank said of the protesters in Prague in September
2000:

> Outside these walls, young people are demonstrating against globalisation. I
> believe deeply that many of them are asking legitimate questions, and I
> embrace the commitment of a new generation to fight poverty. I share their
> passion and their questioning. We live in a world scarred by inequality.
> Something is wrong when the richest 20% of the global population receive
> more than 80% of the annual income.
>
> (*The Guardian*, 27 September 2000)

The protesters did not say anything that was especially new about the role of
global capitalist institutions, but by besieging the summits they were able to place
the arguments of critics such as Susan George and Vandana Shiva on a par with
those defending the global capitalist order. This helped to energise existing criticism of global inequality and the self-interested nature of western development
models from less radical groups such as those involved in Jubilee 2000
campaigns. The Seattle protests in 1999 helped to draw attention to the fact that
the WTO was dominated by the wealthiest nations, throwing the organisation
into a crisis from which, it was widely argued, it could only emerge by giving
more power to the poorer states. Moreover, it was by refusing the tokenistic
forms of access to the 'portals' offered by their opponents, that the protesters
were able to do this. Less radical non-governmental organisations (NGOs) had
been granted consultative status at some of these conferences, but their position
papers were routinely ignored and rarely influenced decisions. By choosing their
own strategy and location for action, protesters were able to express their opposition more effectively and expose the vulnerability of their opponents.

It is both through their critique of the system and through their actions that
movements challenge power. In contesting power social movements also reveal
the relationship between power and knowledge. Social movements show how
forms of knowledge such as technological development, management theory,
marketing, or social policy become part of an epistemology of rule. When left
uncontested these forms of knowledge frame the way that we understand the
world and maintain the actions that reproduce oppressive social structures.
These include the view that new technology is inherently progressive and irre-
sistible, that hierarchical forms of organisation are always necessary and

efficient, and that the marketplace simply reflects consumer choice. When social movements such as the greens say that 'it doesn't have to be this way' they are also changing the relationship between power and knowledge. This does not happen easily, because it requires that movements develop a new grammar outside and in opposition to the mainstream and they have to act without the backing force of institutions and routine practices, which make dominant forms of power seem normal or inevitable. The force of a social movement's challenge depends on the extent and range of collective action that movements can carry out. The factors that shape this will be discussed below.

An ideal type

The definition of the green movement as a social movement is an ideal type that can be used to investigate a broad range of environmental groups. As used by Max Weber (Gerth and Mills 1991) the concept of an ideal type was intended as a way of defining characteristics that would help empirical investigation. An ideal type is therefore a theoretical construction, not in itself a description of reality. Ideal in this sense means an idea, rather than superior, or the best. Its usefulness depends on whether it helps us to understand and explain empirical cases. And those groups that fit the ideal type best are not necessarily those best able to achieve political change, rather they are simply more obviously social movement groups than others.

To be clearly part of the green movement, an environmental group must meet all four criteria mentioned above. They must share some common identity as greens, and this can be expressed in cultural practices as well as texts. Groups involved in green networks are prepared to take protest action as part of their sustained challenge to those with power. There is no single form of green organisation, and no group that leads the movement but green organisations and individuals are linked through informal networks. Finally the movement as a whole must challenge existing forms of power. Not all protest is social movement protest; social movement protest 'challenges society's general orientations' (Touraine 2000: 90). Greens do this through their arguments for an expansion of democracy as an antidote to statist forms of power; through their arguments for greater equality, materially, culturally and globally and through their arguments against productivism and for ecological rationality.

Yet, as the discussion above suggested, the applicability of this ideal type to empirical cases is variable. The identity of groups is itself always heterogeneous, and both identity and strategy are dynamic, changing as a result of the experience of activists. Some have therefore argued that single criteria should take precedence in defining the movement. For instance, Eyerman and Jamison (1991) define a movement in terms of its identity. However, because they regard institutionalisation and a focus on practical policy change as a move away from a social movement's identity, the social movement phase of environmentalism is very brief. Others have given priority to the network ties between actors (Diani 1995; Doyle 2000 and Rootes and Miller 2000) and are less concerned with the content of green identity, and tend to define it as 'environmental concern'. If we make the standard required to have a green identity less demanding than that defined by Jamison *et al.* (1990), then it is possible to identify a radical green identity at least as an ideal type, which fits a large part of the environmental movement. However, it is not useful to define this identity as loosely as 'environmental concern' because that blurs the important differences between those who lack any broader radical commitments, and those whose interpretation of political ecology is rooted in a broader politics in which democracy and equality are also central issues. Making the network ties between different kinds of greens the central criterion would also obscure the differences between greens and less radical environmentalists. Moroever, it limits the analysis of movements too much. For instance, Greenpeace usually appears marginal in analysis of movement networks (Rucht and Roose 2000; Rootes and Miller 2000; Doyle 2000) because of its policy of relative independence from other environmental groups. While the degree of interaction within the network is an important subject of analysis, it is not necessarily more important than the analysis of movement discourse.

The drawback of using four criteria rather than one is that there are more categories on which groups differ. However, untidy though it may seem, groups that do not fit the ideal type unambiguously can still be usefully analysed by comparison with an ideal type of the green movement. Green parties, for instance are less involved in protest and most are now more inside than outside the political system, particularly those in government. But, they retain an ideology which challenges core features of power, they remain supportive of protests and sometimes participate in them; their organisations retain formal links to other parts of the environmental movement, while their activists are linked informally with other environmental activists.

Grassroots environmental groups provide a contrasting case. Many do not share the culture and symbols of other parts of the green movement, nor all of its ideology. The following quote from anti-toxics organiser Lois Gibbs on her attempt to bring together the major US EMOs and representatives of the locally-based anti-toxics groups illustrates the gulf in culture and strategy:

> It was hilarious ... People from the grassroots were at one end of the room, drinking Budweiser and smoking, while the environmentalists were at the other end of the room eating yoghurt. We wanted to talk about victim compensation. They wanted to talk about ten parts per billion benzene and scientific uncertainty. A couple of times it was almost war. We were hoping that by seeing those local folks, the people from the Big Ten would be more apt to support the grassroots position, but it didn't work out that way. They went right on with the status quo position. The Big Ten approach is to ask: What can we support to achieve a legislative victory? Our approach is to ask: what is morally correct? We can't do something in order to win if we think it is morally wrong.
>
> (quoted in Harvey 1999: 158)

As Chapter Seven shows local environmental groups are often radicalised by their campaigning experience, and linked by network ties to other parts of the environmental movement. Moreover, when they develop different agendas to those of existing green ideology, the interaction produces new political debates with the potential to change the identity and ideology of all involved. For instance, environmental justice campaigners in the USA criticised the major EMOs for focusing too much on protecting non-human nature and ignoring the disproportionate amount of environmental pollution suffered by poor and ethnic minority communities. They also highlighted the dangers for greens of underestimating connections between class and ethnicity and environmental politics. Many local environmental groups meet the criteria for being social movements more clearly than they do as part of the green movement.

Also, as noted above, the major environmental movement organisations have also become increasingly institutionalised, increasing the size of their organisation and co-operating more with government and business. EMOs, such as World Wildlife Fund (WWF) and conservationist groups founded in the late nineteenth century, have only weak claims to social movement status because they do not challenge existing forms of power and do not take protest action. Only by sharing a minimal common identity as environmentalists and some organisational and inter-personal ties with the more radical greens could these groups be seen as overlapping with the green movement. Others, such as Friends of the Earth (FoE) in its various national guises, were committed to more radical political change, saw themselves as radical environmentalists, played a major role in the anti-nuclear movements and in some cases in the formation of green parties, and included protest as part of their strategy. But, as the opportunities for political change improved as global environmental issues became of increasing political importance, many of these groups became less radical, at least in their actions, even while retaining radical values. On the other hand conservation groups often broadened their agendas to take account of the systemic causes of environmental degradation and formed coalitions with other EMOs in which they were often prepared to be more critical of the failings of government and business.

The groups that approximate the ideal type most clearly are those like Earth First!, which have a clear identity as radical environmentalists, committed to radical change, carrying out protest action, and interacting with other parts of the green movement.

As this shows, the variety of environmental movements and their changing character means that applying the concept of social movement in order to define 'a green social movement' is difficult. Some have argued that, given the imprecision of definitions of social movements, it would be better to abandon the task and simply treat all environmental groups as interest groups (Jordan and Maloney 1997: 54). The term interest group is most associated with political science, whereas social movement is associated with sociology. Interest group analysis is broader than many sociologists often suppose in that it embraces informally organised groups as well as formally organised pressure groups and can include protest as well as lobbying. It mainly focuses on how groups affect policy-making, but can also deal with how groups assert a broader political cause.

There is no clear analytical distinction between the kinds of groups dealt with in the interest group and pressure group literatures in political science and the social movement literature in political sociology. Rather the difference tends to be one of focus. The political science literature tends to focus on assessment of groups' involvement in policy-making and their interaction with established political institutions,[5] whereas the strength of the social movement concept is in the attention given to the cultural and sociological dimensions of collective action. For instance, a social movement analysis of local environmental groups is likely to assess the extent to which such groups develop a broader political critique of the limits of current democracy and place their own campaign within a framework shared with other environmental groups. It is also more likely to seek to explain this in relation to broader structural changes in society, such as new attitudes to science, or the expansion of higher education. There is also a focus on the forms and scale of action undertaken and how social movement organisations (SMOs) mobilise support. A second difference is that social movement analysis focuses on the network relationships between different groups within a social movement, whereas interest group analysis deals with groups as separate entities with clear boundaries.

Interest group analysis is strong in analysis of policy outcomes, but weak on questions of identity and ideology. Social movements cannot be defined adequately as single-issue groups focusing on particular policy goals. Green ideology and green identity is based upon a general analysis of the political and social system and the relationship between humanity and the natural world. While green groups can agree on short-term goals, green identity and ideology cannot be defined in terms of changes in policy. If analysis of environmental groups were restricted to analysing them as interest groups it would not take sufficient account of green ideology and identity. The ideology of social movements develops and changes as a result of experience and action. This book is concerned with analysing the development of green ideas through the experiences of green activists. This is the most important reason for retaining an analysis based upon the green movement as a social movement.

Debates within social movement theory[6]

Social movement theory is concerned with much more than the question of how to define a social movement. This is not the place for a full review of the theoretical debates within this field, however, it is important to explain two of the traditions that are drawn upon in the rest of this book. The first is the school of 'new social movement' theory developed mainly by European theorists such as Alain Touraine (1981; 1985; 2000) and Alberto Melucci (1989; 1996). They argue that new kinds of social movements began to emerge in the 1960s and 1970s. Where most previous movements had focused on claiming rights such as the vote or material equality, many new movements seemed to give greater attention to the development of new identities, which allowed participants to redefine their experience and provided a framework from which they could seek to challenge features of social life that were seen as natural and unchangeable. In contrast to the earlier movements they did not seek to capture the state and their challenge was as much cultural as material. They did not believe that the revolutionary overthrow of the state would be sufficient to bring about a new society and directed their actions as much at the general public as the state. The environmental movement, the women's movement, and the peace movement were among the core groups included in this category. In the case of the environmental movement it was only the more radical sections of the movement that fitted the ideal type of the NSM.

Although the NSM concept has been the subject of criticism, much of this misses the major point being made by NSM theory. This is the extent to which new movements seemed to be responding to structural changes in society. In this sense, it is not whether previous movements organised or acted in ways that are similar to the so-called 'new social movements' that matters most (Calhoun 1995), but whether the new movements are acting in ways that are only possible in the new kind of society. Thus the newness of the new social movements is not based upon an empirical analysis of all the actions of all movements that call themselves women's movements or environmental movements. What counts as decisive in this debate is whether we are living in a new kind of society and whether this has led to new kinds of collective action, which can be seen in a variety of movements, including actions by feminists and greens. Only some of the action of the NSMs has to be of a new kind to sustain this argument.

What is held to be new by NSM theorists? Common to all NSM theory is an emphasis on the increased importance of culture and information in current society. Melucci (1996) argues that NSMs are new because culture and symbols have now become more important as sources of power, and the actions of NSMs reflect this change. One example is the conflict over the use of bio-technology. The economic value of research in this field lies in the knowledge about genetics and the potential to transform that knowledge into commodities for sale. Arguments over the use of bio-technology to create plants that need fewer pesticides or to control the genes that produce illness or disability raise basic ethical questions the answers to which will shape how cultures evolve. Greens argue that much of bio-technology is seeking to perfect nature and that its effects are impossible to predict and likely to include harmful unanticipated effects. To support their case they argue that scientific research is being misused to pursue aims defined by those most likely to profit from the exploitation of new knowledge and ask why there is so little research and investment in developing renewable energy. Arguments over such questions are a good example of what Melucci (1996) means when he says that battles about which are legitimate forms of knowledge are characteristic of contemporary politics in a way that seems to be new.

Another example of what is new about the conflicts engaged in by greens is provided by Manuel Castells (1997). He sees the environmental movement as engaged in a battle with the dominant political and economic forces over representations of time and space. The dominant forces (major corporations, political elites and media among others) work increasingly as networks, transcending the limits of nation-states. They view time in terms of efficiency and production and space in terms of markets. Environmentalists argue for the importance of local spaces against the homogeneous 'space of flows', based on instantaneous financial transactions and global media. They assert the importance and independence of the evolutionary dimension of nature, which Castells characterises as 'glacial time'.

The debate about the usefulness of the term NSM should revolve around the question of whether we are really living in a new social order, an issue which is too large to be resolved here. Nevertheless, in the criticism of the NSM concept, one significant claim associated with it appears to be relatively unscathed. This is the view that as sites of power become more and more plural, the significance of the nation-state as the principal target of action declines. This occurs as a consequence of processes such as economic and cultural globalisation, the proliferation of new forms of media (and the decentralisation of access to media), the erosion of state sovereignty by both globally integrating and socially fragmenting forces, and the increased power of local government in some countries. It is of course possible to overstate the extent and impact of these developments. Previous social movements have sometimes bypassed the state, as in utopian and religious communes, for example. Also, many of the contemporary NSMs continue to direct much of their activity at national governments. But the real empirical question concerning this aspect of the novelty of NSMs is whether there is activity within the movement that can be explained by these developments? Two examples can illustrate the usefulness of this structural approach.

One activity which is a product of these large-scale social forces is the global action against transnational capitalism exemplified in successive protests against capitalist summits in Cologne in 1998, London and Seattle in 1999, and Washington, Melbourne and Prague in 2000. In these protests, not only did global institutions and processes become the site and focus of protest, but global networking became a principal means of organisation for the movements themselves, of course ironically made possible by the technological changes (especially the emergence of the internet) which are themselves central to globalisation. Although heads of government or finance ministers were present at some of these meetings, they were not targeted as individuals, it was the collective of governments and financial institutions that were targeted.

A second example are actions that target the profit flows of specific companies. In 1995 Shell became a target of green protests because of its failure to intercede with the Nigerian government to prevent the execution of Ken Saro-Wiwa and its plan to dump a redundant oil platform in the North Sea: it changed its plans on the Brent Spar as boycotts in Germany and Scandinavia began to have an effect. British Petroleum and Fidelity investments were targeted by activists because of their involvement in oil exploration that threatened the tribal homeland of the Uw'a people in Colombia and bio-tech companies such as Monsanto saw their share values drop sharply as protests against genetically modified food mounted in Europe. These kinds of action focus on the vulnerability of transnational companies to rapid share decline through loss of brand loyalty.

The literature from NSM theorists attempts to relate movements' ideas to major shifts in recent society such as globalisation and the rise of the information society. As Melucci puts it he is more interested in the 'why' of social movement activism whereas other theories focus more on the 'how' of social movement action.

NSM theory is oriented towards understanding what is at stake in new mobilisations and what this reveals about what is happening in society. It views movements as analysts, interpreters and agents of social change. However, there is very little attention in the NSM tradition to how movements develop and why some develop differently to others. And yet, as we have seen, there are significant differences between the types of groups involved in the green movement. The recognition of a common identity is not in itself enough to constitute a social movement; those involved must also take action together and challenge power. When they do so their action is affected by constraints and opportunities that vary between different states and localities and over time. Assessing and explaining the patterns of movement action has been the main focus of a tradition of movement analysis particularly associated with the 'political process' work of McAdam (1982); Tarrow (1998) and Tilly (1995).

Their main focus has been on patterns of protest or contentious action. However, theirs is not a narrow study of protest events. They explain protest and movements as rooted in much broader processes. For instance, the development of movements is shaped by social changes, such as the changes in employment patterns that encouraged second-wave feminism in the late 1960s or the changes in social structure in the southern USA that encouraged the civil rights movement. They also show how cultural factors shape the emergence of movements. For example, the language of rights and liberation used by the US civil rights movement in the 1960s was taken up and applied to new themes by the New Left and others. Research from this tradition also shows how movements tend to emerge from within existing social networks or out of existing movements. When applied to the green movement this helps to make sense of the common social characteristics shared by greens and the links between early green groups and the New Left.

The resources that groups are able to mobilise also shape the collective action of movements. These are principally of two kinds: time and money (Oliver and Marwell 1992). Groups that concentrate mainly on mobilising activists' time tend to be more participatory in style, as in local environmental groups and direct action groups. Those that tend to focus more on mobilising money tend to be more centralised and less participatory, as is the case with many EMOs such as Greenpeace. But most green groups need to try to mobilise both, and green parties in particular seek both money and activists' time. The emergence of new resources or a decline in previously available resources is likely to affect the level of movement activity.

Movements are also affected by the cyclical nature of protest activity. Tarrow argues that protest cycles begin when 'early risers' take risks that establish the movement as a challenging force (1998: 24). This is most likely to occur:

> when ordinary citizens, sometimes encouraged by counter elites or leaders, respond to opportunities that lower the costs of collective action, reveal potential allies, show where the elites and authorities are most vulnerable, and trigger social networks and collective identities into action around common themes.
>
> (Tarrow 1998: 20)

The focus then is on the 'clues' that changes in the nature of opportunities provide about when mobilisation is likely to arise. As protest cycles develop they inspire and encourage others to act, force authorities and opponents to respond and institutions to adapt. It is at this stage that movements are likely to be at their most innovative. But as states begin to bargain or arguments over strategy within the movement deepen, it becomes harder to maintain protest. Some parts of the movement are likely to want to bargain for reforms; others, often disillusioned, pursue more radical and sometimes extreme strategies. While the general pattern is one of early enthusiasm and later decline, the way in which protest cycles 'end' varies according to movement constituencies, and strategic choices made by movements and opponents.

A strength of this approach is the emphasis on the learning that occurs through protest and movement action. Movements inherit repertoires of action from previous generations, and learn how to adapt or extend them through their practice. But it is important to recognise that movements may continue even when protest appears to be minimal. So we should distinguish between protest activity, which is largely cyclical and other social movement action, which is not. This is consistent with the fact that movements emerge from existing social networks. Movements may not end so much as metamorphose into new movements and the fact that the boundaries between movement networks are often blurred makes this more likely, for example local activist communities share social ties which cross 'movement' boundaries (Carroll and Ratner 1996; Doherty, Plows and Wall 2001). We also know from studies of ex-1960s activists in many countries that most remain both radical and active in other movements. As Tarrow says, 'a high proportion of activists from the 1960s emerged empowered, transformed, and connected to informal networks of other activists. A kind of movement "social capital" was the most durable outcome of that period of contention' (1998: 169).

Tarrow (1998: 175) argues that movements have their greatest impact on political culture rather than in substantive policy gains. They leave a residue of reform, which may include new policy, but also new ideas, values and new forms of action that can be used by political actors and future protest movements – what Welsh (2000) calls 'capacity building'. In contemporary environmental movements what appear to be new environmental movements often include many people with previous activist experience. This is true in the case of grass-roots environmental groups as well as national campaigns. Thus while protest activity is likely to take a cyclical form, the culture of the movement is likely to be more enduring. Moreover, movements diffuse their ideas through wider society. For instance as public authorities respond to movement challenges, some activists find themselves in new posts, as 'experts donated to the system' (Melucci 1996), as for instance with many who worked on Local Agenda 21 projects for local authorities after the Rio UN environment conference of 1992.

Tarrow's cyclical approach helps to explain a common pattern of institution-alisation and radicalism within the environmental movement. There seem to be two generations of green protest, the first mostly emerged from the 1960s New Left, and eventually became institutionalised in various forms in established EMOs and green parties. A second wave reacted against the failures of the first generation. This was evident in the anti-toxics campaigns and environmental justice campaigns in the USA and in the non-violent direct action (NVDA) networks which grew in the USA and Australia in the 1980s and Britain in the 1990s. The assumption of the cyclical model would be that the latter two groups would either institutionalise or decline. However, it seems possible that they will be able to continue without significant institutionalisation. Both types of groups are strongly opposed to institutionalisation and may be able to sustain activity that complements that of the better-resourced EMOs and green parties. This is because they are rooted in communities whose culture is resistant to institutional-isation, but which have the resources and commitment to sustain grassroots activism. Moreover, the lessons of the established groups means that the second wave of green groups is more aware of the benefits of staying outside the system.

The concepts in political process theory help to explain how particular groups in the green movement take action. They are useful in explaining why groups emerge at specific moments, why some groups become institutionalised, or why confrontational action was more common in France than in Sweden in the 1970s and 1980s, for instance. Yet, as Welsh (2000) points out, political process approaches tend to take the existing political framework for granted. This is because their focus is on explaining the pattern of public events rather than because they necessarily support the existing system, but it means that less attention gets paid to how power is exercised to shape what gets on the agenda.

The next chapter combines political process concepts with more structural analysis to explain why green movements emerged in the form that they did. As political process theories would lead us to expect, existing networks, especially those of the New Left, influenced the greens and provided resources and experi-ence for new environmental groups. But greens also developed new ecological ideas and new strategies – based less than those of the New Left on overthrowing the state – that allowed them to define a new identity. New resources were provided by the expansion of higher education and welfare state professions in areas such as health, education and social work. As the argument in the next chapter shows, the social base of green movements can be linked to green ideas through examination of experiences in the New Left, education and welfare state professions. However, this alone does not explain the form of green ideas. For this we need to combine analysis of structural developments, such as the evidence of an ecological crisis, the changed view of the role of the state after the 1970s and the changes to left-wing politics that followed the de-centring of class by a more complex approach to inequality.

Conclusion: the scope of the green agenda

When most people speak of the green movement they often use the term broadly, to cover almost any action based on environmental concern. This might include applying environmental criteria to what we buy in shops, signing a petition in favour of the conservation of endangered species, or sabotaging a digger used to destroy a forest. As the discussion above shows, not all such action is necessarily social movement action. A limitation of the broad use of the term green movement is that the distinctiveness of the more radical groups within this broad spectrum tends to get lost. If we want to understand the actions of those who are prepared to burn diggers we need criteria that separate them from those who are not prepared to do more than drive to the recycling centre. There is also the reverse danger that the ideas of the radicals may be seen as definitive of environmentalism and wrongly ascribed to those whose activism extends only as far as joining a conservation group to gain cheaper admission to stately homes and national monuments. Using a tighter definition of social movements based on the four criteria of shared identity, protest action, network ties, and a challenge to dominant forms of power, avoids both these traps.

This is also consistent with the distinction made by Dobson (2000) between environmentalists and greens.[7] Environmentalists are concerned with tackling environmental problems but believe that they can be dealt with within the existing framework of society. Greens, however, believe that far-reaching changes in social, political and economic structures are necessary to deal with the ecological crisis, and, with other crises, such as those of the over-extended state, and complex forms of inequality. This ideological distinction is also close to that adopted in many definitions of social movements. According to Melucci, a social movement must challenge the limits of the existing system (1996: 28). This means that social movements are radical by definition, and that the green movement is distinct from the broader environmental movement because it challenges existing forms of power. Where movements are able to identify and articulate new ways of looking at power (making political something that had been taken for granted), new ways of acting politically and new structures of feeling about politics, we can say that they have the potential to transform whole social systems (Welsh 2000). This book will examine whether green actions meet this challenge.

2 The emergence and growth of the green movement

In the previous chapter two approaches in social movement theory were introduced. New social movement theory emphasises the importance of structural changes in reshaping social movement action. Political process theory is centred on the analysis of mobilisation and protest. Both are important for explaining the emergence of the green movement. The two concepts that form the basis for the analysis in this chapter are collective learning and a political generation, which include the central questions tackled in both the new social movement and political process approaches. Political generation and collective learning are particularly useful for understanding how experiences are related to the emergence of a new identity.

Mannheim (1952) developed the idea of a political generation to refer to a group that shared the same formative social conditions and a common set of values or political beliefs. This idea is useful in emphasising how particular historical contexts influence the development of worldviews. But Mannheim tended to analyse beliefs as the property of individuals when it is the regular interaction with other activists that shapes how commitment and collective identity develop in social movements. As Whittier says: 'The concrete lived experience of organising a challenge together, not an abstract "spirit of the times" is what gives participants a shared worldview' (1995: 17). Her study of the local feminist community in Columbus Ohio showed how the collective identity of activists differed according to the point at which they entered the movement. The first activists in the late 1960s had been involved in the New Left and retained commitments to other forms of radicalism. For those who got involved later the existence of many practical projects such as rape counselling, music groups or mechanics courses, encouraged confidence in the possibility of a life lived autonomously within a feminist community. Yet, when the funding which sustained many projects ended in the 1980s, new activists who joined faced a more hostile atmosphere in which it was harder to maintain the optimism of the 1970s generation.

Although it involves a much greater level of generalisation than in Whittier's study, the idea of political generations can also be used to explain the experiences of green activists. Two experiences can be regarded as important in the emergence of a green identity. The first was the emerging evidence about the ecological crisis. The second was the rupture with the dominant traditions of the old left made by the New Left in the late 1960s. The impact of both experiences will be discussed below. However, subsequent cohorts have also shaped the development of the movement, responding differently to new contexts in the 1970s, which differed in different countries.

The concept of collective learning helps to explain how green identity developed and changed. Klaus Eder (1985; 1996) has argued that collective learning should be central in the understanding of social change. Building on Habermas's theory of communicative action, Eder argues that evolution can be connected to agency if we see discourse as important in social change. But this does not imply a unilinear or universal pattern of evolution. Instead the nature of social learning varies within societies, between them and temporally. The key point is that agents carry out social action reflexively. Eder argues that by changing the nature of the discourse about nature and the environment, the green movement became an 'epistemic community' which created new forms of knowledge and new moral debates which can only be addressed effectively by collective debate in society. His main concern is with the effects of green discourse on society, but an alternative is to look at collective learning within green movements. Collective learning helps to explain how greens developed their identity over time. Most accounts of identity tend to lack this dynamic dimension. Collective learning can also embrace the green response to the strategic dilemmas which they have faced as social movements.

The collective learning of the green movement did not end with the introduction of new ideas about ecology. The greens were also shaped by their relationship with other social movements such as the women's movement and the peace movement and by the wider alternative milieu. The development of new movements in the wake of the New Left made the idea of a universal left strategy and a singular revolutionary transformation implausible as a basis for mobilisation. Yet, this new form of left politics only evolved through experience in trying to mobilise politically. Moreover, what it meant in practice depended upon national contexts. The nature of the alternative milieu, and the nature of the left differed from country to country, as did the institutional opportunities and constraints provided by the political system.

This chapter will examine the roots of green identity in the emergence of new forms of environmentalism in the early 1970s and the influence of the New Left. It will then provide a brief overview of the development of green movements in western countries, and explain how this was affected by national contexts and transnational trends.

Since green activists are mostly drawn from particular social groups – the young, those with higher education and the new middle-class welfare professions – we might ask how the emphasis on political generation and collective learning connects with this very specific green social base. Rather than the traditional attempt to explain the green social base by analysis of class interests, we can use the concept of generational experience and collective learning to explain why it is that green activists are concentrated in these particular social groups.

How new are the greens?

In the late 1960s and early 1970s new environmental groups emerged in most western countries. These groups appeared new because they linked environmental problems to structural features of western society, such as excessive consumption based upon an economy driven by a productivist logic, according to which material growth was an end in itself. But it was not only formally organised groups, but also in informal networks of those committed to an alternative lifestyle more in harmony with nature and less materialistic, that this new green movement could be identified. But we cannot take the newness of the green movement for granted, particularly as environmental groups and thinkers had long been present in most countries.

The new activism of the late 1960s and early 1970s was the first new cross-national wave of environmentalism since the late nineteenth century when many conservation and animal welfare groups had been formed. The principal aim of the National Trust in Britain and comparable groups in most of northern Europe and the USA formed at the end of the nineteenth century was the preservation and conservation of landscapes and national monuments. In the USA John Muir founded the Sierra Club in 1892 to campaign for the protection of wilderness areas as an important part of national heritage. New nature conservation groups were established at that stage in most countries and the protection of birds killed for the trade in hat feathers was a concern of middle-class and aristocratic campaigners, particularly women, across Europe. This activity reached a peak in the 1900s but began to subside thereafter because most of its immediate legislative aims had been achieved. A wider romantic reaction against industrial society and its excesses had preceded and overlapped with this early environmentalism. While this was extremely diverse in content, a core theme in Romanticism was the view that greater knowledge of the natural world brought by science was making humanity presumptuous and arrogant. Those who lived closest to nature were seen as 'repositories of moral virtue and a valuable culture' (Gould 1988: 3). However, it was only the limited policy aims of the conservation and preservation movements that were translated into public policy, not those of the more wide-ranging and radical critics of modernity. The Romantic calls for a return to nature or to go 'back to the land' were never sufficiently powerful to threaten the existing order.

By the 1900s the counter-cultural critics of industrial society had failed to develop political movements, indeed most had not even tried. Perhaps the only place where a positive alternative vision of an ecological future was developed was in the still nascent socialist movement in Britain. Figures such as William Morris, Edward Carpenter, Prince Peter Kropotkin and Robert Blatchford linked their critique of industrial society to radical social and political ideas and for many contemporary greens they represent the early history of the green movement. Morris and Kropotkin both argued in favour of decentralisation: small self-governing communities were to be preferred for both political and environmental reasons. Both also opposed the capitalist expropriation of the fruits of labour and argued that work could only be made fulfilling when production was for local and community needs and where workers were able to make products that were both useful and beautiful. However, there were also differences that make it hard to see them as greens of the same kind as those who emerged after the 1960s. The idea of nature as having intrinsic value aside from its use for humans is not present in the writings of any of the green socialists or anarchists of this period, nor is the idea of limits to growth. There is no sense in either writer of the kind of diversity of identities which has made the idea of life in homogenous small communities seem potentially oppressive to many contemporary greens.

The question of how new the green movement is has been the subject of much debate. Some stress the continuities in environmental concern (Brand 1990; Sutton 2000) as sufficient to show that there are cycles of particular movements in modernity; others identify many intellectual precursors in a broad tradition of thinking about the relationship between politics and the natural world from which greens can draw an intellectual pedigree (Marshall 1992; Wall 1994b). Moreover, it is not only in the late nineteenth century that we find important precursors of contemporary green thought. In the twentieth century the environmentally driven theories of town planning of Patrick Geddes and his disciples Lewis Mumford and Radhakamal Mukherjee and Gandhi's opposition to 'the indefinite multiplicity of wants' (quoted in Guha and Martinez-Alier 1997: 156), and favouring of a village-based path of self-reliance for an independent India were particularly important.[1] Gottlieb points to the forgotten role of public health campaigners in the USA such as Alice Hamilton and figures such as Robert Marshall who made links between issues of social justice and the protection of nature.[2] Aldo Leopold's *A Sand County Almanac* (1948) built upon the preservationist reverence for nature and provided much of the ground for the later 'deep ecology' of Naess (1973) and Devall and Sessions (1985). Importantly, Leopold moved beyond the emphasis on protecting only special sites of wilderness and focused on the interaction of humans with whole ecosystems, stressing the importance of humble and uncharismatic plant and animal species as much as the great Redwoods or the wolf. In Australia early environmentalists were influenced by Theodore Roosevelt's 'progressivism' and made ethical defences of nature, at least in their private writings (Hutton and Connors 1999: 21). These thinkers and ideas have certainly influenced the cultural context from which contemporary green politics draws its ideas, and in that sense it is clear that they are an important part of green history. Sutton (2000) suggests that existing analyses of the green movement over-emphasise the role of radicals such as those in green parties and underplay the role of the more reformist groups that make up the mainstream of the environmental movement. Viewing the movement from that vantage point there is much more continuity with the conservationism of the late nineteenth century than is usually acknowledged.

But Dobson (2000) gives the best reason for seeing the green movement of the early 1970s as new. What marks the break with earlier environmentalism is the evidence about the finite nature of the earth's resources that emerged in the Club of Rome's *Limits to Growth* Report in 1974 (see Meadows *et al.* 1974).

Whatever the limitations of the evidence presented in that report the idea that the planetary ecosystem is under threat and that this requires radical changes in order to avert a catastrophe was new and still remains a core green principle. Many green ideas, such as commitments to decentralisation or to egalitarianism are not new in themselves but they acquire new meanings in the context of a world of finite resources, as for instance with the idea that justice now requires that we consider the needs of future generations. While Sutton is correct to argue that not all environmentalists accept that radical structural changes are necessary to deal with the environmental crisis, those defined as greens in this book, do. Thus the evidence that the world was facing a potential catastrophe led to an argument (or for some reinforced their view) that major social, political and economic changes in western society were required, and this separated 'greens' from other parts of environmental movement.

The new ecological crisis: an anti-authoritarian response

However, this was not all that separated the greens from other environmentalists. Even before the Club of Rome new environmental activists were making a link between pollution and environmental degradation and structural features of western society. The alternative beginning most often argued for modern environmentalism is the publication of Rachel Carson's *Silent Spring* in 1962. The silence of the spring referred to the poisonous spread of pesticides through the food chain from plants to insects and ultimately to other animals, including birds. One of the central themes of this work was the evidence of the systematic spread of chemicals throughout nature and thus to humans. 'For the first time in the history of the world, every human being is now subjected to contact with dangerous chemicals, from the moment of conception to the moment of death' (Carson 1962). Carson's book had a tremendous impact, provoking great debate in the media, new legislation in the USA and an outraged defensive response from the chemical industry and scientists responsible for pesticide development. Her focus was principally upon the failures of science: its over-specialisation and compartmentalisation had led to a failure to appreciate the inter-dependence of different parts of nature and the complexity of the food chain. Carson's work was published in many countries and was followed by the publication of books by many other critical scientists, who helped to renew a public debate, largely absent in the post-war development boom, about the costs of human exploitation of nature.

Critical science therefore played an important role in the revitalisation of environmentalism in the 1960s. However, this was also balanced by other kinds of knowledge. For instance, Welsh (2000) shows that even before critical scientists began speaking out against the risks of radiation and safety flaws of nuclear energy, there was much evidence of distrust of the reassurances provided by the industry among those most exposed to its risks. Local knowledge often contradicted the narrow laboratory-based science of the nuclear industry and Welsh shows that rather than critical science opening the door to local opposition among lay people, the distrust of the latter may have encouraged critical scientists to ask further questions.

Ulrich Beck (1992; 1996) and Anthony Giddens (1994) have seen the rise of public contestation of expertise as evidence of a shift to a new more reflexive modernity. In a situation in which the experts cannot agree, lay people are forced to make a choice about which evidence to trust. This encourages a more sceptical and critical attitude, which undermines the legitimating authority of science. However, if science itself never had the trust, which Giddens in particular assumes in his description of the shift from an earlier simple (and trusting) modernity to a more reflexive (and sceptical) modernity, it is hard to show that there has been a shift from simple to reflexive modernity. Brian Wynne (1996) argues that we cannot take the absence of overt dissent as sufficient grounds to show that people trusted the experts. Moreover, since there is also evidence of ordinary people's knowledge, such as the evidence of leukaemia clusters close to the Sellafield nuclear plant in Britain in the 1970s, being suppressed or ignored, there are positive reasons to doubt that trust existed.[3] This means that we cannot be sure that critical science preceded and made possible wider popular distrust.[4]

Whatever the status of trust in science in earlier periods, there is widespread agreement that experts are now more routinely challenged than was previously the case. For instance, an ICM poll in 2000 showed that 73 to 75% of UK citizens did not trust their government 'to tell the truth about the safety' of 'food', 'nuclear installations' or 'GM foods and crops'.[5] Why this should be so is clearly relevant to explaining the origins of the new wave of environmentalism in the past three decades. The most obvious explanation is that new evidence emerged of the systematic effects of industrial society's misuse of nature. Thus modernity has now reached a stage where unintended consequences of industrial production and modern patterns of consumption have begun to undermine the material and cultural conditions that made it possible. Ecological problems themselves explain the growth of environmental concern.[6] As evidence of the finite nature of resources, rising human population and the systematic and indeterminate effects of pollution began to grow, this encouraged a rethinking of humanity's relationship with nature.

At one level then it is simply worsening environmental problems, producing a more critical reaction both from affected lay people and experts that explains the new environmentalism, and this feeds back into a new politics of risk through which the structural character of environmental problems as rooted in modern methods of production and consumption are exposed. However, while this works as a general explanation of rising levels of environmental concern it does not provide an explanation of why forms of environmental concern differ. For instance, it does not tell us enough about why the green movement took a particular anti-authoritarian and pro-egalitarian form, in which it engaged with questions of power, democracy and global social justice. After all, among the more apocalyptic experts there were many who presented the environmental problem in terms such as those of Ehrlich, who wrote of *The Population Bomb*

(1968) and Hardin who wrote of the *Tragedy of the Commons* (1968). For them the rise in the world's population due to high rates of growth in the Third World was seen as the principal threat to the planet's ecological stability. And others (Ophuls 1977; Heilbroner 1974) argued that only a strong government would be able to overcome the selfishness and self-interested behaviour of individuals and thereby prevent the worst effects of the ecological crisis.

Concern about population growth is important for greens, but most argue that the reasons for high population growth in the south are principally due to poverty: children provide hope of extra income and security in old age. Second, they point to the fact that western affluence is the principal cause of global ecological crisis: 'roughly 80 per cent of the resources of the planet as well as its sinks are being used by the 20 per cent of the population that lives in Europe, North America, Oceania and Japan' (Anon. quoted in Guha, 2000: 143). Also, while some environmentalists have favoured the kind of authoritarian measures suggested by the eco-survivalists, they have generally been excluded from green movements (an issue which will be tackled further in the next chapter). Greens therefore are optimistic that people can be persuaded to develop an ecologically sustainable lifestyle, and reject the eco-survivalist assumption that people are essentially self-interested and will have to be coerced.[7]

The influence of the New Left[8]

The explanation for green anti-authoritarianism has to go beyond the emerging evidence of ecological crisis and lies mainly in the influence of the New Left on the new ecology movement. This stems from its role as the first major critic of the certainties of post-war politics and the radical democratic ideas which it advanced as the basis of an alternative politics which challenged the limits of representative democracy and inspired the green commitment to grassroots forms of democracy. Also, while ecological themes were not central to the New Left, they were present, and it is possible to establish links between the ecological dimensions of parts of the New Left and contemporary green politics

The New Left was not a single movement and whilst it produced political organisations these never encapsulated the movement as a whole. This means that its boundaries are difficult to establish. Second, there is some uncertainty about when it began. Most commentators identify two phases. The first began in the mid-1950s and was marked by the disillusion of members of the communist parties in Britain and Denmark caused by the Soviet invasion of Hungary. Although new parties were formed in several countries, perhaps the main role of this early New Left was the intellectual inspiration it provided (Kenny 1995). New Left thinkers such as E.P. Thompson and Raymond Williams recovered and extended socialist traditions of thought in which popular culture was central, posing questions for conceptions of an alternative society that were based on centralised policy direction and command planning. The civil rights movement in the USA undermined the claims that democracy had been achieved in the USA and renewed faith in popular protest after a period in which the left had been hounded by state repression. The second phase of the New Left began first in the USA following the expansion of Students for a Democratic Society (SDS) after 1964 from a small group of around 1,000 members, many of whose members had partici- pated in the civil rights movement, to an estimated 100,000 members by 1968. At the same time the shift within the mainly black student organisation the Student Non-violent Co-ordinating Committee (SNCC) from non-violent protest based on claiming equal civil rights to a broader politics of black power and liberation, also marked a decisive and influential shift to a New Left politics. Vietnam was the central object of protest, but the movement was not a single- issue one. For the New Left in the USA and then Europe in the second half of the 1960s the war in Vietnam was evidence of a deeper malaise in society. The political parties and networks of intellectuals which had been the driving force of the first phase of the New Left were now eclipsed by a new generation more concerned with making a direct challenge to the political system. This second phase of the New Left was more international and radical than the first phase and its supporters were principally the first generation of students in the newly expanded universities of the 1960s. It was also primarily a protest movement with no permanent organisations.[9] As Caute (1988: 20) comments: 'The most fundamental characteristic of the New Left was its libertarian distrust of state power, parties, competition, leadership, bureaucracies and, finally, of representa- tive government'.

In sociological terms the most striking feature of the New Left was its genera- tional character and especially its concentration among students. In part this was a result of factors common to many countries such as the expansion of univer- sity education and the reaction against the Vietnam War, most acute for a generation brought up during the Cold War and now confronted with the reality of the meaning of the defence of the free world and democracy. There were also nationally specific explanations, such as the sense of exclusion from politics expe- rienced by a younger German generation, as a result of the Grand Coalition between the (Social Democratic Party (SPD) and the Christian Democratic Union (CDU) of 1966–9, which to the New Left indicated that both parties were the same and that there was no real possibility of achieving change within the system, and the effects of colonial wars on Dutch and French politics. In Spain the breadth of opposition to Francoism and in Italy, the extent to which the movement had begun as an insurgency within the political parties (Tarrow 1990), meant that the protests of the late 1960s were less student dominated than elsewhere. More generally, this period saw a radicalisation of part of the working class with a rise in strikes, factory occupations and more demand for workers to control their own enterprises as well as for more grassroots control of trade unions. Yet, one of the weaknesses of the student movement in most coun- tries was the failure to create effective alliances with working class groups.

Any thesis about the relationship between the greens and the New Left has to proceed cautiously, partly because there is no unity between the New Lefts of different countries, only some very general shared themes, and also because there are many differences as well as similarities between the ideas of the New Left and the greens. To cite some of the differences we need only point to the technological optimism apparent in the writings of theorists such as Gorz (in his early writings) and Marcuse, as well as in New Left parties such as the French Unified Socialist Party (PSU) or groups such as the Dutch *Kabouters* (Lucardie 1980: 262). This shows the importance of the ecological crisis as an influence upon radical movements in the 1970s.[10] Second, the New Left was still largely dominated by Marxism, even if the existing Marxist or socialist parties were rejected, or seen as in need of significant reform. But the most innovative New Left ideas about society and politics came from their radical democratic project,[11] and it is these which have most clearly been inherited by the green parties, even though other features of New Left politics were abandoned as the movement ebbed during the 1970s.

Green movements have taken up several themes first developed in the post-war era by the New Left: particularly, the expressive and convivial dimensions of politics; the idea of a radical democratic project that depends upon initiatives from outside the formal political institutions and also an optimism that change from below was possible. European and North American New Lefts drew inspiration from a wave of post-colonial examples of change from below. Successful guerrilla campaigns such as Castro's in Cuba and the FLN's (Algerian Independence Movement) in Algeria showed that centralised power and capitalism did not always triumph, as had the successes of the civil rights movement in the USA. Globalised consciousness and a transnational sense of solidarity spread through reporting of Third World liberation movements in mainstream, and alternative media helped New Left activists in the west to see themselves as part of a global struggle against imperialism and capitalism.

Another legacy of the 1960s was the importance of direct experience of protest. Participating in demonstrations that broke the rules and angered or challenged the establishment made participants feel their actions mattered. As well as this the experiments with new repertoires of action – such as the creation of the 'free universities' during occupations of university buildings – gave later activists confidence to develop other alternative forms of community, particularly through squatting or other ways of occupying a site such as protest camps.

In explaining the protests of the 1960s Tarrow (1998: 131) emphasises the importance of new resources. The boom years of the 1950s and 1960s had increased the money and the time available to young people. A new youth culture had emerged and there were new markets for youth consumer goods. Although Tarrow does not stress this directly, we can see in this process the preconditions for a stronger sense of generational awareness.

A second new feature of the 1960s was the link between the expansion of media communications, and particularly of television news reporting and decentralised forms of protest and organisation. As Tarrow says, 'If movements could transmit their messages to millions of people across the airwaves, encouraging some to follow their example and others to take sympathetic notice of their claims, it was possible to create a movement without incurring the costs of building and maintaining mass organisations' (1998: 131). This thinking influenced iconic and symbolic actions such as the Yippies showering the New York stock exchange with dollar bills in 1968, and later, other spectacular actions by Greenpeace or the banner that US Earth First! threw over the side of the Grand Canyon dam in 1981 to make it appear that it had a crack running down it, as well as a myriad of other similar actions.

The New Left is often seen as a precursor of contemporary green politics (Finger 1992; Markovits and Gorski 1993: 115). In several green movements the continuity thesis is unarguably justified simply because of the crucial role of activists from the New Left of the 1960s in the formation of new more radical environmental movements, and later green parties. The story of the formation of NOAH in Denmark is a good example: In 1969 the annual meeting of a natural history society was taking place at Copenhagen University when a group of students came in and locked the doors. One of the students recounts the events:

> We locked them all in. We were about twenty people. After we had locked the doors, we cut off the ventilation and started to poison them. It was pretty violent. We got up on the stage and talked about air pollution. We burnt garbage and tobacco in large quantities. We poured waste water from a nearby factory in an aquarium with goldfish who slowly died. On the side walls we showed films about cancer and pollution and we had a loudspeaker with a traffic alarm blasting. We sprayed water in the audience from Emdrup lake. And we had taken along a wild duck which we covered with oil. 'Come and save it', we screamed. 'You talk about pollution. Why don't you do anything about it?' finally we cut off its head to end its suffering and we walked down along the first row of chairs so that all who were sitting there got blood on their clothes. After an hour we opened the doors and said that we wanted to start an environmental movement and that the founding meeting was being held in the next room.
>
> (quoted in Eyerman and Jamison 1990: 66)

While most new environmental groups did not have as dramatic a start as NOAH they did to varying degrees mark a break with existing environmentalism in their anti-authoritarian ethos, in the political character of their analysis of the causes of environmental degradation, their use of protest and their vision of a new kind of society. There were also more conservative currents among the new environmental groups, although these were much less influential. For instance, in Britain, People, founded in 1973, which later became the Ecology Party and then the Green Party, was inspired by the more survivalist writings of Edward Goldsmith. Goldsmith's authoritarian views on questions such as the family and immigration were challenged by leftist activists within People, but it was not until the early 1980s when membership expanded beyond a few hundred, that a clear counter-cultural and anarchist-influenced wing of the party emerged (Rüdig and Lowe 1986: 275; Wall 1993: 40). As well as the new formal organisations in most countries there were also networks of radical environmentalists without any formal organisation, developing alternative technology, establishing green communes, taking part in local environmental campaigns and later in the 1970s, taking part in non-violent direct action against nuclear energy.

In the USA New Left activists took up environmental themes, relating pollution, urban sprawl, traffic and noise to the effects of capitalism on the environment. Pollution was linked to the search of industry for profit and to pressures to consume more. The term ' ecocide' was coined with reference to the use of napalm, used in Vietnam to kill people and the environment, and developed for profit by Dow chemical, one of the firms seen as important in the 'military-industrial complex'. At Berkeley in 1969 a squatted People's Park was created to prevent the bulldozing of a community garden in order to build a university car park. 'Hundreds of young people planted seeds, trees and sod and constructed a swing set, tables and benches. The land was declared "liberated" for the environment' (Gottleib 1993: 102). Gary Snyder, countercultural poet, declared it 'a guerrilla strike on behalf of the "non-negotiable" demands of the Earth and argued that trees were like other exploited minorities, such as blacks, Vietnamese and hippies' (ibid.). When California Governor Ronald Reagan sent in the police to clear the site one demonstrator was killed and hundreds were injured. In London 31 years later, this tradition was revived in the symbolic guerrilla gardening action in Parliament Square on Mayday 2000, in which Parliament Square was dug up and planted with seeds. In the intervening years, however, the idea of liberating defiled space for the environment had become part of the routine practice of the environment movement, with city community gardens, permaculture schemes and many other initiatives in urban ecology.

In 1970 in the USA and in Europe New Left protests began to decline. In the USA the death of six student demonstrators at Jackson State University and Kent State University polarised the anti-war movement and led a small number to turn to violence. In France the collapse of the protests in May 1968 and de Gaulle's overwhelming electoral victory in June 1968 destroyed what had briefly appeared as a possible revolutionary movement. In Italy polarisation followed the 'hot autumn' of 1969 and the bombing campaigns of the far right, aided by the Italian secret services, prompted a turn to political violence and the growth of left-wing terror groups such as the Red Brigades. In Germany too some leftists turned to political violence, partly provoked by repressive and violent police action, and gradually, as their activities forced them underground, to a terror campaign of kidnappings and assassinations. It was only a small number, however, that took this route.

For some New Left activists, the lack of strong organisation and the failure to link effectively with the working class explained the failures of the New Left. As a result they turned to more orthodox forms of Marxism, such as the *K-Gruppen* in Germany or the Ligue Communiste Revolutionaire in France. Others saw possibilities to pursue a more radical reform agenda within the mainstream parties of the left. In Germany, the call by SPD Chancellor Willy Brandt, 'to dare more democracy' led to a large growth in membership of the Young Socialists. The French socialist party also sought to position itself as a party of the New Left in the 1970s, partly as a matter of political strategy, since it helped to distinguish it from the Communist Party, which was still dominant on the left in the early 1970s. Many political elites regarded the extra-parliamentary challenge of the New Left as over by the early 1970s.

Longitudinal research on New Left activists in several countries[12] contradicts the widespread view that activists gave up on their ideals. While many were disappointed, those who had been most active remained committed to a radical politics and many became involved in the social movements that were inspired by the New Left. In that sense, they were the beginning of the political generation from among which the green movement emerged.

The growth of a green movement

In the early 1970s the growth of a green movement was by no means assured. Authoritarian survivalist thinking dominated much of the public debate about environmental crisis and it was only by the end of this decade that the themes of grassroots democracy and global egalitarianism were clearly established alongside ecology as the framework of the green identity. During the 1970s the green movement was shaped by the growth of an alternative milieu, in which the green movement overlapped with other social movements, particularly feminism, anarchism and libertarian socialism. This milieu was linked by the aim of seeking to build communities in which it would be possible to live in a way that broke with the conventions of mainstream society, based on challenging hierarchy and integrating personal lives and political change. It was the influence of this culture along with the heritage of the New Left, which marginalised survivalist currents in the environmental movement. New environmental movement organisations were formed in most countries, although their strategy and ethos varied greatly, they were more confrontational and protest-oriented than the existing conservation groups. This decade also saw the first tentative efforts in elections, although in all countries the movement remained electorally marginal. The principal issue for the developing green movement in this decade was the conflict over nuclear energy, and it was the alliances made between groups involved in this movement that shaped the nature of the green movement during the 1980s. At the beginning of the 1980s peace issues replaced nuclear energy as the main focus of protest for the greens and their supporters in many countries, and in many countries environmental protests declined. This was also the decade when green parties were formed in most countries and had achieved electoral representation in most of western Europe by the end of the decade. Many EMOs gained increased resources during this decade, and membership size increased rapidly at the end of the 1980s as new evidence of global environmental problems emerged. As a consequence they were able to expand their organisations and improve their status as counter-experts. In the USA new grassroots networks began to form on issues such as toxic waste and the threats to human health, which had largely been ignored by mainstream EMOs. In Australia there was a wave of environmental direct action, which succeeded in making threats to wilderness sufficiently controversial as to affect the outcome of the national election in 1983 (helping Labor to victory). In the USA Earth First! was formed in 1980 as a new network using direct action to protect wilderness areas. The defence of wilderness, as earlier in the twentieth century, differentiated part of New World environmentalism from that in western Europe.

During the 1990s the differentiation within the green movement increased further. The 1992 Rio Earth Summit legitimised environmental issues as questions of high politics, and this meant that greens were no longer distinctive in their recognition of environmental problems, although they continued to view them as more serious and with wider social and economic implications than other political groups. Greens were in national governments in five western European states at the end of the decade, albeit as small minority parties, and in cases where they had previously been weak, such as Britain and Australia they were able to elect members to various assemblies. In the USA the campaign of Ralph Nader in 2000 succeeded in making the greens the principal party to the left of the Democrats, gaining enough votes (3%) to be blamed for preventing Gore from beating Bush. There was a sustained wave of environmental direct action protests in Britain in the 1990s and greens across the world joined hybrid protest coalitions to oppose the dominance of neo-liberal principles in international economic institutions, which appeared to threaten environmental regulation, and entrench the power of western corporations. The growth of new international bodies also led to an increased focus among major EMOs on international lobbying, including establishing more transnational networks which included EMOs from the south.

Explaining the development of the green movement

There is so much variety within the green movement both in types of groups and national circumstances that any attempt to provide a general explanation of the factors shaping the movement will inevitably be over-generalised. More of the detail on specific cases can be found in the chapters on the different types of environmental groups. In the section below the development of the French and German green movements is examined. This will be followed by a more general explanation of factors that explain how the green movement developed differently in different countries.

Germany

In Germany the green movement has been marked by three main features: the strength of the alternative left milieu in the wake of the decline of the New Left extra-parliamentary opposition in the late 1960s; the alienation of large parts of the middle-class left from the SPD in the 1970s and strength of green organisations and continued high levels of environmental protest in the 1980s and 1990s.

The alternative milieu was particularly strong in Germany, and by the mid-1970s there were significant alternative left ghettos in districts of the major cities such as Kreuzberg and Schönberg in Berlin Hockenheim in Frankfurt and St Pauli in Hamburg. The self-sufficient character of this culture is captured by the following description:

> The average *Stadtindianer* [urban Indian] wakes up in a commune, buys his rolls at a collective bakery round the corner, and gets his muesli from a macrobiotic shop. While having his breakfast he reads his Pflasterstrand, Info-Bug, zitty [papers of the counter-culture]; he then goes – provided he is not a 'zerowork' supporter – to work in a self-organised business or in some sort of alternative project; every five days he has to take care of children in a *Kinderladen*, his 2CV is only repaired at a left-wing collective garage ... Nearly everything is covered as far as daily life is concerned. At the same time communication is very intense compared with that between the average West Berlin citizens. *Stadtindianer* and *Spontis* talk to this sort of person only when they have to, for instance with policemen during a police raid. In West Berlin and in Frankfurt there are people belonging to this scene who are very proud of the fact that they have not spoken a word to anybody outside their scene for two and a half years.
>
> (quoted in Wasmund 1986: 206)

Leftist groups such as the 'Spontis' along with the growing women's movement played an important part in the anti-nuclear power movement. They were distinct from the *K-Gruppen*, who favoured more hierarchical organisations and a more orthodox Marxism, often based on encouraging militancy among factory workers. Both groups, however, joined anti-nuclear demonstrations at Brokdorf, Kalkar and other sites in the late 1970s. Although the catalyst for anti-nuclear protest had been the peaceful occupation of the site of a proposed reactor at Whyl in south-western Germany in 1975, in which locals and students from Freiburg had co-operated, the presence of the *K-Gruppen* and *Autonomen* (urban anarchists often involved in the squatting movement) meant that at many later demonstrations there were pitched battles with the police. The violence caused divisions between radicals and more mainstream environmental groups. However, by the late 1970s both the alternative left and the *K-Gruppen* faced declining activism and a crisis of orientation.

The German student movement in the 1960s was marked out from others first by the vehemence of the generational conflict in Germany, which centred on criticism of those responsible for Nazism and on a perceived failure to remove legacies of Nazism from post-war Germany, and this partly explains the strength of the alternative culture in Germany. The link with middle-class environmentalism through the anti-nuclear movement, and the search for a new strategy meant that many activists from the *K-Gruppen* and alternative left scene were involved in the formation of Die Grünen in 1980. Through involvement with Die Grünen and through more state funding for alternative projects in the 1980s and 1990s the ghetto culture of the alternative left was broken down to some extent by a process of partial institutionalisation.

The alienation of the German extra-parliamentary left from the SPD is important in explaining the strength of leftism in the German green movement. The actions of the SPD in the mid-1970s pushed large parts of the young and middle-class left to look for a new political movement. One element in this was the treatment of the youth wing of the SPD – the *Jungsozialisten* or *Jusos*. Some on the left, inspired by SPD leader Willy Brandt's call to 'dare more democracy' had sought to pursue New Left ideas within this group. The failure of this project was marked by the expulsion of its leaders from the SPD, and the turn of the SPD to the right under the new leader Helmut Schmidt in 1974 (Padgett and Paterson 1991: 51–3).

Even before Schmidt replaced Willy Brandt the 'reform euphoria' which had arisen in response to Brandt's apparent opening of the SPD to the themes of the Ausserparlamentarische-Öpposition (APÖ) had abated, but Schmidt and the right-wing faction of the SPD made a number of decisions which deepened the gulf between the SPD and large parts of the young middle-class left. The first was the ban on radicals in public service, which created an image of the SPD as undemocratic and a threat to civil liberties.[13] The second was the result of the debate in the party over nuclear energy in 1976. The victory of the advocates of nuclear energy was partly due to pressure from the coalition partners in the Free Democratic Party (FDP) and from the trade unions but as anti-nuclear opposition spread it placed the government in clear confrontation with the alternative left and with the growing network of grassroots environmental campaigns or 'citizens's initiatives'. Third, was the response to the terrorism of the *Röte Armee Fraktion* (RAF) which reached a height in the 'hot autumn' of 1977. The attempts to crush terrorism were particularly directed at the alternative milieu in which the terrorists had their base. Few in this milieu supported terrorism as a strategy but the government's repressive reaction helped to further convince them that they were all seen as a subversive and dangerous part of society.[14]

Citizens' initiatives, (*Bürgerinitiativen*, BIs) had begun to emerge in the 1960s and had been sponsored by the SPD as part of its effort to expand citizen participation in politics. The first BIs focused on issues such as local transport and playgrounds. From the 1970s members of the *Jusos* and *K-Gruppen* were also involved in such campaigns (Fogt 1989: 93; Markovits and Gorski 1993: 96). A co-ordinating body, the *Bundesverband Bürgerinitiativen Umweltschutz* (BBU) was established in 1972 which linked many of the environmental BIs. The 1,000 groups (approximately 300,000 members) linked by the BBU (Langguth 1986: 7) largely shared a commitment to grassroots organisation as an end in itself and increasingly linked their environmental campaigns to a broader social critique, highlighting the political and social obstacles to ecological reforms (Rucht 1989: 78). The BBU therefore represented the most social movement-oriented part of the environmental movement. As well as groups affiliated to the BBU still larger numbers of anti-nuclear groups were organised at local levels.

The most important of the more conventional environmental movement organisations was the *Bund Für Umwelt und Naturschutz Deutschland* (BUND). This occupied an important space between the resolutely anti-political conservation groups such as the *Deutscher Naturschutzring* (DNR) and the WWF and the more radical groups. The willingness of the BUND to support the aims of protests helped in mobilising greater protest than in France, where there was no intermediary between the conservation organisations and the political ecology sector (Rucht 1989). For the BBU the victory of the SPD leadership over anti-nuclear elements in the party in the 1976 debate about nuclear power was a turning point in its strategy. It was most clearly after this that the ecology movement – especially the BBU – developed a broader critical stance demanding wider social change to make ecological changes work.

The influence of contemporaneous events in France was also important in the BBU's shift. The success of the French greens in the 1977 municipal elections influenced the decision of the BBU to propose a green electoral list for the European elections in 1979. Before 1979, however, the BBU had signalled the growing independence and breadth of ecological politics in Germany by publishing a document that argued for the wholesale re-orientation of the economy on ecological principles (Langguth 1986: 8). Yet, the nascent green movement was still very heterogeneous with major regional differences.

These ideological differences, which covered a spectrum from the Marxist *K-Gruppen* to ecological conservatism, were so substantial that the prospects for the formation of a green party seemed poor in 1979. In fact, as will be seen in Chapter Four, the coalition between BBU activists and the various parts of the alternative left which formed Die Grünen was the only basis for a viable green party in Germany. The ecological conservatives had no roots in extra-party social movements and moreover the citizens' intiatives and the extra-parliamentary left were linked by features that the eco-conservatives lacked. Both had sought to develop the experiment in participatory democracy as an end in itself that had begun in the New Left (Mewes 1997: 32). If this is one dimension of a shared collective learning experience a further one which shows the influence of political processes was the effect of successive steps taken by the SPD after 1974 which alienated larger sections of both the alternative milieu and the citizens' initiatives. The gulf that emerged between the SPD and various parts of the middle-class left contrasts markedly with the success with which the French Socialist Party (PS) (Ladrech 1991) or the Dutch Labour Party (PvdA) (Lucardie 1980) were able to accomodate New Left demands.

In Germany, the green movement in the 1980s was characterised by three features. First was the growth in power of Die Grünen, The party succeeded in being elected to the Bundestag in 1983 and improved its position in 1987. However, Die Grünen remained divided between ideological sub-groups, and particularly over how to practice grassroots democracy in a political party and over whether it was justifiable to enter government as a minority party, initially at regional level, and then at national level, with all the compromises that this entailed. German political institutions served to provide opportunities and constraints that shaped the electoral fortunes of Die Grünen. Among the opportunities were: an electoral system based on proportional representation; strong regional parliaments, which allowed the party to build a regional vote and gain elected representatives before entering the national parliament, and to maintain a presence in political life when the western greens failed in national elections in 1990; and state funding for political parties. Among the constraints were the 5 per cent threshold for electoral representation, which provided a pressure for disparate groups to work together electorally: failure to surmount this has excluded greens from regional and national parliaments on occasion.

A second feature has been the vitality of environmental protest and the strength of environmental organisations. Contrary to the perceptions of many commentators, analysis of levels of protest in the 1980s and 1990s (Rucht and Roose 1999a; 1999b) showed that there had been no decline, although there had been fluctuations and changes in the types of protest. Protest against nuclear energy remained particularly high in comparison to other countries where they declined substantially in the 1980s. In Germany continued controversy over nuclear reprocessing plants such as Wackersdorf in Bavaria in the mid-1980s and the transport of nuclear waste in the 1990s led to mass confrontations at Gorleben and Aarhuus in 1997 and 2001. The networks of campaigners against nuclear energy remain decentralised and have not been institutionalised to the same degree as the mainstream EMOs.

The large support for EMOs is reflected in their membership and through the widespread participation in protest of many sectors of society. This, combined with the electoral strength of Die Grünen, put German elites under more pressure than in most countries to respond to environmental issues. Hence, environmental policy and debate about the need for more far-reaching changes went further in Germany than almost any other country. Only Denmark and the Netherlands, each also with large and active green movements, are comparable, although in neither of the latter countries has a green party played as strong a role. In all three countries the more challenging forms of environmentalism associated with the movement in the 1970s and 1980s have been argued to be weaker now (Blühdorn 2001; van der Heijden 2000; Jamison, Eyerman and Cramer 1990) due to the opening up of political elites to environmental policy of a more reformist kind. The degree of deradicalisation of the movement will be examined further in subsequent chapters.

A third factor in Germany was the tradition in policy-making which required that policies be justified in terms of broad political principles (Weale 1992). This meant that there were more opportunities for environmental groups to appeal to ideas such as the precautionary principle, particularly in the courts, as a means of challenging environmental policy. Second, the corporatist traditions of policy-making between government, business and trade unions were also suited to the inclusion of moderate environmentalism, as long as this did not threaten the core interests of any of these groups. The discourse of ecological modernisation (discussed in Chapter Three) provided a means by which this could occur, but which became increasingly dissatisfying to the more radical EMOs such as the BUND and BBU, as its limits became clear during the 1990s.

France

In France, as in Germany, the alternative left was only one of three currents that emerged from the New Left after May 1968, the others being the 'gauchistes' including Maoists and Trotskyist groups such as the Ligue Communiste Revolutionaire (LCR) and the party centred New Left, embracing both the PSU and the New Left wing of the Socialist Party (PS). As in Germany the relationship between them affected the growth of green politics in the 1970s.

The French alternative milieu never established the same scale of alternative projects and institutions as in Germany, but it did achieve a degree of cohesion and political impact through several confrontations with the government. These allowed the movement to develop networks through which groups became linked at the local level. However, rather than expanding, these networks shrank as divisions over strategy increased after 1977.

The beginnings of the alternative milieu lay in an exodus of urban leftists to rural communes in the late 1960s. For this group, according to Prendiville (1993a: 42), the idea of social experimentation was a break with French traditions of political radicalism. The idea of a rapid and cataclysmic social change that had animated May 1968 was rejected for the idea that the alternative society could be created in practical social experimentation from the base upwards. This part of the left earned the epithet 'baba-cool' which referred to hippy style, but also to other aspects of its lifestyle such as wholefoods and the attempt to put into practice the utopian ideals of 1968. The attitude to the state remained anarchist in inspiration. In 1975, for instance, *La Gueule Ouverte* was calling for 'the disappearance of parties, of the vote, of the delegation of power and hierarchies, and thus, of the state'.

The occupation of the site of a proposed extension to the military base at Larzac in the Midi, begun in November 1971 and continued until the abandonment of the project by the government in 1981, was the first symbolic focus of this alternative politics. The protests at Larzac were motivated by concerns that were to become central to greens – the defence of the lifestyle of peasant farmers against outside interference and anti-militarism. The occupiers included the local peasant farmers who faced eviction if the extension went ahead, but was led by leftist radicals including José Bové, later known for his campaign against neo-liberal globalisation. The first demonstration in 1971 attracted 5,000 participants – but the spread of 'Larzac committees' across France meant that the following year 20,000 came. The Larzac campaign was thus the first political manifestation of the grassroots 'baba-cool' left, but it also had the support of the socialist theoreticians of the PSU and the more traditional revolutionary politics of the Trotskyists (Bennahmias and Roche 1992: 24; 33).

Ecology became a political issue in France earlier than elsewhere and although in general environmental consciousness has been lower in France than in other northern European countries from 1970–1976 it was above the European average (Hofrichter and Reif 1990). The first mass membership national conservation organisation, the *Fédération française des sociétés pour la protection de la nature* (FFSPN), was founded in 1968 from existing, mainly local, groups. Like many conservation groups in other countries it was regionally based and did not exert political pressure on central government. The conservationist and non-political character of this organisation meant that it had no links with those parts of the French alternative movement that saw ecological problems as requiring major social and political change (Rucht 1989: 80–3). When political ecology groups did emerge they adopted the organisational principles of the post-1968 alternative milieu. The French branch of Friends of the Earth, *Les amis de la terre* (LADT), was formed in 1972. In the 1970s LADT was the only environmental organisation with branches nation-wide. In contrast to the more centralised and formal organisation of the British Friends of the Earth (FOE) LADT had no national office or organisation until 1977 and even then the centre had no powers over local groups and only the vaguest of statutes (Sainteny 1991: 16).

Further evidence of the early impact of ecological themes in France was the candidature of the agronomist René Dumont in the 1974 presidential elections. This election was the first in which the French Communist Party (PCF) and Socialist Party (PS) had agreed to support Mitterrand as a joint candidate and in these circumstances it puzzled some on the left, including some political ecologists, as to why Dumont should stand. The argument of the Dumont campaign was that the essential ideas of political ecologists were not being taken seriously by either the right or the left. Dumont's platform was radical because it presented issues such as the redistribution of wealth within the context of restricted economic growth. He also related French lifestyles to the exploitation of the Third World and had detailed plans for reducing car use, air and water pollution and conserving energy – criticising the unaccountability of the electricity authority *Électricité de France* (EDF) for choosing centralised nuclear energy instead of decentralised renewable energy (Gurin 1979: 150). The early themes of a distinctive political ecology platform had already been established over the two years preceding the Dumont campaign and although no permanent green party emerged for another ten years, this marked the beginning of the electoral campaigns of the French greens. Although Dumont gained only just over 1% of the vote, one important consequence of his campaign was its effect on relations within the environmental sector. It hardened the split between political ecologists and the conservationists who saw Dumont's campaign as too leftist. It also raised the problem of the electoral relationship with the left for the first time.[15]

The anti-nuclear movement was as significant and as important for the future of the greens as it was in West Germany. The first demonstrations against nuclear energy in France preceded Dumont's presidential candidature, but it was only after 1974 that the nuclear question came to dominate ecological politics. In this year the 'Plan Messmer' was announced under which France was to expand the number of its nuclear plants rapidly – placing its faith in nationally produced nuclear energy in order to overcome the dangers of the dependence on energy imports revealed by the oil crisis of 1973.

During the 1970s the greens were wholly absorbed in the anti-nuclear movement. Yet, this movement remained very heterogeneous and was reluctant to
organise at the national level. Only LADT had a national organisation and
this was only a framework for its largely autonomous local groups (of which
there were around 100 in the late 1970s) (Rucht 1989: 81). A decentralised
form of organisation was seen as an essential alternative to the centralised
and restrictive character of decision-making in the French state. However,
whereas in Germany similar decentralisation proved more effective, since
there were more local and regional legal and legislative opportunities, in
France it made it more difficult to place effective pressure on the government.
With the exception of Plogoff in Brittany there was relatively little support
from the local inhabitants at nuclear sites and this meant that the movement
concentrated upon mobilising its supporters from all over France in confrontational demonstrations. In contrast to West Germany there was no upsurge of
politicised middle-class environmentalism with which the New Left could link
itself. By the time that the anti-nuclear issue had become nationally significant, political ecology in general was already the preserve of the radicals of
the alternative milieu.

Although the movement had grown rapidly there was no real strategic
consensus. One problem was the unrepresentative nature of such national gatherings as occurred. Although there were several hundred anti-nuclear and
Larzac committees in 1975 only 35 groups attended the annual *Assises*.
Communication between groups relied heavily on *La Gueule Ouverte*
(Bennahmias and Roche 1992: 49). The main issue debated was the attitude to
violence at demonstrations, such as the major confontation at Malville in 1977
and not, as Touraine, Hegedus and Wieviorka stressed, what the movement
aimed to achieve and against which opponents (1983: 27). In France the
strategic obstacles facing anti-nuclear movements were especially great since in
contrast to Germany there were no effective judicial or administrative avenues
of influence. For instance, although local inquiries were statutory for new
nuclear plants they were limited to the issue of compensation (Robinson and
Boyle 1987: 79).

This is indicative of a more general problem facing French social movements.
The dominance of the strong central executive under the Fifth Republic has left
radical outsider groups few opportunities to influence policy between elections.[16]
When protests failed it was rational for supporters of the social movements to
attempt to change the government since in the French political system there was
no other obvious avenue of influence. That the PS was the primary beneficiary
of this development reflected a combination of effective strategy and the logic of
the political system. The two-round majoritarian election system made it almost
impossible for a new party to break into the political system, without forming an
alliance with an existing party. Moreover, at this stage there was little support for
the idea of forming a permanent green party. The PS had made promises to
various social movements that helped to persuade many among their supporters
that a left government could make a real difference. The PS's victory in the

national elections of 1981 produced two immediate benefits for the greens: the cancellations of the Larzac project and of the proposed nuclear plant at Plogoff in Brittany, where local resistance had been much stronger than elsewhere – fuelled by Breton regionalism. In taking these steps the socialist government removed the two most significant locations for the protest activities of the alternative milieu – leaving them only with abstract issues on which to seek to mobilise further action. Moreover the promised moratorium on nuclear energy was never instituted.

The PS's victory effectively demobilised the social movements leaving the latter with a choice between absorption into the establishment or counter-cultural retreat. As the PS took power it metamorphosed into a more traditional social democratic party. As Birnbaum (1983: 187) noted of the first years of the Mitterrand government: 'the government governs and the party organises action. One finds oneself very far from the self-management utopias and collective mobilisations'.

In France the green movement remained small whether measured in membership of EMOs or green parties, and with limited capacity to mobilise, with low levels of environmental protest in the 1990s (Fillieule and Ferrier 1999). Yet, Les Verts, the French green party, which was formed as a formal party in 1984, made a major electoral breakthrough in 1989 by winning many seats in municipal elections and gaining 11 per cent of the vote in European elections. Although the party suffered internal crises over strategy in the early 1990s and had to deal with electoral competition from a technocratic and non-green environmental group led by Brice Lalonde, an ex-leader of LADT, it survived and was able to form an effective electoral alliance with the socialist party, entering a coalition government ('la gauche plurielle') with the PS, the Communists and two other left parties in 1997.

Political ecology has remained principally an electoral phenomenon in France, with relatively less concern among French greens with the wider alternative lifestyle that means being a good green in northern European countries, North America and Australia (Faucher 1999). Environmental protest has declined, and tends to be less confrontational than is the norm for protest in France (Fillieule and Ferrier 1999). Membership of EMOs is very low in comparison with other countries in western Europe and so the institutionalisation of the movement has been mainly through the participation of Les Verts in government.

As with Die Grünen, political institutions shaped the development of Les Verts. The two-round majority election system meant that the party was unable to win any seats in the national parliament and its candidates in presidential elections used it as a platform to advertise green ideas. From 1986 to 1993 the party placed emphasis on preserving its autonomy from the PS. While this brought some success, failure to overtake the PS in the 1993 election forced a reconsideration and an acceptance of an alliance with the PS. This caused some divisions in the party, but allowed it to overcome a major crisis and play an increased role in national political life.

A comparative model

Four factors seem to be consistently relevant to explanations of how the green movement developed in Germany and France. These are:

- the institutional characteristics of the political system
- the strategic choices made by the new EMOs
- the nature of the major parties of the Left
- and the nature of the alternative milieu.

These factors also explain many of the differences between green movements in different countries. Moreover, these factors interact and cannot be easily separated. For instance, the greater openness of the PS to new social movement themes and the barriers to new parties posed by the electoral system as well as the ideological commitments to biodegradable forms of organisation all contributed to the reluctance to form a green party in France. Also, these factors need to be seen as dynamic, since the strategies of movement activists and their allies or opponents alter as a result of experience. Thus the resistance to forming a green party in France was reduced after the disappointment of the experience of socialist government and as a result of the demobilisation of anti-nuclear protest in the early 1980s.

Space does not allow for more than a brief discussion of how this model might apply to green movements in other countries. Moreover, green groups are affected differently by these factors. Among the institutional features of the system, electoral systems and the structure of government are most relevant for green parties, policing strategies and levels of welfare benefits are most relevant for direct action networks. These represent the most important political opportunities and constraints on movements. Similarly, the nature of the major party of the left affects the whole movement, but in different ways. Greens have at times been tempted to rally behind what seem to be the greater chances of progressive change represented by left parties, as in France in the late 1970s, and green parties have formed alliances almost exclusively with parties of the left. However, the weaknesses of the major party of the left also provoke much criticism and push greens to develop a separate political identity, as occurred relatively early in Germany. The Green presidential candidature of Ralph Nader was much criticised by supporters of the unsuccessful Democrat Al Gore, including many representatives of US EMOs. The latter argued that Gore had a strong record on environmental issues and Nader had stolen votes from Gore. Nader's reponse was to emphasise the failure of Gore to address the central issues in the green programme, including US over-consumption. Essentially, this was an argument between greens committed to a wide-ranging radical agenda, and environmentalists, stressing incremental reforms.

Socialist and Labour parties have made little move towards the more radical anti-productivist position of greens and are also often seen by greens as weak in practice on egalitarianism and democratisation. In Britain the alternative milieu had largely ignored the Labour Party in the 1970s, particularly when it was in government but during the 1980s the Labour Party underwent a short-term radicalisation in which it appeared open to influence from social movement activists. Activists from the women's movements, anti-racist movements, peace movements and radical community politics joined the party or worked with it, particularly in progressive initiatives by Labour-led local governments (Wainwright 1987). It was only in the late 1980s and 1990s as Labour gradually retreated from radical commitments that there was a reaction against the party. By then, however, a new generation of green activists had emerged, committed to direct action.

In Australia the openness of the Labor Party to moderate environmentalism, more oriented towards wilderness protection, than structural changes in the social or economic order, made it harder for a more autonomous ecological radicalism to develop. In Italy, political violence polarised politics in the 1970s and meant that ecological politics was subordinated to the historic conflict between Marxism and Christian Democracy. It was principally activists from the New Left, disillusioned with the suffocating character of this polarisation, who saw in political ecology a means of developing a new kind of progressive and egalitarian politics (Diani 1990). But, this could only develop when the crisis of political violence abated in the 1980s. Moreover, the greens were unable to develop a strong electoral base. Their emphasis on a new politics in which party organisations were to be weak and on opposition to political corruption, became less distinctive after the political crisis of the early 1990s that led to the demise of the Christian Democrats and the Socialist Party and the ideological transformation of the Communist Party into a social democratic party. The greens were therefore absorbed into the coalition of the left, which, under the new electoral system, provided their only realistic prospect of electoral representation. The reduction of ideological polarisation meant that there was also more space for issue politics, leading to greater attention by EMOs to policy and expertise, and a rise of local environmental campaigns in which, in contrast to past traditions, parties played little part (della Porta and Andretta, 2000; 2001).

In the US, it is the weakness of the left that has most influenced the greens. Although the alternative milieu has fed progressive campaigns such as the anti-nuclear direct action of the 1970s and Earth First! it was only in the late 1990s that networks across parts of the US left, including greens, were formed in order to protest against neo-liberal forms of gobalisation. This in turn gave impetus to the green presidential campaign of Ralph Nader. Prior to this the greens and the left remained largely separate networks, with the US left dominated by social democrats and orthodox Marxists with little interest in green concerns (Ely 1997: 204). Nevertheless, the institutional barriers to green electoral breakthroughs remain major. US EMOs remain split between non-radical professionalised groups oriented to legal action and lobbying in Washington and more grassroots-oriented anti-toxics and environmental justice networks.

The strategic choices made by new EMOs in the 1970s were important in shaping the character of green movements nationally. One important factor in this was the responsiveness of political elites to the new environmentalism. In some countries, such as Sweden (Jamison, Eyerman and Cramer 1990) and Norway (Hunold and Dryzek 2001) radical environmentalism was effectively forestalled as governments provided more opportunities to influence policy and funding in exchange for a co-operative relationship with the new environmental movement. This was conditional on the latter remaining moderate and led to quiescent environmentalism and only a weak green movement. While the logic of these choices can be explained by the degree of inclusiveness of state structures, they were also influenced by discursive considerations in which the ideological traditions of the left and the alternative milieu played a role. For instance, in Denmark EMOs such as NOAH and OOA favoured grassroots democracy and protest, and linked ecological themes to questions of democracy and social justice, much more so than new environmental groups in Sweden. Denmark had a strong new left, and other radical traditions such as in the education system, which differentiated it from Sweden. The leader of the smaller Swedish New Left criticised the Danish as 'petty bourgeois' because they were interested in practical experiments in building an alternative society. To the Swedes, these seemed a diversion from the major structural issues (Jamison, Eyerman and Cramer 1990). Thus we need to add the cultural push of existing traditions of radicalism to the pull of state structures in explaining the choices of environmentalists. Where the alternative milieu was strong and had broken with the major socialist party, alternative green radicalism was stronger, as in Germany, Denmark and the Netherlands.

The concept of an alternative milieu is intended to link some of the strengths of Alberto Melucci's theory of social movement identity with the analysis of the cultural and institutional practices that mark out alternative culture from the mainstream. Melucci concentrates on the cultural and expressive dimension of social movements. For Melucci social movements challenge dominant cultural codes by asking critical questions and suggesting alternatives. By doing so they make forms of power 'visible' (Melucci 1989: 70). Public events and protest may be a part of what a movement does, but equally important are attempts to live differently which may not produce formal organisations. Movements are therefore as much identified through cultural symbols and latent networks as through formal organisations and protests. 'Political activity is but one of a range of practices which question the rationality of states, markets and subjectivity in diverse ways' (Jowers et al. 1999: 100).

Using the term milieu helps to bring out the importance of a certain kind of semi-institutionalised community for the development of the green movement. Melucci stresses the cultural importance of alternative lifestyles as a means of opposition in everyday life to the dominant values in society. Yet, such lifestyles also have an institutional basis which he barely acknowledges. The 'institutions' of the alternative sphere are the co-ops, LETS, community gardens and arts centres, self-help projects, communes, cafes, bookshops, alternative health projects, analysts and other personal services and in particular its media. These exist in most northern European countries and Australia and North America.

They form a milieu because as institutions they share similar cultural assumptions and collectively they have persisted over several decades.

In research on social movements the alternative milieu is much less researched than the overtly political and public activity of movements. One reason for this is its diffuseness and lack of clear boundaries. Since it is a politics based on alternative cultural practices it exists mainly in local activist communities, which themselves may be sub-divided socially and politically. For instance, research on local activist communities in Britain shows that green activism overlaps significantly with a wider alternative milieu, in which feminism, animal rights and radical forms of spirituality such as paganism, and a non-doctrinaire form of anarchism co-exist (Doherty, Plows and Wall 2001). Epstein noted a similar broad range of traditions in the US non-violent direct action movement (1991). The specificity of the green movement is less clear within the alternative milieu than it is when analysing formal organisations.

Collective learning

The concept of collective learning appears more important the more that the evidence suggests that the strategic choices of greens are shaped by a variety of ideological and cultural factors as well as the constraints imposed by political institutions. Even those who are most confident in the ability of political opportunity structures to explain the strategic choices of social movements are keen to stress that these models are not determinist. The strategy of groups cannot be explained only by analysis of the political context. As Kriesi *et al.* (1995) argued what counts as an opportunity varies according to the nature of the movements' identity. Groups whose strategy is focused on changing particular policies are more likely to take a pragmatic attitude to alliances with other political groups or accept compromises with government. For them, the political system appears open if governments involve them in policy-making. This would apply best to the major environmental movement organisations such as Friends of the Earth. These groups are more likely to be prepared to work even with governments that they see as anti-environmental, if by doing so they can produce some positive outcomes on a single issue, such as increasing the amount of renewable energy generated. However, even these groups will retain a critical stance towards overall government policy in such cases and refuse to work with or accept sponsorship from certain businesses such as oil companies or arms traders. More counter-cultural groups, such as those involved in non-violent direct action (NVDA), tend to view such alliances as counter-productive and neither seek nor are offered alliances with mainstream groups.

Dryzek and Hunold (2001), referring to the rise of direct action protests in the UK and environmental justice movements in the US, point out that the 'resurgence of environmental activism in the structurally inhospitable contexts of the USA and the UK suggests that reflexively aware activists and movement may be able to tweak the historical logic' (2001: 23). In effect, their argument emphasises the collective learning experience as an important feature of green praxis.

Greens, while affected by structural constraints, have been able to make choices that showed that they were also effective analysts of their own movement. As Dryzek and Hunold say, 'These developments themselves indicate the waning of structural determinism: state structure may be less important than it once was in determining the form taken by social movements. This in turn indicates height-ened reflexivity. That is, activists can look to the history of inclusion and see what it has and, more to the point, has not accomplished, and draw lessons for their own orientation to the state' (ibid.).

Transnational learning

The collective learning of the green political generation also occurs through the kinds of transnational influences and networks that characterised the movement from the outset. As noted above, major texts such as Carson's *Silent Spring* (1962) and the Club of Rome's report were translated into many languages, and their ideas were also reproduced in domestic debates. The alternative culture of the green movement has also been marked by transnational diffusion. As communes and other experimental forms of community developed in the 1970s there was much inter-change between countries. Welsh comments:

> This dispersed, amorphous body of alternative experiments was integrated by traditional means such as network newsletters, person/worker exchanges, participation in network events such as potlatches, demonstrations and occu-pations. Rather than a myriad isolated local experiments there was an intensely dialogic network which extended throughout mainland Europe.
>
> (2000: 159)

The New York Times estimated that there were over 2,000 communes in 34 states of the US by December 1970, though participants suggested that the real figure was much higher (Gottleib 1993: 102). There was a similar growth in communes in most of northern Europe. Many were only short-lived, based on ad hoc experiments, and a response to the crisis of repression that faced both the hippy counter-culture and the New Left at the end of the 1960s. However, others survived, though often in less radical form than when initiated. Some, such as Larzac, became a focus of national campaigns with local groups organ-ised to take direct action in their defence. Others such as Christiana in Copenhagen a squatted barracks, which became a counter-cultural community that declared itself autonomous from the Danish state, were visited by alternative radicals from many countries.

Communes cannot be usefully assessed in relation to questions of success and failure, not least because they were so diverse in aims and form. By no means all were inspired by green ideas, although, particularly in the 1970s there was a growth of communes based on egalitarianism, self-reliance and living a less consumerist lifestyle. They were not able to transcend the constraints of existing social structures and the dominant culture. But rather than therefore writing them off as naive failures, they can better be understood as experiments in what Sargisson (2000) calls 'transgressive utopianism'. The central point about transgressive utopianism is that it is not based on 'utopia as a blueprint of perfection' (ibid.: 2). Instead, utopia is an attempt to find a space from which it is possible to think differently about the world. In order to do this it is necessary to transgress the norms and paradigms that constrain thinking. 'It is wild, unruly, rule-breaking thought that is politically driven and that expresses a profound discontent with the political present.' Also, 'It is, above all, resistant to political closure and it celebrates process over product' (ibid.: 3). Transgressive utopianism is not restricted to communes or intentional communities but is seen more widely in the extended forms of politics of the wider alternative milieu in the experiments with alternative forms of non-hierarchical organisation, concern with the constitutive work of community such as childcare and the culture of action.

Tactics and forms of organisation also spread transnationally as a result of monitoring movements in other countries. In 1975 British and American anti-nuclear activists visited the site occupation at Whyl, just as activists from other countries visited British anti-roads camps in the 1990s. British anti-nuclear activists were particularly influenced by the forms of action and organisation used by the US Clamshell Alliance at Seabrook in New Hampshire in the late 1970s. While the latter campaign was generally regarded as a failure, many of its practices could be seen as helping to build the capacity of the movement cross-nationally (Welsh 2000). For instance the practice of using a structure based on 'spokes' in which those involved were divided into small groups whose views then were carried by individual spokes to smaller groups, reappeared at many large direct action events, including the demonstrations at Seattle in 1999 and Prague in 2000. And this process of transnational diffusion built on earlier diffusion which helped to spread the New Left: research has shown the importance of personal ties between German and US activists in the spread of New Left ideas from the US to Germany (McAdam and Rucht 1993).

Since the green movement is so heterogeneous its ability to 'learn' collectively might be questioned. Certainly this learning has not been unilinear. For instance, many of the same questions recur in direct action protests in the 1990s as had also characterised the 1970s and 1980s. Among these are divisions over the use of violence and questions of feminist praxis. The US Clamshell Alliance fell apart over divisions between those willing to cut fences as part of their protest and those who opposed them (Epstein 1991). Similar, though less divisive, debates occurred in Britain. *Peace News*, which by the mid-1970s had become the principal paper of the more anarchistic, ecological, peace and animal rights networks in the UK supported the anti-nuclear protests in France and Germany, despite the violence of some demonstrators. This led UK activists to a recognition that direct action was most likely to be non-violent when carried out by small affinity groups who trusted and knew each other well enough to be able to achieve a consensus on how to act. For Epstein (1991) the autonomy of affinity groups was the main reason why the Livermore Action Group (which targeted a nuclear weapons producing plant in California) was able to work more effectively despite having a more socially and ideologically diverse constituency than its predecessors in the Clamshell and Abalone Alliances. This affinity group structure was adopted by peace movement NVDA activists in many countries in the 1980s. Also, in the 1980s cutting the fence at military bases became an accepted part of the repertoire of peace movement non-violent direct action, and activists made a strong distinction between violence against people or animals and property damage. In the 1990s debates over non-violence have remained lively (and are discussed further in Chapter Six), but have been much less divisive than in the 1970s because different groups of activists have learned how to manage differences by working together on some actions and separately on others. Furthermore even the small numbers who are supportive of the use of political violence in mass demonstrations have learned from the experience of urban leftist terrorism in the 1970s that a terrorist strategy leads only to isolation and repression.

Feminist influence

Feminist praxis can also be seen as influential upon the collective learning of the green movement, though again in uneven and unpredictable ways. In the 1970s and 1980s the right of feminist activists to organise autonomously was hotly contested by opponents within the alternative milieu, but by the 1990s women only spaces and gatherings were both widely accepted and understood by activists in green parties and direct action networks. In France, where there had been less linkage between the women's movement and the green party than in most other countries, there was suspicion in the early 1990s of giving priority to feminist demands and adopting practices such as parity in representation within the party organisation (Doherty 1994). However, by the end of the decade increased debate among French activists and the increased contact between green parties that followed the expansion of green representation in the European Parliament led to a convergence in policy. The French green party now make women's issues central in their party programmes. On the other hand, this is not necessarily a story of untroubled incorporation of feminism. In Germany in 1987 a group of women members of Die Grünen published a 'Mothers' Manifesto'[17] which attacked the radical feminist arguments that had dominated the women's section of the 1987 election programme because these seemed to devalue women who failed to conform to radical feminist ideals. In doing so, the backers of the Muttermanifest were not attacking feminism per se,[18] rather this was an argument that was already occurring within feminism between a 'cultural feminism' based upon revaluing women's differences from men and the more materialist concerns of socialist and early radical feminists. In Britain, some green party members such as Penny Kemp and Derek Wall defined green feminism primarily in terms of structural and material inequalities (1990: 94–5), others, such as Sara Parkin used the kind of ecofeminist language that other feminists have seen as suggesting a dangerous biological essentialism:

Whether men are motivated by revenge or a confused notion that they must appear strong in order to be considered worthy of protecting their mothers, it seems likely that only women have the power to absolve them and make it clear that we don't need their sort of defence ... Women certainly need to be able to use the green movement to regain confidence in their considerable power to liberate, not simply women from men, but men from themselves and to re-establish the confidence of all of us in the wisdom of natural instincts and patterns of life.

(Parkin 1988b: 171)

Since feminism was split on these and other issues it was hardly surprising that greens too should differ over which kind of feminist principles they endorsed. Patriarchal practices also remain an issue of contention. In the 1990s activists in the direct action protest camps in Britain and the USA commented that men were patronising women who were poor tree climbers, failing to take responsibility for cooking, cleaning and maintaining the camp fire, and were often aggressive and macho during actions. However, it is a feature of the more limited utopianism that characterises this milieu, that failures of this kind are expected. The evidence of collective learning is that the significance of feminist critique is now much more accepted in the green movement and no longer seen as a distraction from 'more important' issues. Contrary to the view that feminists intimidate their critics into silence through the exploitation of 'political correctness', the opposite seems to be the case. Feminist ideas are part of the taken for granted culture of the movement, in contrast to the 1970s and 1980s when their status was much more contested, as evident in the many arguments then over women-only spaces. But also in contrast to previous decades, when the idea of what constituted 'ideological soundness' was much more dogmatic, there is now greater tolerance of different practices and views of feminism.

Alternative media and the internet

The alternative press provided an important medium through which the collective learning of this milieu could occur. Indeed, such media may be more important in the creation and maintenance of an identity among the contemporary alternative left than they were in the early twentieth century for socialists. As Raschke points out (in Poguntke 1992: 344) even where the organisational networks characteristic of earlier socialism are absent the alternative media can help to secure the symbolic integration of value milieus. Benedict Anderson (1983: 62) has argued that newspapers were important in the emergence of anti-metropolitan nationalism in the Americas. By creating a sense of simultaneity among colonists within a particular territory they expanded the scope of the daily or weekly narrative of events and defined a space within which an imagined community could emerge. The alternative left community was not wholly imagined, it had already defined itself as different by its lifestyle, but because it had no organisations that represented the whole milieu, it was most obviously through forums such as the German daily paper *die tageszeitung*, the French anarcho-green monthly *La Gueule Ouverte* and *Peace News, The Ecologist* and *Green Line* in Britain, that dispersed groups were able to debate and develop ideological positions and to pass on news of related movements in other countries that was ignored or mis-represented in the mainstream media. In recent years, the growth of the internet has greatly expanded the amount of internal debate and compressed the time and space dimensions in such debates.

Alternative news media such as the *www.Indymedia.org* network, oriented towards direct action protest, provide instant reports allowing activists to largely ignore mainstream media. The links between sites in different countries also provide an immediate transnational simultaneity. News from an activist persp-ective from across the world can be available as quickly as the satellite broadcasts of government or corporate owned media. By providing a contin-uous narrative of events from within the transnational direct action community, the expansion of such media creates a new kind of 'imagined community' which partly transcends national boundaries. While such networks would still exist even without the internet, the degree of community which exists through a shared narrative, and debate is much greater. Nevertheless, as yet, these processes have not displaced national particularities altogether. This is most evident in email discussion lists. Most of the activity of greens is still local and dependent on face-to-face ties or national and dependent on national organisational frameworks. However, debates at these levels have also been altered by the internet, and particularly by email. The savings in time and money in distributing information by email are well known, and help to reduce the slope of the pitch for resource-poor groups such as the greens (Doherty 1999c). But email also greatly expands the space for discussion. The myriad of discussion and news lists provide a kind of permanent meeting space, which supplements the traditional monthly meeting, or allows virtual meetings to occur between activists too dispersed to be able to meet. Many of these are issue based, between campaigners on specific issues, such as farming, housing or transport. Others emerge in response to new strategic dilemmas. For instance, in Britain in September 2000 protests against fuel prices by farmers and hauliers, part of a trans-European wave of protest, posed a challenge to green arguments about the need to reduce use of fossil fuels. An email list quickly emerged in which activists from various EMOs such as Friends of the Earth, Transport 2000, and from the Green Party and direct action networks debated how to respond. This debate revealed divisions over whether or not counter-demonstrations were justified, but also created a space in which differing views of the state and social change could be discussed by activists from different types of green group. This kind of cross-movement debate, open to anyone with net access, could not have taken place even a few years earlier. It might have taken weeks for movement newsletters to appear and longer still for communication between different networks about what to do, to take place.

Even local branches of green parties, EMOs or local environmental groups often have their own lists in which practical business and debates occur. Although some of the latter could occur via telephone conversations, it is the potential for greater inclusiveness which marks a qualitative change. News and views can be disseminated much more effectively through email lists. Even those groups with 'telephone trees' used them mainly to alert activists in an emergency, to use them as a discussion tool would have been too demanding. Clearly the greater inclusiveness only applies where activists have access to the internet and the exclusion of those without access has been a subject of debate (Pickerill 2000), but internet access among activists has expanded so fast that this is less of an issue than even a few years ago.

In a period in which identity is widely regarded as fragmented it may be dubious to use terms such as alternative to define this milieu, since it begs the question of alternative to what? Yet, what they share in culture is most clearly based on the idea that they are exploring new forms of practice that seek to extend the egalitarian and democratic traditions most associated with the left. The roots of the milieu lie in the New Left, but the hopes of an immediate evolutionary transformation were undermined by the recognition that the system could not be overthrown by seizing the state. New politics of ecology, feminism and questions of race made the idea of a single strategy and index of progress implausible. Transformation was required in culture and in private lives as much as through political institutions. The most radical parts of this milieu aim to live as autonomously as possible from dominant structures. Yet, a central result of the post-1968 learning experience has been that only relative autonomy is possible. It is not possible to live the life of a pure revolutionary without contradictions.

Thus the greens exist within an alternative milieu which in contrast to the New Left is more pluralist in its concerns and in which progress is based on shifting alliances with diverse constituencies. However, one question that can be asked about this learning process is whether it has been restricted to particular social groups?

Who are the greens?

It has been noted that what separates many contemporary social movements from earlier movements is that for movements such as the greens there is no obvious social interest arising from shared circumstances to explain their activism. Saving the environment, expanding democracy and overcoming inequality is not more in the interest of green activists than of others. However, green movements, and particularly activists in green movements, come from distinct social groups: greens are not socially representative of the population in general. When the first green movements emerged greens were much more likely to be young (though now, greens also include those in their forties and fifties as activists have aged), have higher education and work in public sector professions or 'creative' fields such as journalism. It will be argued that while their politics cannot be explained as a result of social interests, there are experiences common to these groups that can explain their disproportionate support for green movements and parties.

There is a remarkable consistency in surveys and ethnographic analysis, stretching over two decades which show that the generation educated since the 1960s and those in public sector new middle class jobs are more likely than others to support the social movements of the alternative milieu (Parkin 1968; Cotgrove and Duff 1980; Kriesi 1989; Mattausch 1991; Epstein 1991; Bennahmias and Roche 1992; Rüdig, Bennie and Franklin 1991; Poguntke 1993a; Lichterman 1996). This category includes inter alia radical environmentalists and green parties, feminism, and peace movements. However, it is also recognised that it is only sub-sets of each of these social groups that support these movements. Among graduates those who took humanities and social science degrees are more likely to be politically radical than those who took technical degrees (Rootes 1994; Gouldner 1979: 65–70). And within the new middle class it is public sector professionals who are most likely to be egalitarian on both economic and new politics issues (Heath and Evans 1989). Of the German post-1968 generation it is those between the ages of 25 and 44 rather than the first-time voters who are now the group most likely to support the greens (Hoffman 1999: 143).[19] There is much less evidence on the gender divide, but what is available suggests a roughly equal balance between men and women in environmental and peace movements (Byrne 1997: 75–77).

It is notable that these findings are strongest in respect of political activists, rather than the more general category of supporters of movements. They apply most to activists in environmental movements, green parties and to voters for green parties. We have less precise information for radical direct action groups, although what evidence can be gathered gives good grounds for seeing the general pattern as similar. Activists in direct action movements are usually younger than in other environmental movements, not least because of the demanding character of this form of participation. A study of Earth First! (UK) found that none of 28 founders of the movement had children, and half were students (Wall 1999). Epstein (1991), studying US NVDA activists also noted the dominance of well educated activists, living on low incomes, many of whom had chosen not to have children, either because of doubts about bringing children into a world under threat, or because it would make activism difficult. By no means all radical environmental action is carried out by people from these social groups. Grassroots local environmental campaigns tend to be more diverse in their social composition. Yet the limited evidence available about European groups of this kind also supports the contention that activists in such groups are drawn from the new middle class. A survey of activists in ALARM UK, the coalition of local anti-roads group bore out the familiar picture of the social characteristics of green activism (McNeish 2000). In the USA, however, the anti-toxics campaigns that followed the Love Canal campaign in the late 1970s were notable for the leadership role played by women from lower middle class households, most of whom had not had previous political experience (Szasz 1994). In the environmental justice campaigns of African-American and Hispanic communities, existing organisations or networks based on civil rights campaigning, trade unions and community associations have often been important in providing the basis for campaign organisations (Lichterman 1996; Pulido 1996). It was argued in Chapter One, that some grassroots environmental campaigns are less clearly part of the green movement when it is defined as a social movement, but no less important for being partly distinct. Pulido argues that there is a distinct 'subaltern environmentalism' found in the struggles of the poor over health and working conditions. In what follows, therefore, it is important to note that it is not being argued that only the experiences of new middle class professionals, those with higher education and those politicised since the New Left that lead to radical environmentalism. Instead the first aim is to explain why the specific kind of environmentalism that constitutes green identity seems to be most attractive to these specific social groups. The second more positive claim is that experiences in higher education and welfare professions have helped to shape the character of green discourse.

Rather than searching for a particular social interest in order to explain the support for green ideas it seems more fruitful to provide an explanation that begins with core elements of green discourse and examines what in the immediate circumstances of green activists would strengthen such ideas and even help them to develop. If many of the more general values of the greens are not in themselves new, (for instance, participatory democracy, non-violence and redistribution) then it is an understanding of the social and political contexts within which they have been applied that provides the best means of explaining what is new about their politics. Class is still a central issue in understanding why support for green ideas is concentrated among the workers in public sector professions, but rather than searching for class interests the relationship between class and green ideas is better explained in terms of experiences or social circumstances which can plausibly lead to higher than average support for green ideas.

Higher education

The expansion of higher education is often noted as a structural facilitator of the emergence of a constituency for green and other new social movement politics but is it also a determinant in itself of new ideas? Between the 1960s and 1989 the proportion of those passing through higher education quadrupled in western Europe (Alber 1989: 196) and has continued to rise.[20] Although education and political competence do not always go together Inglehart (1990a: 335–71) has convincingly charted the growth in organisational and tactical knowledge of politics among this group. Survey evidence also shows that there is a strong link between education and support for socio-economic changes to deal with the environmental crisis. Among less well-educated groups the environmental crisis is taken seriously but perceived more in terms of personal threats (Rootes 1994). Class rather than education is also a factor here, since for the most part it is the working class that is exposed to the most immediately visible environmental threats in workplaces and in sub-standard housing.

What is less clear is whether expanded higher education has itself been a cause of new political values. There are some reasons for thinking that it has.

First, the evidence suggests that graduates of higher education are more likely than others to be supporters of leftist social movements (Müller-Rommel1990: 22). Also indicative is the evidence that graduates in Britain are more likely to be libertarian and more likely to be egalitarian than all other groups in the population (Heath and Evans 1989). As Rootes (1994) has said, most accounts of the social base of the new politics have tended to ignore the effects of the content of education itself on values. This is despite American research that supports the view that social science and arts students tend to be more left-wing than science students even though the views of the two groups were more similar when they entered university (Lipset and Dobson 1972). A survey of the German alternative projects in the early 1980s found that two-thirds of those involved were students or graduates and, of these, two-thirds came from social sciences or humanities, teaching or social work (Betz 1993: 107).

The fullest attempt at an explanation of the role education may play as a cause of new political values is Alvin Gouldner's (1979) analysis of the 'culture of critical discourse'. By this he means the willingness of those with post-1960s higher education to evaluate the views of speakers independently of their status. According to Gouldner those who shared this culture felt responsible for the whole of society but were split between humanistic intellectuals – more concerned with social and political problems and a problem-solving technical intelligensia. The rapid expansion of higher education also meant that the still mainly middle-class students had less chance than previous generations of students of an elite post at the end of their education.[21] At the same time the content of courses and the culture of universities were altering to fit the needs of a production system that was increasingly reliant on graduates as technocratic managers of information. In general terms, then, the expansion of higher education changed its nature, reducing its elitist character while creating new tensions but, most importantly, it gave a larger proportion of the population, and particularly the first post-war generations, greater opportunities to develop the political skills and language necessary to challenge existing elites on their own terms.

Accepting that education should provide its recipients with more 'cognitive resources' there is still no necessary reason why such values should be articulated in a new ideological form. Beyond the possibility of direct influence from the values of left-wing academics, we can only guess at possible sources of influence. One explanation for the gap between science and non-science students might be the relative rigidity of sciences in comparison with arts and social sciences, and the political themes that arise in social sciences and arts (Rootes 1994). Education has usually been seen by academics as neutral, and this has included those analysing the greens. For the latter, the most important questions have concerned new-middle-class interests, but education is probably itself a source for values which might be further and differently developed in specific occupations. There is no need therefore to see education and occupation as mutually exclusive influences (Cotgrove and Duff 1981: 106; Eckersley 1989: 213). Both education and occupation can be sources of values and contribute to the development of values that might reflect earlier primary socialisation.

Welfare professions

The expansion of professional employment in the welfare and public services was less dramatic than the expansion of higher education but possibly more important for the subsequent character of the green movement. Many explanations of the concentration of green activists and voters in such occupations take note of the expanded number of workers in these fields but they fail to appreciate the importance of the development of emancipatory occupational cultures among radicals working in these fields.

Many of the welfare professions, notably in health and education, predate the welfare state but the numbers employed in these fields and in public administration increased as governments sought to improve public services. Equally importantly, the expectations of their roles were also altered. A second group of professions is a more direct consequence of the establishment of the welfare state. Social and community and youth workers, arts workers, legal workers in law centres, public sector housing workers, welfare-rights workers, educational advisers and others are relatively new fields of employment which result from the more extensive scope of public services.

What do such professions share that might explain their role in the development of green ideologies? As well as the need for graduate and often post-graduate educational qualifications for such work there are three characteristics of the welfare professions which make those in such professions more likely to support green movements and which mark out their relative importance as a source of creative input for green politics.

The first and often noted feature of such work is its autonomy (Gouldner 1979: 34). Most welfare professions have comparatively less structured daily working routines and also leave large areas of their work to the initiative of individual workers. Although not free from hierarchy and management control, the degree of such control is comparatively minimal and clearly demarcated in comparison with most occupations in business or industry. This is also the case in other fields which contain a high proportion of green supporters such as the media, the arts and independent crafts. Work which depends on successful social contact or individual expression cannot be governed wholly bureaucratically. This experience of working autonomy might help to explain the concentration in the greens on a critique of overly restrictive bureaucratic mechanisms. Those accustomed to working independently and depending on their own resources are likely to develop a positive belief in their own and others' ability to assume a more direct role in the self-management of other areas of their lives.

A second and politically perhaps the most important characteristic of these professions is their ambiguity in relation to the dominant imperatives of modern society. It has been argued that the service class can be defined in terms of its reproductive function for capital (Ehenreich and Ehenreich 1979) and yet such an approach is too general in its inclusiveness of the whole service class. What can, however, be drawn from this approach is the importance of the surveillance and disciplinary role of the welfare state professionals. Foucault (1977) stressed the contribution of statistics and the growth of major public institutions such as hospitals to the efficiency of modern institutions of social control: but along with formal disciplinary methods went an ethic of individual responsibility and an emphasis on self-discipline. The disciplines advocated within professions such as teaching, social work and therapeutic health work provide a further extension of the efficiency of such discipline: the cruder and more mechanical regimes of earlier institutions are now replaced by a more psychological approach to the individual's needs and circumstances.

However, it can equally be maintained that viewing these professions only in terms of their disciplinary function is too absolute. Workers in the welfare professions can also be seen as advocates of their clients, pupils or patients, struggling to establish certain gains for them against the weight of institutional inequalities. This is where the ambiguity of these professions lies. They are at once disciplinarians concerned with mechanisms of social control, (the side of the welfare state most often emphasised in radical critiques) and yet also providers of resources aimed at transforming the unequal life chances of those they work with. For instance, midwives have a long tradition of radicalism. Ella Jackson of the Association of Radical Midwives (Arm) comments:

> The Arm does have its roots in feminism. Midwifery is a vocation and we genuinely care about what we are doing. I don't think that you can get into women's issues without being vocal and committed. Being a mid-wife is about empowering women to have the kind of labour they want, it's about giving birth back to women.
>
> (*The Guardian*, 27 September 1999)

This leads to the third characteristic of such professions, the inescapably political nature of their work. Funding for social policies is potentially open-ended. Welfare workers are often hampered by resources inappropriate to the scale of the problems facing them, but in many cases the intervention of institutions in a new area can provoke a progressive expansion in the rate of need at a faster rate than resources are available to deal with it. The pressure for a continuous expansion of health services and the tendency of any final state of 'health' to recede constantly is perhaps the best example. Thus the scale of public provision needed to respond to particular issues can never be finally fixed and the utility of such professions is under continual public scrutiny.

The welfare state professions are usually directly, sometimes indirectly, the representatives of the state charged with intervening in areas previously seen as the sphere of the family or tradition. Yet the debates in these professions are often self-critical. There is a particular concern to combat inequalities of race and gender as well as class and the debates over educational curricula, care of children and citizenship rights are a constant feature of the journals and training for these professions and in marked contrast to the occupational culture of business or industry. The conflicts and continuities with socialism are perhaps most evident here, since the welfare state professionals were in the best position to observe the limits of a social democratic project centred on the welfare state and yet also conscious of the continuing necessity of some forms of universal services.

However, it is not just in relation to issues of equality, but also questions of life politics that the political character of welfare professions is evident. As Sarah Hipperson, mid-wife and one of the last women to leave Greenham Common Peace Camp in 1999 said: 'I have asked myself why there were so many midwives at Greenham. The whole process of bringing life into the world was part of my life, not just as a midwife but as a woman, whereas nuclear weapons were about the destruction of life. And I think that's why I went to Greenham' (*The Guardian*, 27 September 1999).

The evolution in the New Left and green parties from the revolutionary strategies espoused by the New Left in 1968 to a willingness to experiment with building alternative institutions was at least partly due to the sense of political failures in the post-1968 era. A further contributing factor, however, may well have been the experience of the members and activists from the green parties working in the welfare field. Those who had principally been clients of the state were accustomed to having their lives conditioned by the less flexible elements of the welfare state and likely to develop an anti-bureaucratic stance (Offe 1985: 833). But those working in the welfare professions may have produced many of the more positive policy proposals later advanced by green parties which depended on a positive view of the role of the state as an agent of social transformation. The ability to develop emancipatory policy in specific fields may have contributed to a more optimistic view of the importance of an improved welfare system, but one based on more local community structures (Barry and Doherty 2001). Yet, to the social democratic concern with improved universal services were added new themes – the rights of claimants, clients, patients or pupils to greater control over their services and more participation in decision-making, and the failure of policies based on the concept of the general citizen to take account of non-economic forms of inequality based on gender, ethnicity and sexuality.

Political generations, structural change and collective learning

This explanation of why the green movement developed a left-wing form of politics helps to explain why support for green ideas is highest among three social groups. Each of these three – higher education, welfare occupations and the generational experience of those who came of age during or after the second phase of the New Left in the 1960s, can be seen as providing reasons in themselves for an above-average receptiveness to green ideas or conversely why groups with these experiences were the principal sources of collective action which developed the green movement.[22] While this explains the over-representation of green activists and voters in these categories, the experiences of those in these groups can also be seen as a source of input for the democratic and egalitarian ideas of green movements.

While these shared experiences provide predisposing factors for green and other alternative left politics they do not guarantee it. For Mannheim a political generation was defined by its shared beliefs, not only by shared experience of formative conditions. These beliefs are not static, but develop as hybrid movement networks develop, and collectively deal with the core questions of political action: what is wrong with the world? How can it be changed? What might a better world be like? And how to cope with living in the world as it is?

The account here has situated the greens as shaped both by factors internal to the movement as well as its external environment. The idea of the movement itself is an ideal type and when we consider the less formal dimension of movement life, as defined by the cultural practices and extended politics of the alternative milieu, it becomes harder to separate the green movement from others with which it shares both ideas and overlaps in membership and support. However, the nature of the movement is also affected by its interaction with opponents and in their participation in public collective action, the greens are easier to distinguish. In the 1970s the degree of conflict with the state, particularly over nuclear power, and the attitude of the main parties of the left to this issue, and the strength of the alternative milieu differentiated green movements within countries and cross-nationally. The collective learning of the movement embraced the response to opportunities and constraints of national political systems and the transnational spread of green practices. As we will see in subsequent chapters, while the overall experience of the green movement could be defined as that of a political generation, new cohorts have arisen within this generation and developed new strategies which show them to be reflexively aware of the limitations of the past and also shaped by new conditions. For instance, the grassroots environmentalism in the USA that developed since the 1980s, mostly avoided institutionalisation because of the recognition that institutionalisation had limited the major EMOs to lobbying and the development of policy expertise. Direct action groups in Britain in the 1990s made a similar judgement. Also, whereas in the 1970s and 1980s the relationship between ecology and socialism had been a major subject of debate, in the 1990s the crisis of socialism, meant that this issue was much less controversial and greens were mostly prepared to be identified as on the left, without necessarily being socialist.

The collective learning of a political generation is uneven and complex. It does not necessarily lead to progress. Rather it is the process of reflection based on experience, whether derived through direct experience or from learning about other groups' experience, that is at the centre. Moreover, it is also important to stress that collective learning does not always unite the green movement: often it heightens awareness of the differences between groups, particularly over strategy. An awareness and acceptance of the differentiation of the movement, united by shared ideological framework, but divided over strategy, seems to be a feature of its evolution. This Manchester Green Party activist does not regard being in the Green Party as ideologically superior to activism in other green groups, rather his justification is more pragmatic:

I think that there isn't any single way of achieving a socially just and ecologically sustainable society, which I take as my mission to help achieve that. Direct action I think can do certain things and voting in elections can do certain things ... and education and awareness such as Greenpeace and Friends of the Earth do, can do certain other things, but none of them on their own can achieve that goal. So I feel they all need to happen in parallel ... The Green Party had gone into a bit of a decline in the early nineties for various reasons and it was the entity that I felt needed most support. The other things like FoE had a membership of nearly a quarter of a million compared to the Green Party with about five thousand at the time so I think I made a reasonable decision you know.

(interview, 13 July 2000)

In the argument in this chapter priority was given to the emerging evidence of ecological crisis as a structural change that shaped the green movement, but otherwise the explanation largely bypassed the major theoretical debates about modernity and post-modernity in sociology. As noted in Chapter One, in many accounts the rise of new social movements such as the greens is linked to major structural changes in the post-war era, through which the process of rationalisation has colonised increasing areas of social life. For instance the application of market criteria and instrumental judgements of the usefulness to the economy of forms of education, and the expanding territory of social policy and the surveillance of marginal groups, can all be seen as elements provoking the search for radical alternatives among greens and other movements. Here, however, the focus was more specifically on how the greens integrated their radicalism with strategic choices.

Green praxis has reflected a loss of confidence in the idea of a singular revolutionary transformation, even while the aims of the movement remain radical. At the same time the movement, for the most part, has developed a strategy and identity which is based on balancing different commitments to ecology, egalitarianism, and a grassroots form of democracy. This has sometimes been criticised because it is not based on a philosophy in which ecology has any clear priority over other issues. British Green Party activists Irvine and Ponton said of the manifesto of Die Grünen:

For the most part it was founded upon a scarcely reconstructed leftist platform. Revealingly 'ecology' was only one of the 'four pillars' of green politics. In fact, most green parties have been handicapped by the lack of a consistent and rigorous set of values ... the typical green manifesto is a shambles, a dogs dinner, cooked from bits and pieces of demands raised by dissident groups across society.

(Irvine and Ponton 1992: n.p.)

Irvine and Ponton failed to persuade fellow Green Party members to adopt their view and were part of the group that retired from activity in the Green Party as a result of failure to change its organisation in 1992. Their statement provides as good a summary as any of the roots of green party ideology in the broad alternative left. This broad framework, is consistent with the idea that power, inequality and environmental degradation cannot be dealt with effectively by consistently prioritising one issue over the others. As we will see in the next chapter it is the avoidance of ecological reductionism which distinguishes the praxis of green movements from the more ecocentric arguments of green political theorists.

3   Green ideology

Green ideology is new, but it is a new variant within the traditions of the left rather than an alternative to the left/right divide. As we saw in the previous chapter, the green movement emerged as a consequence of changes in the nature of the western left in the late 1960s and still has its roots in the traditions of the left. Greens share a commitment to ecological rationality, egalitarianism and grassroots democracy, the latter two placing them on the left and shaping their view of ecology. None of these three commitments has a priori privileged status in relation to the others and greens seek to achieve a balance between all three. Thus green ideology is not based primarily on ecology, although ecological rationality would never be ignored or rejected as unimportant by greens. Moreover, whenever some greens have tried to establish the superiority of ecology over other green concerns this has caused divisions in green movements, and it is those who want to maintain the three-fold combination of values that have remained within the main green groups. This view of green ideology will be elaborated and defended in the first part of the chapter. The focus here is on the content of green ideology. In the second part of the chapter the consequences of this view of green ideology for how we understand the relationship between green ideology and action will be analysed.

Insofar as this book is concerned with green ideology it approaches ideology as a shifting product of social movement action. Social movement ideology has to be analysed as dynamic; it develops dialectically in debates with opponents and in response to conflicts over ideas within the movement. Green ideology developed through the debates among critical communities of activists and intellectuals, linked by network ties that cut across formal organisational links. It was influenced by the real and anticipated reactions of opponents, and by the learning experience of activists and intellectuals in previous movements and later in the green movement. There is therefore no single point at which we can say green ideology was created, rather it emerged as the movement developed from the late 1960s onwards. The three principles of ecological rationality, egalitarianism and grassroots democracy gradually became more evident as the framework within which green discourse was organised. This more settled framework has developed in response to the need to defend and extend the arguments made by movement groups, and also as greens have faced the problem of who they can work with in practice. This has led some to leave the movement, when they cannot accept this framework, and has also shaped how greens view potential political allies.

Political theorists have carried out much work on these questions, but their focus has been primarily on how green ideology is analytically distinct from other ideologies. Rather than revisit this debate I want instead to change the focus from distinctiveness to internal coherence. This means developing a broader descriptive accuracy framework for green ideology than that developed by most political theorists (Barry 1999; Dobson 2000; Eckersley 1992; Hayward 1995). And, this differs from those who see green ideology in terms of ecologism in that the status of ecology is not privileged above other issues in explaining how greens view the world, although ecologism is one variant of this broader green ideology. Ecological rationality is an essential part of green ideology, and is its most distinctive feature, but it is not enough in itself to explain the views of sufficient numbers of greens. The aim then is to find a version of green ideology that can accommodate as many green activists as possible.

It is in fact not easy to do this. There have been some surveys of the values of green party activists (Bennie, Rüdig and Franklin 1995; Lucardie, Voerman and van Schuur 1993) which support the view that greens are egalitarian, committed to ecological rationality and grassroots democracy. In other green groups there is less explicit attention to ideology. For some it is implicit in their practices, as for instance in the targets chosen by direct action groups or in their culture and everyday practices. In other cases, such as environmental movement organisations, it is evident only in their general mission statements[1] or at the margins of their more issue-specific campaigns, as when they address questions of inner city poverty, for instance. Perhaps the most stable and general statements about green ideology are those made by green parties. More than other green groups they have had to develop a collective view on the central political problems, their causes and remedial actions. In the section below the evidence is based on statements drawn from green party programmes. The relation between these ideas and those of other greens will be addressed later.

Three core themes in green parties' ideology

Green parties' ideology is best defined in terms of three themes:

- The first of these is the ecological crisis. All green movements and parties view the potential for ecological collapse as requiring major changes in the nature of production and consumption. Responding to the ecological crisis also provides new opportunities to develop new post-material values less tainted by greed and the need for ever-greater affluence.
- The second is that greens have also taken up the traditional left-wing concern with the redistribution of power socially as well as politically.

However, the effects of the welfare state on social stratification means that their principal concern is not with the working class *per se*, but also with groups that they see as having been excluded materially and culturally from effective participation in post-war society. However, it is not only exclusion, but also the structural character of inequality that concerns greens.

- The third is that green parties have also identified the restrictive forms of participation in liberal democracies and the closed character of bureaucratic decision-making as obstacles in-themselves to political and social change. In this, however, they are hardly alone. What defines green anti-statism as distinct from that of the New Right is the greens' view that existing state institutions are insufficiently democratic rather than a constraint on the market. Like the New Left, the greens want to change the nature of politics itself so that taking part in effective political decisions becomes a regular part of everyday life.

It will be argued that these are the three essential themes in green party ideology and that all are present to some degree as irreducible elements in their discourse. The sections below will attempt to outline the main features of this type of green ideology as it is found in programmes from three green parties. These are the green parties of Britain, France and Germany. They represent three different positions on the spectrum of pure green and left green identity. The German Greens (henceforth, Die Grünen) are the largest and best known of the green parties and the green party most identified with the left. The French Greens (Les Verts) include many ex-combatants of the New Left but have varied in their choice of an 'alternative left' or a pure green identity at different stages in their history. They therefore represent a mid-way case. The British Greens (the Green Party) have had probably the weakest links with the left, either biographically or in terms of their own view of their identity, although opposition to a left identity seems to have decreased in the British Greens since the Labour Party's ideological renovation. Despite these differences, importantly, they share an ideological framework that allows them to take very similar positions on a wide range of issues.

Green party programmes

The similarity of green policies on these three areas can be illustrated through a comparison of their programmes.[2] The summary below is based on analysis of programmes from the German Greens in 1983, 1994 and 1998, the French Greens in 1989, and 1997, and recent drafts of the British Greens' ever-evolving meta-programme, known as the *Manifesto for a Sustainable Society*. The examination of party programmes[3] is essential to understanding the ideology of a political party, but, of course, it cannot be sufficient in itself. We also need to know about the context in which decisions were taken. These issues will receive more attention in the later chapters on the green parties. For now, programmes provide a means to illustrate shared features in green party ideology. As noted, this can be divided into three areas: ecology, a pluralist egalitarianism and an argument for a more participatory democracy. Yet, wherever possible greens present these as inter-related and the obstacles to their realisation are similarly identified with a single crisis tendency in modern industrial societies.[4] Thus the first point in common between these programmes is their perception of a system-level crisis. Single policy changes will not be sufficient. What is required is a new way of thinking and radical social, political and economic changes.

Green programmes on the ecological crisis

The evidence of a global ecological crisis provides the first reason for change. All three parties argue that since natural resources are finite and the stability of ecosystems is under threat from pollution, urban growth and soil erosion, economic growth must be curtailed and if necessary reversed. However, a society based upon such an equilibrium can still be a prosperous one, yet here non-material concerns to do with the quality of life come to the fore. Ending the productivist logic of a growth and consumerist society depends on changes in values as much as on changes achieved through governments. Thus the inhabitants of the First World states are encouraged to accept the necessity for less consumption by making the equation between existing forms of production and the long-term consequences of worsening ecological crisis. At the same time, green parties want to encourage this new culture by supporting qualitative changes in the economy, breaking up large and unaccountable private and state enterprises, expanding the state's regulatory role in environmental matters, supporting the informal sector, and altering the patterns and nature of work. They want to reduce overall work-time and divide the available work more evenly. Work should no longer serve as the exclusive means of self-realisation; but within the workplace emphasis is placed upon reducing formal hierarchies and increasing workers' rights.

On two areas concerning the relationship between economics and ecology the greens are particularly careful to spell out their position. They are critical of capitalism, but do not advocate classical socialist alternatives such as nationalisation or full common ownership. They criticise socialists for adopting the same productivist priorities as defenders of capitalism, and offer the alternative of a regulated market in which private ownership of goods is defended, but production is constrained by ecological and social goals. On the second area, science and technology, they also have a position that tries to avoid absolutism. In the view of greens, science and technology is invested too straightforwardly with an inevitably superior logic when it in fact reflects the priorities of the existing social system. Developments such as nuclear energy and genetic engineering represent a Faustian logic that ignores the risks of grandiose attempts to control basic natural processes. Science no longer has any yardstick by which to understand its own limitations. Greens therefore want to see the redirection of scientific knowledge to help deal with ecological problems, without seeking to redesign nature. At the same time they are anxious to stress the benefits of 'soft technology,' which includes renewable energy and computers.

Green programmes on the politics of inequality

The green parties' perception of a social crisis is apparent throughout these programmes, but the causes of this crisis are diverse. First, because of their critique of economic growth the greens are particularly concerned to emphasise the importance of the redistribution of wealth. They are also critical of the limits of current welfare provisions which are seen as bureaucratic and alienating. Green parties would replace existing means-tested benefits with a guaranteed minimum income available to all regardless of whether they were in or out of formal employment. This is evidence of their anti-materialism and their communitarianism. Wages are not seen as the primary or even a necessary motivation for work, but a level of income sufficient to guarantee a minimum standard of living, relative to the community, is seen as a human right. One consequence of the latter view is a strong concern with global equality. Greens give a major priority to ending what is seen as the western exploitation of the Third World and to increasing the levels of aid there with the purpose of allowing the Third World to choose its own path of development, albeit within the constraints of global ecological stability. The green parties are well aware of the need to take action which will have an impact on a global level. However, as First World parties they see the primary responsibility for action as lying with the richer countries. Ecological justice also requires the extension of special consideration to future generations and to the natural world.

The green concern is with reducing the concentrations of economic power and redistributing wealth globally rather than with collective ownership of the means of production in a national framework. As a result they do not accord any privileged status to the working class, preferring to speak of the poor, the marginalised, or the excluded. This also means that they are able to address other forms of inequality not based on class such as racism and sexism as central problems of modern industrial society without the same worries experienced by Marxists over their priority relative to the class struggle. The influence of the ideas of the new social movements is evident in the strength of green commitments to challenging legal and material inequalities and cultural practices that they see as excluding from effective citizenship women, ethnic minorities, lesbians and gays, Romany and New Age 'travellers', and the disabled. Yet their aims go beyond inclusion of these groups within the existing framework of citizenship. Green parties also want to transform the existing forms of citizenship in order to create a more participatory public sphere. They see such change as dependent as much on changes in political culture as formal changes in political institutions.

Green programmes on grassroots democracy and the limits of statist politics

Establishing the social preconditions for effective citizenship at local, national, and transnational levels is one goal of green social policy. However, what the greens regard as real democracy also requires major institutional changes. Here the main measures are directed against the power of the central state and against restrictive forms of bureaucracy. Greens want to radicalise democracy by expanding the arenas of decision-making through increased referenda and public access to policy-makers, and most importantly, by substantial decentralisation of representative decision-making. They want to see more extensive powers for local and regional authorities whilst, somewhat uncomfortably, accepting the need for some transnational sites of decision-making, although these would be based on bodies with constitutionally delimited competence and therefore sovereignty would vary issue by issue. Their vision of democracy is therefore participatory with a strong emphasis on the question of appropriate scale: decisions should be taken at the lowest level possible – hence the widespread use of 'grassroots'.

A further dimension of their challenge to statism concerns the state's use of external and internal violence. On the former, they regarded the Superpower confrontations of the early 1980s as evidence of the dangers of secrecy and lack of accountability. On internal violence, they have tended to criticise the form of policing but moved away from more utopian positions that rejected the use of force in support of the political order. However, greens continue to argue that organised violence should not simply be accepted as inevitable, and that violence by the state, social groups or individuals depends on ideological assumptions that can be gainsaid. Thus they do not believe that military defence is necessary, and are prepared to experiment with alternative forms of defence, and aim also to reduce the cultural and structural causes of non-state violence.

The dominant issue in the domestic politics of liberal democracies in the post-war era – the role of the state in the capitalist economy – is not the most divisive issue in the green parties. With the exception of a debate in the German Greens during 1986 and 1987[5] the green parties have not been seriously divided over the balance between the economy and the state. This is not because this issue is irrelevant to their politics (it clearly is not) but, because they do not see in it a sufficient solution to other problems, notably the ecological crisis; the limits of current forms of democracy and the complex forms of inequality that characterise advanced industrial society.

Their politics can justifiably be called innovative, simply because they direct attention towards new problems and therefore stand outside the dominant traditions of post-war politics. Their arguments are sufficiently coherent to identify inter-relations between different areas: for instance, decentralisation can be justified as enabling democracy, preserving regional identities and as less wasteful on ecological grounds. At the same time there is no single idea sufficient to embrace the whole of the above agenda. Ecology provides a justification for a radical reconstitution of politics, but it does not explain why the greens should be so egalitarian or so concerned about democracy.

Critics might say that even if these views are held in common across European green parties they are policy positions rather than ideology. However, green parties are especially programmatic parties. They use programmes as ideological statements more than attempts to maximise their vote. And, their programmes do the work that we demand of a functional definition of ideology. That is, they describe the nature of the world, they prescribe alternatives and they seek to motivate us to action. So, while it is true that party programmes are special kinds of documents with a strategic purpose and shaped by a variety of contextual factors, they cannot be ignored as resources for assessing the content of green ideology. Other sources such as the writings of green activists and ideologists, debates within green movements and parties, and the actions of greens also contribute to the shaping of green ideology. But green party discourse is the product of tests that are less common for other sources of green ideology. Green parties are required to apply green ideology to all political problems, and they do so collectively, through debates among activists, and they then have to develop and maintain a base of support among voters. This does not make green party ideology analytically superior, but does give it a certain legitimacy.

A shared ideology?

Importantly, these core themes of green party ideology provide a basis for constructing an ideal type of green ideology that applies as well to greens outside the green parties. Losing any one of the three core themes means that the ideology is no longer green. This means that some forms of ecological radicalism cannot be called green ideology. For instance, many of those whom John Dryzek terms 'green romantics' do not have a green ideology because they 'are generally uninterested in the social structures or institutions of industrial society, or indeed its alternatives' (Dryzek 1997: 164) and are not committed to expanding democracy or egalitarianism. This would include some 'deep ecologists', some 'bio-regionalists', some ecofeminists and lifestyle ecologists. When advocates of the latter concentrate on changing individuals' consciousness and do not regard social structures as major constraints on actions by individuals, they stand outside the ideology developed by the green movement. However, neither deep ecology nor ecofeminism are homogeneous traditions and both contain strands that are compatible with green ideology. Deep ecology is associated with the defence of biocentric egalitarianism, or ecocentrism, in which nature is recognised as having intrinsic value and human interests are not necessarily superior to those of the natural world. As the discussion below shows, when this leads to anti-human misanthropic arguments, which reject ideas of democracy and social justice, deep ecologists are not tolerated by greens. Most deep ecologists (Devall and Sessions 1985; Taylor 1995), however, seek to combine ecocentrism with commitments to social justice and democracy.[6] Similarly, most ecofeminists emphasise the need to challenge male-dominated views of nature, which they see as the source of ecological problems, and replace them with a feminist sensibility based on women's experience of being closer to nature, through childbirth, and the historically developed virtues through women's role as nurturers. Again, there is no necessary conflict between this and commitments to equality and democracy even though many other feminists criticise what they see as biological essentialism in many ecofeminist arguments. Both deep ecology and ecofeminism provide important sources of motivation for many green activists in a variety of radical and more mainstream groups. As Dryzek notes, their main impact is likely to be cultural. This is because the sensibilities they advocate do not provide answers to the kinds of political dilemmas that are most central for greens who seek to engage with social and political institutions. This suggests that the three core commitments remain the best litmus test for adherence to a green ideology.

Greens of various kinds seek to balance these three commitments and do not always give an a priori priority to ecological goals over others. For instance, green commitments to decentralisation have been defended on ecological grounds, as providing the best means of organising low-impact and self-reliant communities and, less instrumentally, as more consistent with nature insofar as communities are allowed to be constructed in ways that are shaped by geography and a sense of place; but, also on democratic grounds, because decentralisation increases the opportunities for individuals and groups to participate in decision-making and participation helps to develop citizenship. Decentralisation is also advocated because it breaks up the power of state elites and central bureaucracies. For most greens most of the time these are compatible and complementary commitments, part of what greens often call a 'holistic view'. Sometimes, though, these principles clash and this requires that judgements be made about which principle has priority; the broad framework ideology, however, provides no guidance on how this is to be done. For instance, when faced with the realities of the expansion of political union within the European Community in the early 1990s, greens in the European Parliament had to choose between outright rejection of the Maastricht Treaty or an alternative vision within the existing institutions. They chose to argue for a stronger European Parliament as a counter-balance to the bureaucrats of the European Commission and the narrower national interests of the Council of Ministers which was made up of representatives of national governments. Although they criticised the Maastricht Treaty because they saw its main purpose as supporting productivist expansion of European economies, they did not want to reject it outright because of the fear that it would encourage a return to narrow nationalism (Bomberg 1998: 77). It was non-ecological factors that determined the decision that reform from within was better than opposition from without. These factors were greens' anti-nationalism and the view that it was possible to make progress in democratising European Union institutions. But other green parties in Sweden, Britain and Ireland opposed the Treaty, not because they were less concerned with combating nationalism or more strongly opposed to economic growth than the other green parties, but because they saw opposition as the best means to achieve change.

Another example comes from a discussion within a local British Earth First! group. This group was discussing 'what we stand for' and used the UK Green Party manifesto as a starting point. The Earth Firsters! found little to disagree with in the document. As would be expected, decentralisation was defended on the grounds of conduciveness with sustainability but also because it allowed communities to govern themselves. When concern was expressed about the danger that some of these communities might pursue repressive policies even those most anarchistic among green activists were not necessarily confident in the ability of decentralised communities to defend human rights. As one said: 'You can't decentralise basic moral laws about rape or murder. And having a non-hierarchical system won't protect minorities that are being picked on'. On this and other issues there was much in common with the kinds of debates that occur in green parties. The point here is not to show that anarchist greens should therefore support green parties. In fact, there was no clear agreement in the Earth First! group on how to resolve the classic anarchist dilemma. However, what was clear was that what was being sought was a balance between the same three principles identifiable in the green parties' programmes and ideological debates.

Environmental movement organisations (EMOs) have more variable agendas, as is shown in more detail in Chapter Five. Their narrower focus on particular issues in their campaign work means that they seem closer to the idea of a public interest group than green parties or direct action groups. Yet it is still possible to find statements of ideological commitment from EMOs that are consistent with the general commitment to balancing ecological rationality, egalitarianism and grassroots democracy. For instance the mission statement of the Australian Conservation Foundation (ACF), not the most radical of Australian EMOs, includes in its philosophy both ecocentric statements such as the ACF 'Recognises that we share the earth with many other living things that have intrinsic value and warrant our respect, whether or not they are of benefit to us,' with social and political statements such as the ACF 'Believes that social equity and justice are fundamental to sound environmental outcomes' and 'Values participatory democracy and will work to defend the rights and enhance the role of all people in protecting the environment'. A pamphlet published by British Friends of the Earth (FOE) Boardman, Bullock and McLaren (1999) makes a strong argument that environmental sustainability and social equity are interdependent and sets out a range of policies to meet these aims. Although it might be argued that we would hardly expect environmental groups to argue in favour of inequality or against democracy, their commitment to interpreting environmentalism in these terms distinguishes groups such as the ACF and FOE[7] from the others such as the National Trust and the Royal Society for the Protection of Birds (RSPB) which do not see their environmentalism as requiring the same commitments. Perhaps the strongest test is whether the actions of EMOs are consistent with their philosophy. This issue is examined further in Chapter Five. The major point to emphasise is that not all green groups will act in the same way on every issue. Some such as EMOs are likely to focus more narrowly on issues of environmental policy. However, if they do so in a way which reflects commitments to social equity and participatory democracy, they share an ideological framework with more radical greens.[8]

Local environmental campaigns are too heterogeneous and often too short-lived to be able to make generalisations about their ideology. As Chapter Seven shows, although there is evidence that some of these groups become radicalised as a result of their experience as campaigners, most groups cannot be said to be green in ideology in the sense in which green ideology has been defined here. Yet, this is not to say that these groups do not develop ideology. Many have previous experience in other movements, others find themselves becoming more aware of the weakness of citizens' rights in existing democracies. It is from the environmental justice movement in the USA that new discourses of environmental racism and 'subaltern environmentalism' of the poor (previously seen as the preserve of non-western environmental groups) have emerged. These have posed an important challenge to green groups and have had an important impact on green ideology in general.

Excluding those who break with the green framework

When some eco-activists have given absolute priority to ecology at the expense of equality and democracy they have been regarded as extremists both inside the green movement and by outside observers, and often have been unable to work with other greens. One of the best-known instances of this was the dispute in US Earth First! over whether the movement was concerned with wilderness preservation to the exclusion of social justice and democracy. The infamous column in the Earth First! journal, in which a 'Miss Ann Thropy' (Chris Manes) argued that Aids and famine in Third World countries could be beneficial in reducing human population, caused strong criticism within Earth First! Arguments also raged over the status of feminism, seen as irrelevant to green concerns by wilderess defenders such as 'Miss Ann Thropy'. Underlying this was a basic disagreement about whether humanity was redeemable (Lee 1995). Many of the supporters of Miss Ann Thropy argued that human nature could not be changed and that the only way to save the world was to preserve as much wilderness as possible and prepare for the final collapse of civilisation. The disputes resulted in the departure of the misanthropes, led by Dave Foreman. They could not work with others in Earth First! such as Mike Roselle and Judi Bari who were concerned not only with ecology but also with egalitarianism and expanding democracy as ends of green action.

Other cases of ecological reductionism include Kirkpatrick Sale's (1985) claim that bio-regions should not necessarily be democratic and like nature ought to be as diverse as possible in their forms of government. Rudolf Bahro is another example of an eco-fundamentalist who found it impossible to work with other greens. After leaving the German Greens in 1984 over what he saw as the party's equivocation over vivisection, Bahro (1994) developed a more spiritual and anti-political stance. He argued that only complete withdrawal from the industrial system will provide the spiritual resources necessary for the transformation to a new society and that the charismatic leadership of a 'Green Hitler' might be what is needed to break with the current system. Herbert Gruhl, a founder member for the German Greens soon split from the party to form his own right-wing ecological party because the majority of the founders of the German Greens favoured egalitarian social policies and grassroots democracy over his support for an authoritarian and strong state. And the Unabomber, seen by some as a green because of his strong ecological sentiments and critique of the damage done by technology, derided what he saw as symptoms of psychological weakness evidence in current leftism. He believed that these weaknesses were manifested in feminism and arguments for egalitarianism. His strongly naturalistic assumptions about the instinctual character of human nature were used to support anti-egalitarian arguments, based on the traditional conservative idea that inequality is natural.

The key point about all these cases is that each of them remained committed to a politics based on ecological crisis, yet they were shunned by green activists insofar as they attacked other greens' commitments to democracy and egalitarianism. The degree of reaction depended on the circumstances. Sale remained within the ambit of the green movement because in his other writings he related ecological struggles to struggles for equality and democracy (Sale 1996). Dave Foreman, also sought to make his peace with his critics in a relatively friendly debate with Murray Bookchin (Chase 1991). However, he did not return to Earth First! and his legacy is now rejected by most of its activists. And although others left with him, the Earth First! network survived without them.

Bahro's case is more ambiguous. It is not clear that he has rejected egalitarian and democratic commitments so much as moved away from politics and political action. In Dryzek's terms Bahro is a green romantic. As Dryzek points out (1997: 163) green rationalists differ from green romantics first because the rationalists do not reject rationality and second because they believe that social structures do matter. Thus the causes and solution of the ecological crisis depend on changes in social structure and rationality can provide the basis for green strategy by 'open-ended and critical questioning of values, principles and ways of life – which opens the door to critical ecological questioning' (Dryzek 1997: 172). It is possible to combine some elements of green romanticism with green ideology, for instance, the romantic emphasis on the importance of emotion and experiencing the natural world, do not necessarily contradict rational green argument. But, as noted above, when romantics attack reason itself and seek to replace it with instinct and fail to support equality and democracy as green goals, they place themselves outside green ideology.

Each of the three themes of ecology, egalitarianism and participatory democracy has independent and equal status in green ideology, and without any one of them, it is impossible to explain the character of green ideology. However, while they are all therefore essential, they are not all equally distinctive in relation to other ideologies. As Andrew Dobson (2000) shows, the politics of nature, provoked by the evidence of a global ecological crisis, provide the most distinctive and novel element of green politics. It is not surprising therefore, that this should also lead commentators to seek to base all green ideology on the politics of nature.

Green leftism or ecocentrism?

Dobson (2000) and Eckersley (1992) argue that ecology provides the base on which the green superstructure of egalitarianism and democracy can rest. Robyn Eckersley defines ecocentrism as 'a picture of reality … in which there are no absolutely discrete entities and no absolute dividing lines between the living and the nonliving, the animate and the inanimate, or the human and the nonhuman' (1992: 49); whereas anthropocentrism, the basis of all other ideologies, views humans as having superior interests to the rest of nature. Extending this point, she argues that green politics includes emancipation but extends it to the non-human world. Radical green politics can be 'grounded in a broader defence of autonomy (let us say for the moment, the autonomy of human and non-human beings to 'unfold in their own ways and according to their "species-life") and by association, a broader critique of domination (of humans and other species)' (Eckersley 1996: 226). Ecocentrism can be seen as the third wave of emancipatory politics, extending emancipation to nature, following the liberal championing of civil and political freedoms against feudalism, and the socialist drive for greater equality between humans. But, while ecocentrism is new and distinctive, it might not be the only basis on which the greens can claim to be new.

There are occasions when actions by greens can only be explained as motivated by ecocentrism. For instance, when US direct activist Julia Hill endured a year on a platform in northern California to protect a giant redwood tree she did so not because of the impact of her action on global warming or to challenge a promethean approach to science – both of which could be justified on the basis of human-centred concerns – but because she believed that the forests had a right to protection and that they should be able to 'unfold in their own ways and according to their "species-life"' (in Eckersley's terms). However, the important question is not are such actions green, but are non-ecocentric ecological radicals insufficiently green?

First, even in their ecological politics, it is not so clear that all greens must necessarily be ecocentric. It can be argued that several themes in greens' ecological discourse are potentially independent of ecocentrism, such as the need to be cautious about technology or interventions in nature whose effects are unpredictable, the argument that a sustainable society must be based on reduced rates of consumption, and the importance of the idea of sufficiency as a basis on which to develop a less materialist and less productivist vision of the good life. Of course, ecocentrism, as defined by Eckersley, does not contradict these aims, but nor is it necessary to them. All can also be defended on the grounds that they are better for humans.

Greens' egalitarianism and green commitments to democracy have also been argued to stem from ecocentrism, but they need not do so. The French ecosocialists such as René Dumont, André Gorz, Jean-Paul Deléage and Alain Lipietz, all of whom have been significant figures with the French Green movement, have developed distinctive theories linked by the themes that the new conditions in society, of which the ecological crisis is one, require radical political and social changes towards greater democracy and global redistribution of wealth. None could be regarded as ecocentrics and ecocentric language is almost non-existent within the French green movement (Faucher 1999). Yet, they are characterised as much by their break with traditional socialist ideas as by continuities (see Whiteside 1997), and all share a commitment to the ideas outlined as green above.

A different example comes from activists involved in environmental direct action networks in Britain. Many said that the use of Earth First! as a label for their network was inappropriate. They rejected the idea that the interests of the earth could be placed above those of people and were worried that Earth First! meant people last. However, changing the label was difficult because even if there was little support in Britain for the kind of deep ecological ideas that had shaped the commitments of the early founders of the US Earth First! network, the name itself was still part of their common history.

This also seems to have changed in Australia. As Doyle says:

> Until as recently as five years ago, the environment movement in Australia reflected a particularly narrow definition of 'the environment' ... In the more radical groups 'wilderness' agendas dominated, with their concomitant 'ecocentric' arguments about the rights of 'other nature' ... More recently, the environment movement in Australia has been challenged to reshape its agenda into something more reminiscent of broad-ranging green movements elsewhere ... Reflecting the change in the social movement's agenda is the inclusion of issues relating to urban and rural environments, the rights of indigenous peoples, other issues of social equity, non-violence and democracy, while still advocating the rights of the non-human.
>
> (Doyle 2000: 215–6)

While ecocentrism might provide the strongest basis for a green claim to be distinctive, it is not the only basis for such a claim. There are other ways to be green and radical without being simply 'environmentalist', or a hybrid of existing ideologies. This also means that while green ideology can include ecocentrism, it is not essential to it. [9]

One of the concerns about ecocentrism then is that if seen as foundational it might be privileged over egalitarianism and democracy when conflicts arise between the interests of humans and those of nature. Of course, not privileging one principle over another does not mean that conflicts between them can be avoided. But if, when such conflicts arise, they are kept distinct, it will at least be clear which principle is being favoured.

What are the consequences of making egalitarianism and democracy essential features of green ideology, along with ecological rationality? The first is that green authoritarianism and right-wing green ideology become oxymorons. Conservative inclinations attributed to green thought include the suspicion of progress and the association of history with the growth of destructive human rationality. Yet, on the question of progress, there is little evidence from green movements of hostility to progress, beyond the advocacy of caution with regard to developments that might threaten environmental stability. This can be defended as much on rational grounds as any other given the evidence of ecological crisis and green arguments that scientific research and particularly technological development reflect the interests of those with most power. Although there may be other logical and theoretically interesting possibilities for a politics based on ecology to proceed in a conservative or authoritarian direction this is not green politics as it exists, or is likely to exist, within green movements.

A final alternative means of defining green ideology as based on ecology is Mary Mellor's suggestion that the logics of different ideologies can be kept distinct. She says that a feminist green socialism should be:

> Feminist because it acknowledges the centrality of women's life-producing and life-sustaining work and focuses upon the predominance of men in destructive institutions. Green, because it argues that we should act and think globally to regain a balance between the needs of humanity and the ability of the planet to sustain them. Socialist, because it recognises the right of all the people's of the world to live in a socially just and equitable community.
>
> (Mellor 1992a: 279)

Commenting on this, Freeden (1996) points out that in cases of conflict it might be difficult to establish which of feminism, ecology and socialism has priority. This point also applies to the relationship between emancipation and the interests of nature in Eckerlsley's (1992) definition of ecocentrism. For Eckersley, ecocentrism is based on the assumption that questions of humanity's relationship with nature are logically prior to questions of social organisation. This suggests that in cases of conflict, the interests of nature will be decisive. However, I am not convinced that many greens would go this far. In some cases, such as experiments on animals for medical research, greens might be prepared to run the risk of failing to alleviate human suffering by refusing to make animals suffer. However, there are also cases where nature may lose out to human interests. For instance, few greens would back the shoot-to-kill policy used against poachers in African wildlife parks[10] if poverty is a major cause of poaching. Nor would they want to punish poor farmers who burn rainforests. If egalitarianism, democracy and ecology come into conflict, the priority given to each will have to be negotiated politically. In such conflicts greens will take different positions, while remaining identifiably green.

Mellor's position is perhaps the most persuasive alternative to the definition argued for thus far in this chapter. It avoids the problem that might arise in too broad a definition of green ideology, that of subsuming different ideological positions under one general heading. For instance, greens do not all agree on every form of equality. The forms of feminist discourse differ within and between different green parties, even though their policies on civil rights and male violence and measures to reduce inequalities of gender are similar. And few greens would want to be defined as socialist, even if many would be prepared to be defined as anti-capitalist. Yet, while grouping diverse green discourses under equality, democracy and ecology hides much disagreement, this disagreement is not greater than that in other ideologies. Dobson says (2000: 4) that while we use terms such as social liberalism to signal diversity within ideologies, we still want to retain a sense of the differences between ideologies. Thus greens' distinctiveness rests on their ecological politics. If greens based their other egalitarian and democratic arguments on ecological rationales, such as the view that nature can provide us with guidance about the social world, it would be possible to show that green ideology was ecologically based. Although a few greens do this, most do not, not least because it is so unconvincing to see in nature guidance about the social world, given that nature has also been used to support inequality and even racial superiority and since the interdependence of species in nature does not prove that social equality between humans is necessary. Interdependence might also be said to characterise hierarchical social relationships such as that between master and slave. Thus, it seems that if we are to define green ideology in a way that includes both the most ecocentric greens and a larger number of non-ecocentric green activists, who themselves feel ideologically distinct from other political groups, we need to see ecology, egalitarianism and grassroots democracy as all essential elements in green ideology. And, even if the ecological politics of the greens are their most distinctive feature, they do not define green ideology in full. Thus, as Barry says (1999) green politics is ecologically centred but not ecologically based. Moreover, it is distinctive because it combines ecology, egalitarianism and democracy in a radical form that challenges taken-for-granted features of current society in ways that are not done by any existing ideology.

Green ideology as left-wing

This treatment of green ideology has been very broad. There has been no space to do justice to the rich literature on the connections between green ideology and liberalism (Wissenburg 1998); conservatism (Gray 1993; Freeden 1996) or socialism (Pepper 1993a; Ryle 1988; Williams 1989). Similarly, the complexity of the debate about ecological values in political theory (on which see Barry 1999; Dobson 2000; Eckersley 1992; Goodin 1992 and Hayward 1995) has also been over-simplified or bypassed. The justification for this is the aim of defining green ideology so that it embraced as many green activists as possible in a way that was also compatible with the second aim of showing that green ideology was distinctive. This meant challenging the minimalist definition of green ideology as ecologism and ecocentrism as the basis upon which other green commitments must rest.

Seeing green ideology as based on three principles: ecology, egalitarianism and democracy has the strength of applying to larger numbers of greens than definitions that privilege ecology. It does not prevent us from distinguishing green radicals from environmentalists, or from showing that green ideology is new. However, it also allows us to show why green ideology partly overlaps with other ideologies, particularly anarchism, feminism and socialism and is an ideology of the left. The fact that greens have been unsure about whether or not they are on the left is indicative of the wider uncertainties about what the left means in contemporary politics as well as the problems of classifying the diversity of themes in green party agendas. Their own view of their leftness or their rejection of a left-identity depends mainly on their own perceptions of the national traditions of the left in their own countries (Doherty 1994: ch. 2). However, as socialism has weakened in the 1990s, the resistance among some greens to defining themselves as on the left has all but disappeared. Almost all EDA activists interviewed in a UK study (Doherty, Plows and Wall 2001) defined themselves as on the left. No one defined themselves as on the right. Green parties in western Europe now overwhelmingly define themselves on the left. Lee Rhiannon a member of the Australian Greens regrets the fact that some other parts of the left in Australia do not readily identify the Greens as part of the left (Arena 51, February 2001). The uncertainty about the meaning of the left stems from the fact that socialism as an ideology appears to be in crisis. A number of writers have seen the crisis of socialism as providing an opportunity to distinguish between socialism and the left (Bull 1993: 413–23). For instance, Stephen Lukes (1990: 571–8) has said that whilst the demise of actually existing socialist régimes in eastern Europe has meant the end of socialism as an economic idea and that any future socialism has to occur within a capitalist economy, the core ideals of the left remain relevant. As a historical phenomenon the left has had two principles in combination: the Republican aim of making formal equalities more real and the egalitarian aim of making social inequalities more equal. If we follow this definition of the left, then the core features of green ideology outlined here are clearly left-wing whilst also sufficiently different from the ideology of socialism to be classified as a new variant within the traditions of the left, one that raises ecology to an equivalent status to democracy and egalitarianism. Analysis of the debate over ecological modernisation helps to show from another direction why greens' ideology cannot be explained only by their ecological commitments.

Ecological modernisation

In recent environmental sociology ecological modernisation has emerged as a new paradigm for understanding developments in ecological discourse. It reflects changes in the nature of the environmental debate which followed as the seriousness of environmental problems has increasingly been accepted by mainstream political parties and business. Central to ecological modernisation is the view that:

a capitalist or rather market-based system of production and consumption does not necessarily contradict significant environmental improvements in any fundamental way. More production and consumption in economic terms (GNP, purchase power, employment) do not have to imply more environmental devastation (pollution, energy loss, loss of biodiversity). Within principally the same modern institutional lay-out (a market economy, an industrial system, modern science and technology, a system of welfare states, etc.) we can thus look for – and design – radical environmental reforms.

(Mol and Spaargaren 2000: 36)

Ecological modernisation theory is complex and much more sophisticated than a simple advocacy that everything is all right with capitalism. Much must change for ecological modernisation to be feasible, and how much must change separates weak versions from the stronger versions that would be most compatible with a green ideological framework (Christoff 1996). For the most part greens have been excluded from the discourse coalitions of policy-makers that have formed to pursue environmental reforms based around the idea of ecological modernisation (Hajer 1995). However, as green parties have entered government they have been forced to confront these questions more directly. The record of greens in government will be examined further in the next chapter, but both the Swedish greens (Elander 2000) and the German Greens (Rüdig 2002) have increasingly talked about economic growth within the framework of ecological sustainability and the anti-capitalism that characterised the German Greens' economic programmes in the 1980s is gone. Thus there is evidence that ecological modernisation has had an impact on these green parties.

As a framework for analysing the changing nature of the ecological debate ecological modernisation has some advantages. It does not assume that capitalism is environmentally benign, but clearly it is in conflict with those greens who see capitalism as the motor of productivism. While many greens remain anti-capitalist, as noted above, there is no clear consensus within the green movement on whether capitalism in some form is compatible with ecological sustainability. There is a consensus, however, on the view that current forms of capitalism are incompatible with ecological sustainability. Ecological modernisation arises principally out of a concern with policy studies geared towards 'actual transformations and designs of institutions and social practices' (Mol and Spaargaren 2000: 37). Green parties and many green EMOs are also engaged in this field and as they have had to move from emphasising the problems to deciding what can be done immediately; they have had to develop more incremental policies in response. The central question is whether by doing so they 'make capitalism less in need of a green critique' (Mol and Spaargaren 2000: 22). The dilemma is not dissimilar to that faced by socialist parties regarding the welfare state. For many socialists the welfare state brought important, though insufficient, advances. For many Marxists, the welfare state represented a structural adaptation by capitalism under threat. However, in distinction from the latter, few radical greens would want to see the ecological crisis worsen.

Ecological modernisation is less relevant to understanding the broader nature of green ideology. Greens have been shaped by a broader left discourse on egalitarianism and democratisation and it is not only for ecological reasons that greens are critical of current society. Even if the ecological crisis was solvable within existing institutions this would probably not satisfy greens. Along with neo-Marxist critics of ecological modernisation theory greens share a concern with the new kinds of social inequality that might result from ecological modernisation. For instance subsidies for organic farming might be directed primarily towards the most economically efficient operations, reinforcing the trend towards larger farms owned by financial institutions and further weakening smaller family farms.

Conflicts over environmental issues often cut across traditional class and other social divides. Ecological risks such as climate change can affect all groups irrespective of class (Beck 1992), although it is important to add a qualification to this, since even global risks most affect those with least resources (farmers in Bangladesh for instance, or those with poor quality housing in central America are much more vulnerable to the floods and cyclones that result from global warming, than northern Europeans). Yet there are still some risks such as BSE or genetic modification that are less class-specific and all social groups have to deal with the increased uncertainty that these create. A number of writers (Giddens 1994; Cotgrove 1982; Paehlke 1989) have seen the divides over environmentalism as separate from those between left and right, but while this may be true of environmental debate in general it is not empirically true as regards the green movement. There are certainly conservative forms of environmentalism, and divisions within the environmental movement, however, the green movement as analysed here is linked by its commitment to egalitarianism and a radicalisation of democracy. There is a danger of confusing the engagement of the greens in institutional politics with a turn to the right. Although greens in government are forced to address more difficult decisions, such as the trade-offs between high fuel taxes and the fact that these hit the poor hardest, these decisions reflect the dilemmas of policy-makers. Moreover, despite their marginal position, greens have pursued more radical social policy within government, particularly on the rights of migrants, gay and lesbian rights and reducing work time. In general it has been the greater conservatism of their larger socialist partners that has limited their achievements on these issues. However, this does not mean that the greens are either ideologically homogeneous or immune from deradicalisation. In the next chapter the relative deradicalisation of the German Green Party in the 1990s will be analysed.

Linking ideology and action

The version of green ideology that has been outlined above is essentially a framework. As analysed above, the green framework provides no answers on the question of how to decide on which commitment has priority in cases of conflict: the answers to such dilemmas are worked out by activists in specific contexts. The content of green ideology develops in relation to this kind of ideological work and in the collective action of green movements. As Bevir says:

> Ideologies are not fixed entities that we can take as given. They are ideas and practices that people produce through their activities. They are contingent and changing traditions in which no value has a fixed, central or defining place. We can describe an ideology adequately only by tracing how it develops over time as its exponents inherit beliefs and actions, modify them and pass them on to others.
>
> (Bevir 2000: 280)

If we follow Bevir, then even the three core commitments that I have argued as currently constitutive of green ideology may not remain so. Thus there is no permanently fixed meaning of 'green ideology' if by that we mean the ideological framework shared by most activists within the green movement.

Second, the shared green framework does not tell greens what action to take, although it does limit their choices. As Freeden says:

> The tightly assumed flow between given theory and advocated practice no longer obtains ... ideological morphology dictates the existence of multiple routes from a given theory to a range of given practice and from a given practice to a range of possible theories.
>
> (Freeden 2000: 320)

For Freeden ideology[11] is a distinct thought practice, one which is central to politics because it provides a language of meaning and a set of conventions that shape political discourse and 'enable us to choose to become what we want to become' (1996: 553). Yet, while it provides the tools which make a certain kind of political action possible, it does not resolve the questions of how best to achieve goals. Thus while strategic choices are ideological questions, ideology does not provide the solution to strategic dilemmas. Implicit in this model is a view of ideology as a condition of collective action – one that creates a 'we'. Ideology is therefore related to identity, but not wholly equivalent. Identity includes the culture and ethos of the movement as well as its shared beliefs. Identity must be consciously held and practised, whereas ideology may be discerned by commentators without necessarily being acknowledged as such by activists. Identities can be strong or weak, inclusive or exclusive. Ideologies cannot be any of these. Identities also always define a political relationship in terms of them and us. We would expect adherents to green ideology to have a green identity, but it is possible for some who see themselves as greens to not support a green ideology. This might be the case, for instance, with some environmentalists, who reject the view that major social and political changes are necessary to deal with the ecological crisis. But, an assessment of green ideology that was wholly divorced from green identity would be implausible. An account of green ideology should be based on what distinguishes it from other ideologies and will also be stronger where it can show that this distinctive position is held by a large number of those who identify themselves as greens. The content of ideology must overlap to some extent with real green identities, otherwise it is of only academic interest.

How then do we explain how movements create ideology? One way to do so is to start from the position that social movements' ideas develop in response to the experience of their activists: and new ideas emerge when actors need new tools to make sense of their experience and alter through debate with opponents.

MacIntyre's notion of 'tradition' is consistent with this emphasis on the changing and contextual approach to ideology. MacIntyre regards a tradition as the site of basic epistemological crises (MacIntyre 1985: ch.15). He describes the human condition as a form of dramatic narrative in which problems of truth and intelligibility continually arise. When we overcome such epistemological crises we do so through a reconstruction of the tradition making previously contradictory theses intelligible. This is achieved not so much through an appeal to the facts as by new ways of showing what an appeal to the facts could mean. However, there must be some continuity between earlier and later traditions because otherwise we cannot understand the nature of our task. In terms of social movements' ideology this provides an epistemological basis for seeing ideologies as generated by people in concrete situations working within an existing tradition (in this case that provided by the left) but reformulating it when confronted by new problems.

The ecological crisis provides one example of such an epistemological crisis. The emerging evidence about the effects of human actions on nature called many of the core beliefs of the modern era into question such as the equation between material economic growth and progress, and the status of science as superior to other forms of knowledge. This crisis was resolved by arguing that instrumental attitudes to nature needed to be controlled and balanced by ecological rationality and production constrained by ecological limits. Greens were also influenced by other epistemological crises, including the reaction of the New Left against the statist forms of existing socialism. The experience in anti-authoritarian movements of the late 1960s had a profound influence on many of the generation that formed green movements. They resolved the crisis by rejecting many of the core ideas of the old left such as the need for a disciplined hierarchical party. By acknowledging that there were other 'progressive' forces as well as (or for some instead of) the western working class they, along with other movements, opened up space for issues that had previously been defined as 'diversionary' such as racism, patriarchy and consumerism.

Approaching ideology as a form of dramatic narrative fits well with the view that the greens have been tackling what they see as the inability of the classical ideologies to respond to new crises. Thus greens believe that a reconstitution of theory is needed and yet, whether they accept it or not, this must also draw to a large degree on elements of the existing political ideologies.

Although the analysis of ideology tends to be dominated by political theory, recent debates among social movement theorists have highlighted the importance of ideology in studying movements. Ideologies were often treated as simply an element of movements that needed description, but not explanation. Nor were ideologies necessarily accepted as causal of movement action. However, since Dalton's (1994) analysis of survey material suggested that ideology was a better predictor of the actions of environmental groups than the factors stressed by resource mobilisation (access to resources) or new social movement theory (the social groups taking action and the level of postmaterial values) the concept of ideologically structured action has become a focus for debate. Perhaps most notably, Zald (2000), the originator of resource mobilisation theory, for which a prime aim was to seek explanations other than ideology for collective action, has argued for a shift to examine the origins of movement ideologies in socialisation and the impact of social movement ideas on general publics and state actors.[12]

Eyerman and Jamison (1991) argue that the emergence of new ideas from social movements can be explained as a process of 'cognitive praxis'[13]: social movements exist when a group creates a new 'knowledge interest'. In the case of the environmental movement the three dimensions of its knowledge interest are an ecological worldview, small-scale alternative technology and science for the people, calling into question the ways in which knowledge is produced (Jamison, Eyerman and Cramer 1990: 5–6). These emerged in the debates of critical scientists and the actions of newly formed environmental groups in the late 1960s.

Eyerman and Jamison (1991) also argue that when the new knowledge interest becomes institutionalised, the movement is no longer a social movement. In the case of environmentalism, once the major parties begin to adopt environmental ideas, alternative scientists get posts in universities and the environmental movements begin to become involved in the policy process and develop more bureaucratic structures, the movement is no longer engaged in the creation of a new form of knowledge. As movements fragment into specialised sub-units their common project dissolves. This account of the origins of the new environmental consciousness has the strength of being based on a historical analysis of environmental movements in Sweden, Denmark and the Netherlands. Jamison, Eyerman and Cramer (1990) are describing the origins of modern environmentalism, rather than the origins of a distinctive green ideology. Indeed, radical green ideology would be more difficult to absorb than environmentalist arguments. However, they also show that more radical green ideas were marginalised in these countries as opportunities opened up for more moderate environmentalists. Thus the actions and discourse of the new movement were both affected by changing contexts.

Although the concept of cognitive praxis is useful as a way of explaining the emergence of new social movement ideas, it should not be used in such a way that it limits the definition of social movements too much. Movements do not end because some of their ideas become more widespread and although movements and their organisations are distinct, we need not define movements as necessarily anti-institutional, for the reasons argued in Chapter One. Eyerman and Jamison's (1991) definition of a social movement limits it to a brief moment of knowledge creation, but this means excluding from analysis many of the strategic questions that social movements face, such as the tension between confrontational and consensus styles of action and grassroots and profession- alised campaigning. As the recurrent appearance of more radical groups within the green movement shows, there is still a more radical presence within the green movement, and, as the analysis in the next two chapters shows, even the more institutionalised groups are not necessarily de-radicalised.

Much of the analysis of the relationship between ideas and actions in social movements is based on the concept of framing (Snow *et al.* 1986; Snow and Benford 1992). Frames can be more specific than ideology, but are similar in their function in that they identify a problem, who or what is to blame, and specify the action necessary to resolve it. But whereas ideologies must be distinctive from one another, a frame, because it is more specific, can be shared by activists from different ideological traditions. For instance, anarchists, social- ists, liberals, conservatives and greens could justify direct action against nuclear power as necessary and effective and share the view that it was a dangerous, and economically non-viable technology, without sharing an ideology. The scope of frames can be more specific than the usual use of ideology in political theory – focusing on particular issues rather than delineating an all-embracing worldview. Framing is also sometimes viewed as a strategic process through which movements seek to communicate their ideas, mobilise support and encourage action.

The use of framing has several advantages as a way of analysing movement ideologies. First, it does not treat ideologies as simply given, rather frames have to be developed as a process of interaction; within a movement and between movements, opponents and third parties. This means that frames are likely to develop and change over time.[14] Second, it concentrates attention on how move- ments work to create new meanings. It is persuasive to argue that the impact of movements depends in part on their ability to create frames that have popular resonance.

Among the criticisms made of framing is that it tends to be too instrumental. Framing is presented as something that movements, and usually movement leaders, do in order to persuade people to support them. In fact, Snow and Benford do not see framing exclusively in this way. They also say it is about how people make sense of the world and develop an interpretative map to make sense of events.[15] However, they do leave the relationship between frames and ideology unclear. As Steinberg notes:

> Left ambiguous are (a) whether frames are composed of elements from some systematic belief system that is situated outside of social movement processes, as their references to values and beliefs imply, and (b) whether framing is therefore derivative and dependent upon ideological processes that operate outside the ken of social movements. Alternatively it is possible that ideology is an emergent and interactional product of framing and is essentially produced in framing.
>
> (Steinberg 1998: 847).

Steinberg (1998) argues that this ambiguity can be avoided by analysing framing as a form of discourse. Discourse theorists such as Bakhtin argue that discourse is dialogic, dynamic and riven with contradictions. Discourse is an interactive process of producing meaning within specific historical situations. People do not (usually) talk in ways that match the coherence of ideology understood as a systematic worldview. And for social movements engaged in conflict, the wider discursive context always includes the meanings generated by opponents as well as the multiple voices from within the movement. But discourse is also ideological in that it does not convey neutral meanings. Steinberg suggests that:

> As opposed to the often implicit assumption in framing theory that cognitive order and focus are imported externally from a belief system to a collective action frame, a discursive perspective suggests the reverse: the processes of collective action often impress a conscious and explicit order on ideological discourse. The dialogic and often counter-hegemonic process of structuring meaning can be seen as requiring a specificity and clarity beyond the pragmatic and commonsensical use of the everyday generic and hegemonic discourses within a field.
>
> (Steinberg 1998: 856–7)

Ideology develops as a more coherent and abstract statement of principles because movement activists have to justify their own actions to themselves and respond to their critics. Steinberg argues that frames can be seen as repertoires. They are not static ideas belonging to individuals so much as a product of collective action. The repertoire will be shaped by the discursive traditions of networks of activists – and also by the targets of their action, the point in a cycle of protest and the precise character of the issues involved.

This brings us back to the question of the static picture of ideology presented in the first part of this chapter. As noted, the three core commitments denote a framework which is shared by green activists. However, major differences remain within this framework, such as the differences between the strong anti-statism of anarchistic greens and the participation in state institutions of green parties and EMOs. These are also ideological and reflect central disputes over strategy, which, as will be argued in the final chapter, are probably not resolvable. Moreover, the life of such arguments is less to do with the logic of green ideology per se, than the experience of different parts of the green movement in different spheres.

In order to understand how frames are produced we need to examine the use of frames in action. Frames are a product of the need to make sense of the world, driven by the desire for ontological security. Frames are not necessarily devices to achieve changes within the existing structures of society. Movements can also create frames which reject the institutions and practices that constitute the dominant order. The core principles of Green ideology might therefore be seen as meta frames – based on general values shared and reaffirmed by various different types of green movement. Within these meta frames there are what might be called meso and micro frames (Welsh 2001) which reflect different kinds of actions and strategies developed in pursuit of green goals. Meso frames refer to the engagement with institutional contexts and the kinds of action repertoires deemed appropriate. At this level greens disagree greatly over strategy. Some give priority to more formal and institutional strategies, others to more informal and extra-institutional methods. Micro frames reflect 'an individual's awareness of affect, the quality or intensity of feeling evoked within a particular strip of activity' (ibid.: 15). Welsh argues with reference to evidence from the Prague anti-IMF/World Bank protests, that 'meta frames do not establish an overarching set of meanings that orchestrate other levels of framing at all times. We would suggest that the framing process can also be driven from the micro-level up' (ibid.). As individuals strive to make sense of a 'strip of activity' they draw on both their abstract commitments to general values and their experience of emotion or affect. In informally organised networks such as those in Prague from the ecological direct action groups, weak ties to many groups based on networked word of mouth, help frames to diffuse in ways that are hard to predict, but which given the growth of this movement, may have considerable potential to achieve social transformation.

Conclusion

I have argued that green ideology is best defined by shared commitments to ecological rationality, egalitarianism and grassroots democracy. The concrete content of green alternatives varies but greens of different kinds use these three commitments to define boundaries between them and other political groups. As analysed here ideology is seen as reflected in the discourse of activists and developed through their actions. In comparison with the search for coherence and distinctiveness in the study of ideology in political theory this more sociological approach can only produce a weak framework, within which greens will differ considerably, as for instance in the differences between the anti-statism of anarchistic direct action groups and the more limited grassroots democratic project of green parties or EMOs. But, here the distinction between a concrete utopia as a blueprint, and a utopian orientation towards an alternative kind of society seems apposite. Greens share something of the latter, even while they are less sure about what precise form it will take and the best means of bringing it about. In the next four chapters the problems of green praxis will be examined through case studies of the experience of different kinds of green groups.

4   Green parties

Green parties, once regarded as anti-system parties (Bürklin 1987; Fogt 1989), are now part of coalition governments in several western European states. Since the early 1980s they have grown in electoral strength and representation in most countries, and are now a familiar part of the political scene in most of western Europe, and Australasia as well as Brazil, the USA and the Ukraine.

The first green parties were founded in Tasmania, (known as the United Tasmania Group) New Zealand (the Values Party) in 1972 and Britain ('People') in 1973. But these parties remained small and had little political impact. It was elsewhere in Europe that the major breakthroughs for greens in elections occurred. In France, beginning with the presidential candidature of René Dumont in 1974, and in Switzerland, Belgium and Germany by the end of the 1970s, greens were standing in elections, new parties were being formed and by the early 1980s MPs were being elected. There was an upsurge of green support in the late 1980s, with new parties in Italy and Sweden benefiting from the resurgence of opposition to nuclear energy in the wake of the Chernobyl reactor disaster and a wave of green votes in many countries in the 1989 European elections. Despite some volatility in elections green parties with existing representation tended to consolidate or improve their vote in the 1990s, while in other countries greens gained representation for the first time. In New Zealand a new Green Party was formed in 1990 and gained 7 per cent of the vote in its first election, gaining its first MPs at the next election in 1996. Greens in Ireland, and Australia[1] gained national electoral representation, and in the UK greens were elected to the European Parliament, the Scottish Parliament and the Greater London Assembly, and doubled their previous highest percentage in the general election of 2001, gaining 3 per cent. In the USA the 3 per cent of votes, and particularly the 97,000 votes won in the linchpin state of Florida, by the green presidential campaign of Ralph Nader, in 2000 led to the greens being blamed by Democrats for the defeat of Al Gore by George W. Bush. In eastern Europe, despite the importance of environmental campaigning in opposition to the communist regimes in the 1980s, green parties have been weak in the post-communist era. There have been temporary upsurges, such as the 1.39 per cent won by the Russian greens in 1995 and the more significant 1.5 million votes (and 17 MPs) won by the Ukrainian greens in 1998, but, as yet, no sustained green presence. In southern Europe green parties have also been weak. The Greek greens imploded soon after gaining their first MP in 1989, and there were doubts over the independent status of the Portuguese green MPs in the 1980s and 1990s, who were seen as largely a front of the Communist Party. While there have been some recent successes for Spanish greens, gaining two MPs in national elections in 2000, it is too early to say whether this means that previous regional divisions have been overcome. Table 4.1 below summarises recent electoral results for some of the more enduring green parties in western Europe:

Despite the ubiquity of green parties, it is notable that in national parliamentary elections they tend to achieve only between three and ten per cent of the vote. Only in Belgium, where the two green parties, the Flemish Agalev and Wallon Ecolo gained 14.4 per cent in 1999 and entered a coalition government, have greens broken the 10 per cent barrier. In majoritarian electoral systems such as France, or semi-majoritarian systems such as Italy, greens have only been able to gain national representation as part of a left electoral alliance. Also, green parties have been unable to gain national representation even in PR systems where a New Left party already existed, as in Denmark, Norway and the Netherlands. In these countries the New Left parties were close ideologically to the greens, and already had a vote based on the social groups which constituted the greens' core constituency in other countries. This meant that there was no basis for an independent green party, although it did not prevent some from

Table 4.1 Electoral performance of selected European green parties in recent national elections

Year	Election type	Country									
		Belgium	Finland	France	Germany	Italy	Sweden	UK	Neths	Austria
1994	Parl	–	–	–	7.3	2.7	5.0	–	–	–
1994	*Euro*	11.1	–	2.9[1]	10.1	3.2	–	3.2	3.7	–
1995	Parl	8.4	–	–	–	–	–	–	–	4.8
1995	*Euro*	–	–	–	–	–	17.2	–	–	–
1996	Parl	–	–	–	–	2.5	–	–	–	–
1996	*Euro*	–	7.6	–	–	–	–	–	–	6.8
1997	Parl	–	–	5.16*[2]	–	–	–	1.4*	–	–
1998	Parl	–	–	–	6.7	–	4.5	–	7.3	–
1999	Parl	14.4[3]	7.3	–	–	–	–	–	–	7.4
1999	*Euro*	15.9[4]	13.4	9.8[5]	6.4	1.8	9.5	6.3	11.9	9.2
2001	Parl	–	–	–	–	2.2	–	2.9*	–	–

Sources: green party websites; (Carter 1999 and Hooghe and Rihoux 2000).

Notes:
* Represents the average percentage gained in seats where the party stood a candidate.
1 Génération Ecologie also gained 2% – running in competition with Les Verts.
2 Other green candidates (mainly from Génération Ecologie and Mouvement des Ecologistes Indépendent)   gained 2.7%.
3 Ecolo gained 18.3% in the Walloon region; Agalev 11% in the Flemish region.
4 Ecolo gained a remarkable 22% in Wallonia; Agalev 12% in the Flemish region.
5 Waechter's Mouvement des Ecologistes Indépendent also gained 1.5%.

trying. In Denmark and in the Netherlands unsuccessful pure green parties co-existed with the Socialist Peoples' Party and the Green Left respectively.

Thus while greens were in government in Finland, Belgium, France, Germany and (until 2001) in Italy, they were always minority parties, representing a specific segment of the electorate. Votes for green parties have generally been higher in European elections, probably because since these elections do not decide a government, the stakes are lower, allowing voters to express sympathy with green ideology or to make a protest, usually against the major left party. For instance, the Swedish greens gained 17.2 per cent of the vote in the 1995 European elections, capitalising on opposition among Social Democrat voters to their party's support for membership of the European Union. However, while this may indicate a higher potential vote for green parties, most seem likely to be at best 10 per cent parties for the foreseeable future. One of the main dilemmas they have faced, given their limited electoral constituency, is how to maximise their political impact. How are parties who want to make radical changes in society, politics and the economy to find what the French greens have called 'social majorities', referring to diverse constituencies of support on different issues, rather than a mass of fully converted greens?

This chapter examines the nature of green parties, focusing in particular on the question of whether they have been transformed from movement parties to parties that are no longer substantially different from the others. Greens defined themselves as different, but this rested on several separate claims. First, they were parties based in extra-parliamentary movements. Second they were parties that organised differently, based on principles of grassroots democracy. Third, they were parties with a new ideology. The interpretation of each of these varied between green parties, as did their importance relative to other goals. These variances reflected different institutional contexts, different ideological traditions, and cultural settings within the alternative milieu and the environmental movement in these countries. It is impossible to encapsulate the detail of this experience in one chapter, instead the main focus will be on case studies from two green parties, those of France, and Germany. These two parties have contrasting traditions, electoral strength and operate in very different contexts. Three aspects of their experience will be examined: party formation; the dilemmas of growth, and being in government. In conclusion the question of whether the greens remain radical parties will be examined.

Green party formation

Germany

The German greens were the creation of a heterogeneous coalition with its roots in the anti-nuclear protests of the late 1970s. The formation of the party[2] was preceded by several electoral experiments, which reflected the full range of groups involved in the anti-nuclear movement. The first electoral lists had been formed in 1978 at *Land* (regional) level by conservative ecologists who sought to balance the influence of the far left on anti-nuclear campaigning. Herbert Grühl, who had been a CDU MP was the leading figure among this tendency. He had written a bestseller *Ein Planet Wird Geplundert* (1975) which presented a survivalist view of the ecological crisis, and advocated strong government, and increased spending on defence to deal with the social collapse and the increased threat from the east that was likely to follow the ecological crisis (Hülsberg 1988: 87; Rüdig 1986: 405). The conservative founders became less dominant as these groups drew in more participants from the citizens initiatives. In the 'city states' of Berlin, Bremen and Hamburg various alternative left 'rainbow' electoral lists had also been established.[3] In 1978 Rudi Dutschke, the most charismatic figure of the German New Left in the late 1960s, had called for a new party of the left, but the alternative left was too small and isolated to be able to build a national party on its own.

Between the alternative left and eco-conservatives was a third group, represented by activists from the BBU, some of whom were also SPD members, who had played a major role in integrating the broad networks of the anti-nuclear movement. Noting the relative electoral successes of the French ecologists and the failure of anti-nuclear protests in Germany to make any impact on government policy, the BBU advocated the formation of an electoral list for the European elections of 1979. Electoral law for European elections was less restrictive and this allowed the 'Other Political Association' (SPV) to stand without too much internal negotiation, gaining 3.2 per cent of the vote. This convinced the various currents that had been exploring electoral avenues for ecological politics that a constituency existed and because of the funds provided by reimbursement of electoral expenses, provided important funds to invest in such a venture (Rüdig 1986: 383). A green list gained 5 per cent in Bremen in October 1979 surpassing the threshold for representation, providing further support for the new party project.

The founding congress was held at Karlsruhe in January 1980. With the exception of Gruhl's party, all those who had been involved in the European election campaign supported the effort to include the alternative left groups in the new party. Gruhl's argument that the commitment to a redistributive welfare state was incompatible with reducing economic growth was also rejected at this conference, as was his argument for strong defence. When redistributive, anti-bureacratic and anti-militarist policies were reaffirmed in the party's first full programme agreed at Saarbrücken in March 1980, Gruhl left the party with around 1,000 of his followers. This group formed a new eco-conservative party the Ecological Democratic Party (ÖDP) but this failed to make an electoral impact (Mewes 1997: 38).

The new party's programme was, as the alternative daily *Taz* commented 'a listing of all the demands which the left, including its previously dogmatic and Marxist-Leninist factions, had developed over the last ten years and propagated in public' (quoted in Frankland and Schoonmaker 1992: 129). Yet this was not a leftist takeover of an ecological party. It was the growing gulf between the agenda of the ecological conservatives and the broader social orientation of the

BBU and other social movement groups which shaped the new party as one more oriented to the alternative left than the eco-right. The only feasible basis for an ideological common project was in the shared framework of grassroots democracy, non-violence, social justice and ecology that constituted the four 'pillars' of the first green programme. Yet, even this covered a broad range of ideological positions and reflected a considerable process of collective learning in which the citizens' initiatives had become more oriented towards new forms of democracy and the linkage between ecological politics and social themes, and those parts of the alternative left that joined the greens had come to regard counter-cultural projects as too introverted and saw in ecological themes a new ideological challenge.

The German greens performed badly in their first national election in October 1980, gaining only 1.5 per cent of the vote, but this was at least partly due to concern on the left at the prospect that the right-wing Bavarian leader of the CSU Franz-Josef Strauss, could become Chancellor, encouraging a rallying of the SPD vote. Between 1980 and March 1983 the greens were able to build their votes, credibility and visibility through regional elections. Further, they benefited from the strength of the new peace movement which grew in opposition to the heightened Cold War tension and the plans for deployment of Cruise and Pershing II missiles in Germany. The backing of SPD Chancellor Helmut Schmidt for the new missiles was to the advantage of the greens, and leading greens such as Petra Kelly were prominent in the mass protests of these years. When the greens surmounted the 5 per cent threshold in March 1983 they became the first new party to do so in the post-war era. The threshold had been designed in part to keep 'extremist' parties out of the political system, but now a new radical party had overcome this obstacle. Many commentators thought that this represented only a momentary phenomenon of protest, but they failed to recognise that the greens represented a generation and ideological constituency that had been politically excluded. They also failed to anticipate the determination of activists in the new party to make it succeed, notwithstanding the major ideological crises over strategy that the party faced during the following years.

France

Although the French greens had fought electoral contests throughout the 1970s they did not form a national party until 1984. The political character of the ecology movement was very varied regionally, with supporters of a more purely ecological politics predominant in Alsace, leftists in Paris and Normandy, and regionalists in Brittany. One of the few points of agreement was that the greens' great strength was in their roots in particular localities, often due to involvement in local environmental campaigns. Diverse political ecology lists had their greatest impact in municipal elections, but national campaigns were much harder to coordinate. An important part of the political ecology networks argued that since greens could not compete effectively against the major parties they should pursue a locally based strategy, one which would avoid their energies being consumed by a political system that was inimical to ecological goals and methods of acting.

Thus national electoral initiatives tended to be only supported by parts of the political ecology network. As noted in Chapter Two, the Dumont Presidential campaign in 1974 provoked divisions, and the ecology list that stood in the European elections of 1979 was not supported by Friends of the Earth (LADT), which at this point was the only nationally organised political ecology group, because LADT was opposed to the EEC. Although this list gained a creditable 4.4 per cent of the vote, by failing to reach 5 per cent they not only failed to elect any representatives but also lost any rights to the reimbursement of election expenses. Individuals from the list had to cover their own expenses, which contrasts with the importance for the German greens of reimbursements from a proportionately lower vote in the same year.

As in Germany the failure of protest to make an impact on policy on nuclear energy and disquiet about violence at demonstrations encouraged some to advocate the formation of a political party. This group formed the Political Ecology Movement (MdEP) which had greatest support among the more purely ecological activists of Alsace. Other greens, combining leftists and those still sceptical of party organisation were organised in the Ecological Confederation. LADT represented the third most important grouping. LADT was led by Brice Lalonde, the best known figure in the movement, but after his relatively successful candidature in the 1980 presidential election (3.8%), the party moved from a broad political role to a greater concentration on issues of environmental policy. Lalonde was to become increasingly impatient with what he saw as the utopianism of the more radical greens and refused to join the new party when it was formed in January 1984.

Discussions about forming a party had been given greater impetus by what the greens saw as a betrayal by the new government of the left led by the Socialist François Mitterrand. The failure to institute a promised moratorium against nuclear energy, part of the price for the support from the ecologists in the second round of the presidential elections in 1981, and the lack of commitment to environmental policy by the new government helped to emphasise the importance of a challenge to the left. The inclusion of the leaders of the Movement of the Radical Left (MRG) and the Unified Socialist Party (PSU) in the new administration also reduced the options among existing parties. By 1984 there was sufficient consensus within the green networks to form a party. The new party, Les Verts, Confédération Ecologiste – Parti Ecologiste, was tiny, with only around 1,000 members and very few resources. Its programme was very similar to that of Dumont ten years earlier and its activists differed in strategy, particularly over the attitude to co-operation with parties and movements of the left. At this stage it seemed to be the last throw of the dice by a declining network, but within five years the greens had achieved a major breakthrough in gaining 10.6 per cent at the 1989 European elections. However, the question of relations with the left continued to dog the new party and dominated its internal life for its first decade.[4]

As well as the differences outlined above there are also many points of simi-larity in the formation of these two parties. Both had their roots in the new environmentalism of the 1970s, and both grew in part from the anti-nuclear campaigns of that decade. In Germany (as also in Britain)[5] a conservative eco-survivalism influenced the early stages of the party, although both the British and the German green parties later developed a more left-influenced political programme. The greater strength of the German greens in comparison with the French was due both to institutional advantages in the form of proportional representation, state funding for political parties and regional government and also to cultural factors, in particular the larger anti-nuclear networks and the exclusion of the German alternative left from political institutions (as discussed in Chapter Two).

Dilemmas of growth

Germany

The years between the entry of Die Grünen in the Bundestag in 1983 and their first participation in national government in 1998 were anything but a process of boring and steady evolution. At many points, but especially after the failure of the western greens to surmount the 5 per cent threshold in the 1990 national election, the party seemed on the point of collapse. Factional struggles seemed likely to tear the party apart in the late 1980s and there were continual battles over the practice of grassroots democracy in the party organisation and the degree to which the greens could play an instrumental role as policy-makers and partners in government within the political system. Although green parties were growing in strength across northern Europe, Die Grünen was for many years the most important party for observers because it was the only relatively successful green party in a large country. As a result it came to be seen as representing the fate of green parties everywhere. Yet, while there are features that the Germans shared with other green parties, there were also many nationally specific factors which shaped the party, making it unwise to make Die Grünen the basis for generalisations about all green parties.

The influence of the greens on German political life was particularly dramatic in their first parliamentary period (1983–7). These years coincided with continued strong peace movement protests, high levels of public debate and concern about pollution, reinforced by environmental disasters, and financial scandals concerning the government, which the greens played an important role in exposing. German parliamentary politics was staid, and the greens were the first new party to enter the Bundestag in the post-war era. The radicalism of green ideology and a refusal to accept parliamentary conventions, from wearing suits, to challenging officials and ministers through assiduous parliamentary work, had a major impact.

When the greens appeared to be in the ascendant, with continued mass protests against NATO strategy and against the nuclear reprocessing plant at Wackersdorf in Bavaria, and were making news as a novel and subversive force in German political life, Petra Kelly expressed the dominant view within the green party:

> We are, and I hope we will remain, half-party and half-local action group – we shall go on being an anti-party party. The learning place that takes place on the streets, on construction sites, at nuclear bases, must be carried into parliament.
>
> (Kelly 1984: 21)

The question was, how? Die Grünen were, like all green parties, experimenting with organisational structures developed in extra-parliamentary protest movements and with ways of maintaining a link with their diffuse constituency; some of the more radical ways of doing politics differently proved unworkable.

The first problems for the greens became evident by the mid-1980s. Mid-term rotation removed the MPs with the highest media profile; the decline in peace movement activity increased tensions over the 'two leg' strategy in which the party was supposed to be the representative of the extra-parliamentary social movements, and the first divisions between '*Realos*' and '*Fundis*' began to emerge. *Realos*, such as Joschka Fischer argued that the phase of the 'anti-party' party was now over and that the greens should accept that their role was distinct from that of the social movements and seek to exploit coalition opportunities available to them.[6] Fundamentalist opponents were split between the Marxists from the K-Groups led by Rainer Trampert and Thomas Ebermann and more anarchist-influenced ecological fundamentalists such as Jutta Ditfurth (Raschke 1993: 154–63; Roth and Murphy 1997). These leading fundamentalists had all left the party by 1991. Although there was no organised fundamentalist faction in the party and there were important splits between the Marxists and *eco-fundis*, over ideology and grassroots democracy, party activists, particularly at national level, provided regular support for various 'Fundi' platforms (Kitschelt 1989: 178). Between 1980 and 1988 green conferences sought to maintain the unity of the party by balancing out the gains of fundamentalists and realists. The problem for the party electorally was that this did not correspond to the support evident among green voters for realist strategies.[7]

The idea of a dual parliamentary and extra-parliamentary strategy had a long history in socialist parties but it described rather than resolved the problem of strategy. Even when the working class was no longer seen as the main source of extra-parliamentary pressure and the strongest links were with predominantly middle-class protest movements, as in the Dutch and Scandinavian New Left parties in the 1970s, there were still major conflicts over strategy (Lucardie 1980; Logue 1982). In the first half of the 1980s the main division was over the strategic consequences of defining the greens in terms of left and right. Rudolf Bahro (1986: 46; 64) had come to the view that all deference to a politics grounded in industrialism was dangerous and saw any link with the left as counter-productive. Bahro and others such as the eco-libertarian Thomas Schmid argued that the greens should seek allies in all parts of the political spectrum. The non-leftists were in a minority in the German greens, however. Both *Realos* such as Fischer and the eco-Marxists, plus many non-aligned party activists, saw the party as on the left and supported economic and social policies that reinforced this.

The conflicts between fundamentalists and realists grew more intense after the 1987 election in which the greens had increased their vote to 8.3 per cent. By this stage the greens had also taken part in the first *Land* coalition government with the SPD in Hesse between 1985 and 1987. In the teeth of fierce opposition from fundamentalists on the national party executive, greens in Hesse had finally agreed to a coalition in which Joschka Fischer became a green minister. The coalition ended in discord, however, when Fischer was sacked (or resigned, according to the SPD) for refusing to accept the licensing of a new nuclear power plant. At this stage *Fundis* were in the ascendant and used their domi-nance of the party executive over the predominantly Realo MPs in the Bundestag. A growing part of the party was increasingly hostile to the paralysis that factionalism was creating (Roth and Murphy 1997: 50). Two developments in 1988 changed the balance of power in the party against the *Fundis*. The first was a revolt by a group of activists led by Antje Vollmer who were fed up with factional infighting and stormed a party meeting in protest (Raschke 1993: 172). Although this began as a mediating group, it later adopted positions closer to the *Realos*. Second, was a series of financial scandals concerning the party executive. Fundamentalists were accused of misusing party funds in their management of work on a party research centre, and the next party conference voted out the old executive (Markovits and Gorski 1993: 230). Although the new executive also contained *Fundis*, henceforth it also included more *Realos*.

Although *Realos* had the initiative until the 1990 'unification' election, they were still at times reprimanded by activists when they seemed to be going too far, as occurred to Fischer in 1990 when he said that fundamentalists had no place in the party, or Vollmer when she argued that accepting the priority of ecology and civil rights meant that the greens should no longer define themselves as on the left.[8] In November 1990 the second red/green *Land* government in Berlin collapsed acrimoniously, reinforcing perceptions that the greens were not up to the task of government. Thus, the effect of the '*Fundi/Realo*' disputes cannot be excluded from explanation of the western greens' failure to gain 5 per cent in the election of 1990. A major factor, however, was the greens' opposition to reunifi-cation. Much of their identity had been built around opposition to militarism and the effect on Germany of bloc politics. They continued to support the prin-ciple that there should be two German republics for several months into 1990 long after it had become evident that East Germans wanted unity with the west and that most green voters also expected it (Poguntke 1993a: 50). The leaders of the November 1989 demonstrations, who mostly joined the eastern greens, also favoured separate development and, unlike the other parties, the western greens refused to absorb their eastern neighbours into the western party. Fearing that the East German greens would not manage to achieve 5 per cent of the votes, Die Grünen argued for the separate application of the 5 per cent threshold in East and West. Had the threshold been applied across the whole of Germany, Die Grünen would have been saved from failure by the East Germans.

The acceptance of unification provoked the departure of the ex-K-Group Marxists, while organisational reforms in 1991 which brought the party organisation and *Land* and Federal MPs closer together, and ended term limits of party leaders, led to the departure of the ecological fundamentalists, who denounced them (exaggeratedly) as the end of control of the MPs by the 'grassroots'. These departures did not constitute a real split, since only a few hundred activists left the party. Nor did it signal the failure of the integration of the extra-parliamentary left. The greens remained an alternative left party, but with more consensus on strategy and organisation than in the 1980s.

The shock effect of the 1990 electoral setback was important in these developments, convincing a majority of party activists to back some reforms in the party structure and to be more tolerant of the compromises needed in coalitions. In the 1990s greens took part in eleven *Land* coalition governments, including three renewals of existing coalitions. These coalitions enhanced the power of leading figures in *Land* parties and the learning experiences which they involved are argued by Lees (1999) to have been important in preparing the ground for the national level coalition in 1998. The merger between eastern and western greens was formalised in 1993 with the creation of a nominally new party, Bündnis 90 – Die Grünen (B90/DG). The merger strengthened the *Realos* slightly, but since easterners remain a small minority within the party membership and in recent years the green vote has largely collapsed in the east, the party remains overwhelmingly west German.

To what extent it remains the same party in other respects, however, is open to debate. The experience of the greens in government will be examined further below. However, ideological shifts began to be noticeable even before 1998. In particular, the greens have developed a neo-liberal profile on economic issues that has undermined their egalitarianism to some extent (Rüdig 2001). The relatively high social costs borne by German employers have been a subject of increasing debate in the 1990s. The greens sought to make an innovative contribution on this issue by identifying new ways of financing social policy by shifting taxation towards measures such as energy taxes. This agenda was pushed by the so-called 'New Greens', neo-liberal green MPs such as Oswald Metzger and Christine Scheel, and resulted in the greens advocating cuts in income taxes in their 1998 party programme and supporting the austerity programme of the SPD finance minister in 1999. The New Greens have tried to persuade the rest of the party to accept economic globalisation and seek active co-operation with economic corporations rather than focus on state regulation (Mittler 1999a). In government the greens have been associated with cuts in welfare state services, pension payments[9] and reduction of taxes for small businesses. While this is balanced by their advocacy of measures such as basic income and their strong emphasis on anti-racism, gay and lesbian rights and feminism, the leftism of the greens is now at least questionable. There are still leftists in the party, and the conference still seeks to balance leftists and *Realos* in appointing party speakers but the left has as yet made little challenge to the neo-liberal economic turn in the greens. As Rüdig notes (2002) this may be the most significant change of all for the greens in the 1990s.

France

Although by forming a political party in 1984 the French greens had in part come to terms with the limits imposed on grassroots movements by the political system, for the next decade the new party remained divided over another strategic and ideological question. This was whether the party should define itself as wholly autonomous from existing parties or, on the other hand, as part of the left, even if distinct from the existing left.

The leading partisan of green autonomy was Antoine Waechter from Alsace. He won a majority for his motion entitled 'Ecology is not available for marriage' at the party conference in 1986, and this shaped the strategy of the party until 1993. On the other side were veteran anti-nuclear campaigners such as Didier Anger from Normandy and Yves Cochet from Britanny. They favoured exploring alliances with other groupings on the left with whom they argued the greens shared much ideologically.

The tide swung in favour of Waechter when the hostility to the Socialist Party (PS) was greatest, as it was in the wake of the sinking of the Greenpeace boat the Rainbow Warrior in 1985 and the suppression of information about the effects of the Chernobyl reactor accident on France, both acts of a Socialist government. The greens achieved a major surprise breakthrough in the 1989 European elections, winning 10.6 per cent of the vote and gaining 8 members of the European Parliament. This provided the party with new resources and a new credibility as national political actors. In contrast to the UK greens, who also gained a surprise 15 per cent vote in these elections, but no MEPs due to the non-proportional UK election system, the French were able to consolidate their vote, maintaining a high vote in the 1992 regional elections and gaining 7.6 per cent in the 1993 legislative elections.[10] Rather than a success, however, the 1993 election seemed more a failure as earlier the green list (in combination with the environmentalist list of Brice Lalonde) had threatened to surpass that of the PS. Moreover, the green list failed to win any seats under the two round electoral system, throwing the emphasis of Waechter on autonomy into doubt.

Competition from Brice Lalonde, the best known environmental leader in France, had been an influence on green strategy as soon as Les Verts were formed as a separate party. He refused to join the new party because its members had refused to accept his demand that they support the possession of nuclear weapons by France and moderate their opposition to French nuclear energy. Thus Les Verts faced a rival list led by Lalonde in the 1984 European elections. Lalonde joined the Socialist government as Minister of the Environment in 1988 and formed a new movement Génération Écologie (GE) in 1990. The latter was not a green party and was rejected by green parties in other countries because it lacked the essential characteristics shared by the green party family, although it was still treated by the French media as a rival to Les Verts. Its ideology was environmentalist rather than green in that it took a reparative position towards environmental problems, did not attack productivism and supported both nuclear power and French nuclear weapons. Lalonde rejected the vision of a Europe of the Regions advocated by greens and green policies on social issues such as granting immigrants the right to vote. The organisation of GE also differed from green practice in that it remained under the central control of Lalonde and made few concessions to internal democracy. In short, there was no commitment to an alternative form of politics (Cole and Doherty 1995). However, even if it cannot be viewed as a green party GE had one vital area of similarity with the greens, it attracted essentially the same voters, the young, educated and new middle class (*Libération* 23 March 1992). In the 1992 regional elections GE slightly outstripped Les Verts, and although the two green formations ran a joint list in the 1993 legislative elections ideological and strategic differences made this an uneasy collaboration. Lalonde negotiated with other parties without informing the greens, and when it came to the European elections the following year GE and Les Verts ran rival lists, scoring 2 per cent and 2.94 per cent respectively.

The strategy of green autonomy associated with Antoine Waechter had been under pressure since the greens became more successful electorally. There were divisions in 1990 and 1991 about how to respond to the resurgence of support for the far-right Front National (FN). Cochet and others associated with a leftist position had said that the greens should moderate their strategy of always trying to retain a candidate in the second round of the elections, if doing so meant making the election of an FN candidate more likely. This was rejected by Waechter and others, because it suggested some kind of co-operation with the parties of the left.[11] The Waechteriens were also split over their attitude to the Gulf War, with many rejecting the view of Waechter that the war should be opposed[12] and over ratification of the Maastricht Treaty, which Waechter supported but many greens opposed, preventing the party from adopting a clear line in the referendum of 1992.

In the wake of the disappointing election result in the 1993 legislative elections the party all but fell apart. Party membership dropped to around 4,000, less than ten per cent of the membership of Die Grünen. A new relatively left-oriented but pragmatic current led by Dominque Voynet gained more support than Waechter's list at the 1993 party congress and Waechter and his followers launched a new rival Mouvement d'Écologiste Indépendant (MEI). Waechter, Lalonde and Voynet all announced their intention to stand in the 1995 presidential elections, but only Voynet as the candidate of Les Verts was able to secure the required number of nominations from elected officials. Although her performance was modest, a mere 3.3 per cent, this established Les Verts rather than the other groups as still the major representatives of green politics in France.

The new more pragmatic mood among Les Verts was evidenced by new discussions with the Socialist Party (PS). While the PS was in crisis in 1992 Waechter had rejected its overtures. The PS itself was still weak after the

1993 elections, but the strong performance by Lionel Jospin in the 1995 presidential election, which centred around traditional left themes of social justice, helped the party to rebuild. This, and the acceptance by Voynet and other party leaders that the greens could not make an electoral breakthrough on their own, led to a new strategy. In the 1997 parliamentary elections the greens in alliance with the PS gained 6 MPs and Dominique Voynet became the Minister for the Environment in a broad left government coalition.

Comparing the development of these two parties, the most striking feature, and one matched by accounts of other green parties, is the pacification of the strategic disputes that were previously so dominant. The main issues at stake in these disputes differed: for the French it was the question of political autonomy, while for the Germans it was both control of political leaders and the question of participation in government. In other parties, such as the British it was the question of party organisation that most divided activists (Doherty 1992b; Evans 1993). The intensity of these arguments was affected by the nature of political institutions, the barriers to political representation (particularly the electoral system), the availability of alliances with a left party, and also by the ideological make-up of the party. In comparing the Belgian and West German green parties, for instance, Kitschelt (1989) found that while there were similar disputes in green parties in both countries, the conflicts were more intense in Germany, mainly because the social movement milieu was stronger. Yet, underlying these arguments within green parties was a common dilemma. If radical change was needed and if the ecological crisis made this more urgent, did compromises damage the prospects for change or enhance them? In effect, as will be discussed further below, many of the more radical approaches to this problem were less appropriate for a political party in the current context, but this process of adaptation first required that each party adapt and learn the limits of its role.

It is also notable that strategic disputes have led to departures of leaders of key currents in many green parties. Both wings of the *Fundis* had departed Die Grünen by 1991. In France Brice Lalonde rejected the alternative left radicalism of Les Verts and has continued to pursue his own non-left environmentalist agenda; Antoine Waechter also left the party after the turn towards an alliance with the left. In Britain, leftists departed from the Ecology Party to form the Socialist Research and Environmental Association (SERA) in the mid-1970s and in a more significant split the party's best-known names, Sara Parkin and Jonathon Porritt gave up national activity in despair in 1993. This followed a major internal dispute over party organisation.

Porritt and Parkin were part of a group called Green 2000 that tried to reform the party structures to make them more efficient in fighting elections, but their opponents saw this as centralising power in a leadership elite. The dispute was also about party strategy more generally. Green 2000 argued that the Green Party should concentrate on the aim of achieving a green government by the early years of the twenty-first century (*Green Activist* September 1990). Their opponents argued that by accepting the principle of leadership, even without a single party leader, the 'electoralists' of Green 2000 were accepting the inevitability of hierarchy and oligarchic control. They argued that it would be impossible to change society in any fundamental way by using hierarchical structures, and argued instead that the party should aim to build a green culture, recognise the obstacles to green electoral success and not concentrate on elections to the exclusion of other aspects of green praxis (*Green Activist* June 1990). Although Green 2000 were initially successful, after mobilising the postal votes of many members through direct mail, many of the Green Party's core activists remained opposed to the new structures, and this resistance disillusioned Parkin, Porrit and some of the other supporters of Green 2000. Following the departure of the latter, the Green Party adopted a new strategy which was less exclusively focused on elections. In the UK 'first past the post' constituency electoral system the barriers to new parties were too great for government to be a realistic short-term ambition. Some supporters of Green 2000 remained in the party, including Caroline Lucas and Jean Lambert who became the party's first MEPs in 1999[13], but all sides recognised that the survival of the party depended upon the maintenance of its sense of community.

Each time after such disputes the green parties have recovered and while many of the traditional strategic issues remain divisive, the disputes have become much less intense. In part, this is due to a learning process. The intensity of strategic disputes reflected the period of party formation and development, when the apparent urgency of making a major breakthrough raised the stakes of every decision about the party's future. It is perhaps depressing to point out that the problems are no less urgent, but that the greens have reduced their expectations. However, green parties are perhaps less central to the green project than they appeared in the 1980s. Then they were the ideological standard bearers for a broad new politics and it was hoped that their radicalism might infuse the left as a whole with new politics. Although this hope has not gone, particularly in countries such as France, where there is stronger resistance to neo-liberalism on the left, it now seems to be only one part of a much more multifarious green project, rather than its main plank.

In the late 1990s the major influence on the development of the green parties was the entry of greens into national government in five western European states. Accepting a role in government was consistent with the strategic evolution analysed above, but for some this was the end of the greens as a party of the green movement. To assess whether this is the case it is necessary to examine the evidence from the greens' experience in government to date.

Greens in government

The first greens to enter a national government were the Finns in 1995, gaining the environmental ministry. The greens increased their vote and re-entered government in 1999 as part of a broad coalition of social democrats, conservatives and other parties. In Italy the greens were part of the Olive Tree coalition government led by the (ex-Communist) Party of the Democratic Left from 1996–2001; and in France, the greens entered government in 1997 gaining the environment ministry in the plural left coalition of the PS, Communists (PCF) and two other small left parties. The two Belgian green parties also joined a broad coalition government in 1999 gaining in particular from their campaign against the governing parties in the Dutroux scandal, concerning government corruption and incompetence in the case of a serial child murderer.

Participation in national government had been preceded in several countries by local and regional governmental participation. This was of most significance in Germany where the Länder had more power than regional governments in non-Federal systems. In all but one of the 13 instances of green participation in *Land* governments (up to mid-2001), the greens governed alone with the SPD.

Die Grünen

When the German greens entered government at the end of October 1998 they were in a relatively weak position. After several years of consolidation in which Greens had participated in *Land* coalition governments with the SPD, Joschka Fischer had become recognised as the most effective critic of the Kohl government in the Bundestag and the most popular politician in Germany, and internal disagreement had been relatively peaceful; the greens appeared to have entered a new phase. However, the pre-election congress at Magdeburg in March 1998 had reaffirmed some green radical commitments, calling, though narrowly, for German withdrawal from participation in peacekeeping operations in Bosnia, and arguing for higher taxes on fuel, including the imposition of tax on aircraft fuel. As Rüdig (2002) says, greens appeared to be threatening the core pleasures of consumer society, the car and the foreign holiday. In fact, the green vote only fell slightly (to 6.7 per cent) but critical reaction seems to have undermined the confidence of the party's leadership in pushing for their goals within government. The second reason for weakness was the structural weakness of the greens' dependence on the SPD. Whereas the FDP had been able to choose between the SPD and CDU and thus needed to be courted and wooed, the greens had no real prospect of making an alliance with the CDU. Since the SPD could dominate the greens in a way that it could not do with the other parties, a coalition with the greens was a 'rational' choice in terms of models of coalition formation (Lees 1999), but it also reflected the greens' lack of alternatives. The impetus for the coalition also came from the learning experience of red/green co-operation at *Land* level (Lees 1999; Roberts 2001). After the debacle of the Berlin red–green city government in 1989–90, which collapsed after numerous conflicts between the two parties (Doherty 1994), there were 11 further coalitions at *Land* level, which were increasingly stable.

B90/Die Grünen therefore entered the government with an agenda of incremental reforms recognising their limited mandate. However, while the days of non-negotiable fundamentals were gone, there were core commitments that represented the minimum acceptable achievements for the greens. These included an end to nuclear power, a reform of German citizenship law to extend it to include long-term 'foreign' residents and their German-born children, and ecological taxation. Commitments to all these were included in the coalition agreement with the SPD which was negotiated quickly and without acrimony.[14] In the green conference which ratified the agreement the main criticism concerned the failure to gain four Cabinet posts and the resulting loss of male–female parity. In general, however, the mood of activists was optimistic.

The greens faced major difficulties in the first year with all three of the core policy commitments, and also faced a further unanticipated crisis over German participation in the NATO intervention in Kosovo in 1999. The reform of German citizenship law was considerably curtailed due to the effective mobilisation of popular protest by the CDU. The greens argued that long-term foreign residents, principally Turkish guest workers, should be entitled to dual citizenship. German nationality law gave priority to those who had ethnic German descent rather than long-term residents. In the early 1990s this produced the bizarre situation where 'ethnic Germans' from Russia, many of whom spoke no German, were able to gain citizenship, where long-term guestworkers were not. For the greens, reforming this law was part of their long-term commitment to multi-culturalism and a means of combating right-wing extremism by providing resident foreigners with more civil and political rights. However, during the Hesse *Land* election campaign in early 1999 the CDU launched a petition against the new law, which successfully mobilised CDU voters and damaged the greens. Second, polls showed that voters for all parties except the greens were hostile to reform (Green 2001). Worse, the replacement of the SPD–Green coalition with a CDU–FDP coalition in Hesse meant the red–green government lost its overall majority in the Bundesrat, the second chamber. This was sufficient to persuade the SPD Chancellor Gerhard Schröder to accept a compromise with the FDP in which dual nationality would only be allowed to the age of 23 after which a choice would have to be made. Although the nationality laws were liberalised, few foreign residents seemed willing to choose German citizenship, not least because the process of divesting themselves of their other nationality remained slow and cumbersome. The greens had effectively been marginalised and the reform agreed had little practical consequence (Green 2000).

The phasing out of nuclear power also saw the greens disappointed. Schröder blocked the commitment to a new energy law that would have ended nuclear reprocessing on legal grounds, partly under threat from British and French reprocessing companies to sue for breach of contract and partly under pressure from the German nuclear industry which pointed out that without waste storage facilities, nuclear reactors would have to close for technical reasons. Prior to entering government the greens had also broken with more radical parts of the anti-nuclear movement over whether it was feasible to close down the nuclear industry immediately. After several years of harassing the nuclear industry through a *Land* government, greens from Hesse were particularly experienced in the legal obstacles to fast closure. While the coalition agreement had included a period in which a consensus was sought with industry after which, if no agreement was possible, a new law would be enacted, the negotiations with industry dragged on, and the Green

Environment Minister Jürgen Trittin, was increasingly isolated (Rüdig 2000). Schröder was determined to avoid legal action for compensation by the nuclear industry and he put Werner Müller the Economics Minister who was close to the electricity industry in joint charge of negotiations. Since environmental groups were excluded from the negotiations, and the Ministry was traditionally pro-nuclear, and had until 1998 helped override anti-nuclear measures of *Land* governments, Trittin had to carry the green agenda alone in a very hostile context. While a policy to phase out nuclear power was finally imposed in June 2000, it was so long-term and flexible that the protests of the energy industry seemed largely rhetorical. In practice, no nuclear power stations were to be phased out for at least 20 years. The pyrrhic nature of this victory was reinforced when greens found themselves opposed by a mass protest against the transport of nuclear fuel at Gorleben in 2001 (Hunold 2001). For a party that had its roots in the anti-nuclear movement this was a major threat to its identity. While it is arguable that the greens made mistakes in their negotiating strategy, they faced major structural obstacles in the form of the German nuclear industry, not only due to its lobbying power, but principally due to the very measures which had best served the anti-nuclear movement, the complex network of laws governing the operation of the industry.

The use of eco-taxes to meet social and environmental objectives had been a major part of the green manifesto (B90/DG 1998). The greens had argued that by taxing ecologically damaging consumption, particularly energy, money could be raised to invest in jobs and social welfare, reducing the component of social security contributions from employers. But, since Schröder had committed himself to allowing only tiny increases in petrol tax, it was clear from the outset that the tax would fall far short of the greens' plans. Business gained an 80 per cent reduction in the tax and the coal industry was exempted altogether (Mittler 1999a). As a representative of German FOE (BUND) commented 'almost the only gain was the rhetorical one that if Germany could do it, eco-taxes elsewhere might not be so dangerous' (Mittler 1999b). The wave of protests against high petrol taxes across Europe in September 2000[15] made it even harder for the greens to press this issue further.

The involvement of the German military in the NATO bombing of Serbia was also a turning point for Die Grünen. The commitment to non-violence had been one of the four founding principles of the movement. By supporting the intervention the B90/DG seemed to have broken radically with their past. At a dramatic party congress at Bielefeld in May 1999 green leaders required police protection, there were fights and Fischer was hit by a paint bomb. Although split, the delegates backed Fischer in his search for a diplomatic settlement and urged a limited suspension of the bombing. Those most opposed to Fischer left the party, but as in 1991, their numbers were small. In contrast to the other dilemmas faced by the greens this was a question which went to the core of green ideology.[16] The new green foreign policy which Fischer had worked hard to develop did reflect some concerns specific to German greens. With the end of the Cold War many of the reasons for green anti-militarism also disappeared.

The new dilemma was how to use foreign policy to pursue human rights. Greens had always been particularly concerned about the Nazi legacy. The German New Left of the 1960s had regarded the failure to confront the Nazi legacy as central to the weakness of German post-war democracy. Greens were particularly concerned that Germany should not act differently from other countries in defending human rights. This was important in persuading party activists to accept German involvement in what seemed to many at the time to be a necessary action to prevent the massacre of Kosovan Albanians. The leftist leader of the parliamentary group Kerstin Müller, spoke for many when she said 'We all have doubts and we all are torn' (*The Guardian*, 14 May 1999), but she backed the NATO action.

In every *Land* election since the coalition the green vote has declined by 30–50 per cent (up to the time of writing, Summer 2001), and the party has all but imploded in the eastern Länder. Fischer remains popular, but has alienated more radical greens with his support for military action. Furthermore, while Fischer was engaged in foreign affairs he failed to support other greens in their battles over domestic policy. These weaknesses have led to much speculation about another end for B90/Die Grünen. But, the more relevant question is about its continuing status as a party of the green movement. This question can be divided into two. First, is the party no longer a party of grassroots democracy? Second, has it changed so much ideologically that it is no longer within the green ideological framework. These questions will be examined below with reference to other green parties as well as the German case. First, however, it is necessary to examine the experience of other green parties in government.

French greens in government

The experience of the French greens in government is in some contrast with the German example. Thus far, the greens have improved their vote while in government, scoring well in European elections in 1999 and municipal elections early in 2001. As a result the greens gained a second cabinet minister in March 2000 when Guy Hascoët became Minister for Economic Solidarity in March 2000, joining Dominique Voynet who was Minister of the Environment. Polls rate the greens' performance in government highly and party members are overwhelmingly in favour of continued participation in government (Boy 2002). Part of the explanation for this more positive experience lies in the weakness of the French greens early in the 1990s. The divisions in the party after 1993 had seriously threatened its future and thus not only to have survived but achieved parliamentary and government representation seems a major victory. This conclusion is reinforced by the evidence that green party members surveyed in 1999 had mixed views on the party's success in promoting policy. The greatest perceived successes were regarding the environment, equal opportunities for men and women in politics and work-sharing (ibid.). The latter three fields had produced demonstrable policy gains. The greens had been able to achieve the cancellation of the Rhine–Rhône canal, and the closure of the landmark Supérphenix nuclear reactor at Malville, a site of huge symbolic significance for the French green movement as it was the target for the landmark demonstration in 1977 (see Chapter Two). Voynet enforced long dormant European directives to limit hunting, despite strong protests by hunters, which included them wrecking her office. Furthermore a moratorium was achieved on new planting of GM crops, although this was as a result of court action more than the greens' weight within the government. Although less influenced originally by feminist discourse than Die Grünen, the French greens have in recent years made women's political and civil rights a central part of their platform. They were therefore strong supporters of the law on parity in political representation instituted in 1999. The greens had also been the principal initial advocates of a shorter working week and took some satisfaction from legislation to establish a 35-hour working week.

Yet, while this suggests that the greens achieved much, the picture needs to be balanced by acknowledgement of their setbacks. There is as yet no sign of the moratorium the greens hoped for on new nuclear plants (although no new plants have yet been begun) and the greens have been arguing for the kinds of measures committing the government to longer-term nuclear closure which are seen as inadequate by many German greens. The greater dependence of France on nuclear generated electricity (70 per cent, as compared with 35 per cent in Germany) is a major difference. The greens have not been able to stop motorway building and Voynet has reluctantly supported the expansion of airports around Paris. She was also forced to accept the temporary re-opening of Supérphenix in order to carry out work on waste disposal. The hunting lobby remains strong, gaining 6 MEPs in the 1999 European elections. The policy on working hours was instituted in a way that the greens disapproved of, giving too much flexibility to employers to interpret how it should be carried out. Finally, the greens have been the most critical participants in the coalition. They were publicly rebuked by Lionel Jospin for their attack on government policy on the regularisation of the status of long-term migrant workers who lack formal social and civil rights. The greens have always been more strongly in support of granting rights to immigrants as part of an anti-racist strategy but the PS and other parties in the coalition favour a case-by-case approach.

Discursively, there is much less sign of change in Les Verts than there is in the German greens. Party programmes have remained much more consistent, with a strong emphasis on international solidarity with the Third World, the redistribution of work, giving greater priority to the social economy, and decentralisation of political institutions based on a principle of subsidiarity.[17] The party's (provisional) candidate for the 2002 presidential elections, the economist Alain Lipietz, defined the core principles of the movement as follows: 'between the traditional soul of the left, the support for democracy and solidarity and the essential originality of ecology: the defence of the conditions of life on our planet, the rights of future generations'; between autonomy for the greens and need to construct majority alliances with the forces of progress; between the engagement in protest and campaigns on the ground to transform relations of force and the entry into the institutions' (Lipietz 2001).

It is notable how little experience Les Verts had to draw on. In contrast to the Germans, who had extensive experience in *Land* governments, the French greens had taken part in only one, much less powerful, regional government (in Nord-Pas de Calais), but this may also mean that their expectations were less high. It is not only factors within the greens that explain their more positive attitude to government. One factor, already noted is the 'miracle effect' of the alliance with the PS. This, of course, also means that the greens may be vulnerable if the PS decides not to renew the agreement. Ideologically, it is perhaps easier for Les Verts to work with the PS than for Die Grünen and the SPD to work together. Under Jospin the PS has tried to reposition itself as a party based on more traditional left themes, whereas Schröder has pushed the SPD further towards Third Way ideas. And while the PS is the dominant force in the coalition it is in a less strong position than the SPD, because as well as the greens it must also work with a diverse coalition of the PCF, which has changed ideologically much less than most other communist parties; the nationalist and left-wing Mouvement des Citoyens (MDC) led by the ex-socialist Jean Pierre Chévenement; and the relatively weaker Mouvement Radicaux de Gauche (MRG). The union of the left is really a coalition of distinct ideological groups with distinct electorates. It was 'an attempt to stabilise a crumbling electoral base, rather than a realignment of the French left' (Szarka 1999: 35). The greens have been able to work with the communists on issues such as working rights and briefly on migrants' rights, but were opposed by them on nuclear power. Thus while united by an identification with the left, this coalition also illustrates the variety of forms of leftism in contemporary politics.

Thus far the French experience of green participation in government appears more typical than the German. The Finnish greens are now in their second spell in government, and ' the old ambivalence about political power and the compromises necessary in political decision-making have ceased to be a problem to the greens as a party' (Konttinnen 2000: 132). However, on issues such as nuclear power the greens find themselves marginalised within a largely pro-nuclear coalition.

The first stages of the two (Walloon and Flemish) Belgian greens parties' participation in government in 1999 were relatively consensual. There was some surprise that the greens were able to participate in a coalition that included the right-wing Liberals, as well as the Social Democrats. However, both the Liberals and greens were linked by their concern to challenge political corruption and a desire to exclude the Christian Democrats from government for the first time since 1958. The Belgian greens soon found themselves facing an almost identical range of dilemmas as the French and German greens. The greens and Liberals clashed over the greens' desire to relax asylum regulations and the greens were criticised by some human rights groups for making compromises; the green secretary of state for energy Olivier Deleuze, an ex-chair of Greenpeace Belgium, 'was strongly criticized by Greenpeace for what it perceived as far too slow a phasing out plan for nuclear plants' (Hooghe and Rihoux 2000: 136). ECOLO (the green party of the French speaking Walloons) was in a more difficult position than Agalev (the Flemish green party) because the Socialists and Liberals had enough seats in Wallonia to govern without them, but a national coalition involving both green parties was made conditional on green inclusion in both regional governments. Thus, the Belgian greens were able to exploit their greater bargaining power in comparison to the Germans.

Being in government has brought new difficulties for these green parties, particularly criticism from EMOs and green radicals. As well as facing demonstrations from more predictable sources such as hunters and the CDU the greens have also had to face the ire of their allies. And even when still demonstrating, greens could find themselves in trouble. Claudia Roth a speaker for Die Grünen had to face the hostility of previous supporters when she attended the anti-nuclear demonstrations at Gorleben in 2000. At the Vallée de l'Aspe a green spokesperson was the victim of a 'pieing' (being hit in the face with a custard pie) by anarchists during a march against the Somport tunnel. This is simply a cost of green participation in government, but it also shows that the arguments about strategy are now less within the green parties, as they were in the 1980s, than between them and other parts of the green movement.

Even when greens have not being pursuing very radical policy they have found that the institutional culture of public administration places obstacles in their path. The French green representatives lacked the education in the Grandes Ecoles which defines the elite of the French state 'Thus in inter-ministerial arbitration, ignorance of the customs, the techniques and even the language common to the state management class was a severe handicap to those representing the Ministry of the Environment, especially over the first few months before these skills were acquired' (Boy 2002: n.p.). The German greens faced similar cultural difficulties. In the negotiations to phase out nuclear power Jürgen Trittin was reliant on Ministry officials who had previously been responsible for the defence of the industry against the anti-nuclear movement and so was forced to rely on a small group of outsiders from the green movement (Rüdig 2000). He was then criticised for excessive secrecy. The greens were also weakened by the relative lack of expertise within the party. Whereas the SPD could replace staff from a ministry from within party networks, B90/Die Grünen lacked equivalent networks (Lees 1999: 182).

One institution in which greens have been able to build expertise and political capital more effectively is the European Parliament. Greens have been represented in the Parliament since 1984 and after 1999 they were the fourth largest parliamentary group (in a joint group along with a smaller regionalist group) giving them greater weight on committees. Moreover, environmental issues are one area where the EP has been particularly influential. Although the right-wing dominance of the parliament and its environmental committee is likely to curtail advances from a green point of view (Bomberg and Burns 1999), the role of the greens in European level politics is likely to grow. Fischer's support for greater European integration and greater democratisation of EU institutions has been the focus of much debate; there is now a green Commissioner and European green environment ministers led by Dominique Voynet played a crucial role in preventing agreement between the EU and the USA at the climate change talks in the Hague in October 2000. While critics saw this as a blameworthy failure (Lean 2001), greens argued that a deal which allowed the USA to avoid meeting its commitments, which from a green point of view were already wholly inadequate, was counter-productive. The major international debate which followed the decision of the Bush administration not to ratify the Kyoto agreement also refocused attention on the 60–80 per cent cuts in 1990 levels of emissions that scientific consensus now suggests would be required, rather than the five per cent cuts for industrialised countries agreed in Kyoto.

In domestic governments, however, it is notable that the major achievements of the greens have been in areas that do not directly threaten dominant economic interests. Greens have achieved legislative change on gay and lesbian rights, the rights of migrants, and animal protection in several countries and these are all vital to green ideology. Even on these questions, they have faced strong opposition. But efforts to institute more radical changes in pursuit of sustainability or to end nuclear power have produced much less success and seen effective counter-mobilisation by business interests. Moreover, the major socialist parties have generally sided with business in these conflicts.

While the greens have not achieved the radical changes they would desire through participation in government, they have achieved some positive policy changes and thus far have mostly been able to show that policy for which they are responsible is consistent with longer-term green goals. Where conditions allow and where there is a real prospect of achieving positive change green parties in all countries are likely now to view participation in government as normal in a way that they would not have done in the 1980s. What does this tell us about how far the green parties have been transformed?

The transformation of green parties

One of the major features of the collective learning of green parties has been the relative decline of more fundamentalist positions within green parties. While the limits of green electoral strength were evident from the beginning, this was not always accepted by green activists, as the unrealistic aspirations of the Green 2000 group in the UK Green Party showed. Given the limits of the green vote the debate about whether to seek governmental power which took place in most green parties was much less the issue than the question of how to function effectively as minority parties, something which greens rarely addressed directly. The green position on the limits of change via the state should have prepared them for the limits of being small minority parties in government. The political role played by the green parties now depends on how well they adapt to this role. At best this means a dual combination of exploiting the opportunities for internal opposition within governments, as the French have done, and thereby creating space for wider debates about core green ideological principles, while at the same time making the incremental reforms which the system allows and which are compatible with advancing further green goals. The opportunities for the latter are now ironically probably greater in more centralised systems such as France than in the more decentralised systems such as Germany. Given the changes in green parties, however, some have argued that there is nothing left now of the original green radical project. To what extent are the greens now parties just like the others? This will be examined by discussing changes in three crucial areas, green political identity, grassroots democracy and green strategy.

Political Identity

One of the major changes in the political identity of greens is that the fissures between green leftists and non-leftist greens are no longer a major source of division either within or between green parties. In the 1980s there were many arguments within green parties over the priority of ecology relative to other themes and the question of whether the greens should identify with the left. There were also disputes between green parties over whether parties such as the Dutch Groen Links should be recognised as members of the European Federation of Green Parties. Groen Links was an amalgam of existing New Left parties such as the Pacifist Socialists, the Radical Party, (which had Catholic roots), a small leftist Protestant party and the Communist Party (which had broken with much of its heritage and become particularly influenced by feminism). As these parties became more affected by the changing intellectual debates on the left they sought to redefine their ideology (Voerman 1995). By the late 1980s, there seemed to be less reason to maintain separate parties, so they merged as a single party. There was, and still is, also a smaller more ecocentric green party known as De Groenen. The latter was supported by some of the less leftist green parties such as the British, as the only proper 'green' party in the Netherlands. However, De Groenen were much less successful in elections than Groen Links and the latter were seen as ideologically similar by parties such as the Germans that saw ecology as part of a broad social movement leftist politics. In the end both Dutch parties were accepted as members of the Federation.[18]

The diminution of the red–green conflict is part of the general shift in green parties away from more dogmatic positions towards a broad framework ideology. This does not mean that the greens have moved closer to the socialist parties. In part the heat has gone out of the dispute because of the major changes in identity undergone by most major socialist and communist parties, but also due to the acceptance by green activists that they share more with the parties of the left, which came through the accumulation of experience in representative institutions. Greens have regularly argued for more radical left positions than socialist parties on welfare benefits, women's rights, race and sexuality, and disarmament.

The fear that the ecological message would be diminished if greens were absorbed into larger left parties has also dissipated because ecological issues have become more mainstream in international politics and new frames such as ecological modernisation have, despite their limitations, opened up a space for dialogue about sustainability with the major socialist parties. Conversely, the failure to convert the more radical of the remaining 'old left' movements such as the PDS in Germany, the LCR in France and the Socialist Alliance in Britain to make ecological and other green themes central to their politics, has helped to reinforce the greens' sense of distinctiveness. Thus greens can now recognise that their principal allies will be on the left, but do not expect to merge with these parties or that this means that they will be seen by voters as the same as the other parties of the left.[19]

Another major feature of green identity was the view that green parties were in some respects a representative of other movements within the system. This commitment remains, but greens are now less concerned with integrating the full range of social movements into a single radical force. The example of ECOLO, previously a party less oriented towards the social movements, (Kitschelt and Hellmans 1990) is indicative of current green attitudes to their relationship with social movements. In the late 1990s ECOLO organised a series of Etats Généraux de L'ECOLOgie Politique, a series of forums involving thousands of activists outside the party (intellectuals, trade union activists, and different social movement leaders) (Hooghe and Rihoux 2000: 133) which broadened the range of contacts with sympathetic groups. ECOLO also gained the respect of the 'White Protest' movement against the corruption and incompetence of the two major governing parties, the Christian Democrats and Socialists. Finally, ECOLO, like other green parties, included candidates from other movements on its lists. While other Belgian parties also did this, ECOLO seemed less opportunistic because the '*candidats de l'ouverture*' on its list had a long history of activism in movements which had a close relationship with the greens (Hooghe and Rihoux 2000).

Green parties in most countries continue to take part in protests, but usually as part of a coalition of movement organisations and parties and have not sought to gain control of protest movements or integrate them into the green party. When they have aimed to work more closely with groups outside their party greens have sought principally to use such groups' expertise and knowledge in policy formulation. This reflects the continuing dominant green view that a party organisation cannot itself have all the answers.

Grassroots democracy

Green parties identified the bureaucratic organisations of the existing parties as an obstacle to democracy. They agreed with Michels (1911) that if the political system was to be democratic its parties should be organised democratically. They also shared Michels' suspicion of the self-interested nature of party leaderships. But they disagreed with Michels' pessimistic conclusion that these problems could not be overcome in the conditions of modern mass politics and that therefore oligarchy would make democracy impossible. Michels, writing in the early period of mass working-class male enfranchisement, had assumed the need for parties to form a single collective will, whereas the greens have tried to develop a more pluralist party culture. Michels also assumed that only leaders would have sufficient technical expertise to take decisions for the party, greens have challenged this and tried to broaden the category of relevant knowledge and expertise in decision-making. The form of organisation they developed was therefore considerably shaped by their vision of a decentralised and participatory democracy for society as a whole. None of the organisational features of green parties were absolutely new. For instance, the idea that party representatives should be mandated by the grassroots members was widespread in early socialist parties (Ware 1987: 180). It is the context in which the greens have tried to develop a different kind of party (the decline of the class milieu, the rise of new kinds of social movements, and the rise of media and personality dominated political campaigns) that most separates the greens from earlier cases.

To be sure, the greens failed in practice to live up to all the ideals of grassroots democracy. They have mostly[20] abandoned rotation of MPs because they accepted that the skills that MPs developed through experience were wasted if there was too much turnover. Their party organisations have become more professional, employing more staff where the resources allow, and they have even begun to dress more like the other parties. Perhaps the greatest problem, however, is the relatively low level of activism among party members. Instead of having a positive effect, participatory forms have been at best neutral in facilitating involvement – though the suspicion of many greens about party activity may also have had an effect. As Kitschelt showed (1989), despite the attempts to overcome hierarchy within the Belgian and West German green parties, informal networks of power tended to operate based on factional identities and with factional leaders. These 'stratarchies' operated in parallel to the formal structures of accountability, and partly undermined the claim to have developed a new form of participatory party. The power of factions tended to be greater than their weight of support among mostly less factionalised ordinary members.

Yet, Kitschelt also showed that green structure had prevented the emergence of a single power centre, and this still remains the case in most green parties. First, green party members mostly continue to resist the accumulation of power by individuals through limits on the amount of posts individuals can hold, or by banning the dual holding of party office and membership of parliament.[21] Informally, green leaders have not been able to impose their will against the wishes of party members. In one other major area of internal party organisation the greens have been consistent. Greens have higher representation of women at senior levels of the party, among MPs and cabinet posts. Green parties are strongly committed to the principle of parity and even in those parties such as the Belgians that have not institutionalised this in party rules more than half the MPs in Federal and regional parliaments are women (Hooghe and Rihoux 2000: 136).

In most green parties there has been relatively little formal change in organisational structures in the past decade (Rihoux 2000). In Germany, *Realos*, including Joschka Fischer, have tried (thus far) unsuccessfully to overturn the principle by which those elected to parliamentary offices are barred from holding senior positions in the party.[22] The two party speakers elected in 2000 Renate Kunast and Fritz Kuhn had to resign their posts as parliamentary leaders of the green group in *Land* parliaments before taking up their posts. Then, when Kunast was appointed agriculture minister in early 2001, she resigned her post as party speaker. Party activists may have been more determined to prevent any erosion of the grassroots control of party organisation because of the painful experience of compromise over policy in government (Rüdig 2002). As a result green conferences remain unpredictable and lively, with a measure of real power in the hands of grassroots activists. Other features of green culture also remain intact, such as the practice in Germany of ensuring that those elected to party posts represent a balance of opinion within the party. For the two party speakers' posts for instance, having satisfied the formal requirement that one be female and the other male, and one an easterner and one a westerner, the conference generally elects a leftist and a Realo.

It can be argued that small parties such as the greens, with relatively few resources, and no major institutional backers, need to nurture their activists and maintain the communal character of the party organisation more than larger parties. While those green parties with state funding are less dependent on their members' financial support, the role of activists is broader than this. Small parties still need organisers, campaigners and candidates for office. A shortage of activists has posed problems even for financially wealthy (in green terms) parties such as Die Grünen (Frankland and Schoonmaker 1992). In smaller parties such as the British, it is even more important to maintain the communal ethos of the party. Green party culture is such that activists expect to have an influence over decisions and given the social characteristics of green party activists this is unlikely to change.

However, grassroots democracy is undoubtedly not the same in practice as intended originally. Then, amateurism was not seen as a serious problem compared to the attention that ought to be attached to the ideas in party programmes, but now greens have accepted the importance of effective communication and professional expertise and largely caught up with rival parties in this respect, even though they lack the resources to compete on equal terms.

For both the French and German green parties involvement in government has undoubtedly reduced the power of party activists. While party members are consulted on key issues,[23] and can reject or approve major policies, such as the compromise over nuclear power in Germany, their role is necessarily *post hoc*. The original hopes of a wider and more open participation of the movement at every stage of policy negotiation have disappeared (Demirovi 1997), and were not really operable even at relatively local levels, as studies of green involvement in local government have shown (Holliday 1993; Doherty 1994; Scharf 1994).

Further, there is much less concern now about exploiting the appeal of informal party leaders (Hooghe and Rihoux 2000) and the *de facto* existence of informal leaders has become widely accepted. And yet, while these 'leaders' now have greater freedom to speak on behalf of the party, they are not leaders in a more traditional sense. Neither Fischer, Voynet or green leaders in other countries have the same level of insulation from grassroots dissent as leaders of most other parties, and green leaders do not have any power to impose their will on their party. One of the principal reasons for strong measures of accountability was the distrust of leaders. Experience has taught the greens that their leaders do not, or cannot, use the party in the way Michels predicted, to pursue personal ambition.

The only green party that has moved formally to the more common concentration of power in the leadership has been the Italian greens. Collective leadership was abolished in 1993 during the crisis of the Italian party system and after the party failed badly in the 1999 European elections the then leader Luigi Manconi was replaced by Grazia Francescato, who had been the European spokesperson of WWF. She appointed many WWF staff to party posts and centralised policy-making (Biorcio 2000). In part this reflected long-standing particularities of the Italian interpretation of grassroots democracy. Rather than instituting strong forms of accountability the initial structure of the Verdi had given priority to avoiding strong party structures. The party was initially a federation made up from the 'archipelago' of green groups, involving a wide range of EMOs such as Italian Nostra and WWF, which in other countries kept their distance from the greens, along with more New Left eco-activists and more leftist EMOs such as Lega per l'Ambiente. The weak party organisation was a deliberate contrast to the strength of party organisations in Italian politics before the political crises of 1992–3 (Diani 1990). However, once the old corrupt parties were swept away the opposition to party organisation seemed less relevant.

It has been easier for more marginal parties such as the British greens to retain grassroots democratic practices than for the green parties in government. The costs and limitations of grassroots democracy have become more apparent as the greens have had to face new tasks such as negotiating day-to-day policy with a larger coalition partner. The excessive optimism that was part of the early stage of experimenting with alternative party forms has been tempered by the reality of relatively low rates of participation by party members. And yet, even those green parties in government have maintained many of the formal features of grassroots democracy. Ordinary members of green parties are unlikely to seek or accept a more hierarchical party structure because practising politics in an egalitarian and participatory way is too bound up with green culture. As Faucher's (1999) study of green activists in Oxford and Aix-en-Provence showed, for greens party activism is part of a green way of life in which egalitarian culture is taken for granted and part of the established tradition of a wider activist community. For parties in government this produces the slightly paradoxical situation in which greens hold on to the formal mechanisms of grassroots democracy while the reality of decision-making is that it is increasingly concentrated among party leaders. All this is a long way from the vision of using the party organisation as a model to inspire a wider democratisation of society.

Green strategy

Strøm's (1990) analysis of parties as rational actors defines them as pursuing three goals: vote-seeking, office-seeking and policy-seeking. Green parties, however, do not wholly conform to this model. For greens a balance has to be sought between the three goals outlined by Strøm and a fourth goal which might be called ideological representation.[24] Like other radical forces in the past, the greens seek to balance their gains inside the political system with maintaining a critique of its limits. They regard existing liberal democracy as failing to provide citizens with adequate opportunity to affect decision-making. Voters have to choose between the packages put together by hierarchical and inaccessible parties, which are often influenced by special interests. And parties often compromise with each other and in particular with business interests in ways over which citizens have no control.

What distinguishes the situation of the greens from that of socialists in the 1900s, and this applies as much to the wider green movement as to the green parties, is the decline of the centrality of class interests as the yardstick of progress. Early socialists were divided between revolutionaries and reformists, and superficially green arguments over strategy look similar to this, but socialists in the early twentieth century could agree on the priority of class exploitation and the need to gain control of the means of production as their principal objective (Keane 1988). Greens and others on the left are often now unsure which of the many issues on their agenda has priority and what the consequences for action are. For instance, is dealing with the ecological crisis more urgent than preventing a global war? Should action concentrate on the state or on building autonomous centres of resistance? Does dealing with these issues depend on overcoming sexism, racism and capitalism? Greens vary on all these questions, as do others on the left. Rowbotham, Segal and Wainwright summed up this new ideological field well when they said in relation to the relationship between socialism and feminism that: 'the old scale of measuring consciousness becomes ungainly because you are moving in several directions at the same time. People can be so backward and so forward at the same time that the scale won't work anymore' (1981: 106).

This pluralism means that green parties tended to view a wide range of left and alternative social movements as their allies. But since these movements were themselves often divided, as feminism was for instance between those who gave priority to equality and those who emphasised difference, there was no easy way in which greens could be the party of or for the movements. The greens were informed by other social movements, and overlapped with them, but they had to earn the support of a changing and evolving social movement milieu and could not claim to be its automatic representatives, not least because there was no homogenous social movement constituency or agenda. The green party agenda was therefore one of many possible agendas based on a wide range of social movement demands. This left the greens with the problem of how to relate these diverse demands to action. The principal strategic division among greens reflected this core dilemma.

Green fundamentalists wanted to avoid recuperation, the tendency of the political system to absorb those parts of radical politics which pose no threat to the system, and by doing so undermine the radical and critical edge of the movement. As Petra Kelly said:

The environment, peace, society and the economy now pose such a threat to human survival that they can only be resolved by structural change, not by crisis management and cosmetic adjustments. The Greens can make no compromises on the fundamental questions of the environment, peace, sexual equality and the economy.

(Kelly 1984: 18)

Yet, as realists pointed out, knowing that minor measures were inadequate did not help to show how green parties could best play a role in bringing about more radical change, and this was a major reason for the decline in fundamentalism among activists in green parties. Furthermore, the logic of green pluralism was that it was unlikely that there could be a mass green public. Those who supported the greens on one issue might not do so on another. Recognising this, Wiesenthal, an influential German Realo, defined 'social change as a product of asynchronous, particularised social conflicts involving varying sets of non-radical protagonists,' but said this was 'hard to square with the current concepts of what the relationship should be between movement and party' (Wiesenthal and Ferris (eds) 1993: 34). Wiesenthal was referring to the view of Die Grünen as an empty shell to be filled with ideas and activism from a variety of social movements. One of the major consequences of the green party's collective learning experience has been an awareness that parties are functionally different from extra-parliamentary movements. Green parties can join protests and even initiate them, but they also have to carry out tasks that differentiate them from other parts of the green movement, such as developing a comprehensive and integrated policy platform, working within parliamentary institutions, fighting regular electoral campaigns and building organisational structures capable of sustaining these roles. Once in government, these roles expand to include taking some responsibility for all government action and its concomitant failures.

Thus, whereas in the period of green party formation it was common to see green parties as an extension of the alternative left milieu, this looks less plausible now. The green movement itself is more differentiated. Many Green parties have executive power and much experience within political institutions; EMOs now have much greater support and resources than two decades ago, and have become much more professionalised, and there is a new wave of young direct action networks in many countries. Although green parties share network ties with these groups and a common ideological framework with most, the green parties play a specialised role within the political system. The principal role they might play within the green movement is as avant-garde parties, arguing for more radical alternatives on a wide range of issues, and seeking to use the mechanisms of legislatures and government incrementally. This may include using their power within government to implement existing legislation more effectively, as the French greens did on hunting. But, greens are at best 10 per cent parties, reliant on distinct social groups and a political generation for their support. They have a much more self-limiting role than the vanguard parties (whether Leninist or social democratic) of early twentieth-century socialism, and do not expect to achieve transformations on their own. However, this limited position makes it difficult to point to clear evidence of green party achievements, and for some other greens the green parties appear to have been captured by the system they set out to change.

If greens had given up on their core values in order to gain power then we could certainly speak of a process of de-radicalisation due to dominance of a logic of voter maximisation. The position clearly differs between parties, and it is those in government that have given up most. But even then, the process of ideological transformation is not complete or predetermined. Significant ideological divisions remain and attempts by some greens – such as the neo-liberal New Greens in Die Grünen – to break with their core ideological commitments have been strongly resisted by others.

Other changes are less clear cut. The old utopianism and confrontational critique has gone from most green parties, and with it much of the excitement. Many of the changes are a product of adaptation to the party's limited role. Green parties have chosen to take the small opportunities offered by government office to make positive changes in preference to remaining on the political margins. Since there is good evidence that this is what most green voters wanted this is a defensible choice. Moreover, being in government does not have to mean abandoning long-term commitments to green goals. Yet, there are also dangers evident in the way that green parties have acted since joining governments. For instance, environmental policy has become just one of a number of specialist policy areas for many green parties. As the German Naturschutzbund commented in 1999, 'Environmental Policy within the Greens now has the same status as in other parties. It is a specialist field dealt with by experts' (quoted in Mittler 1999a). Since green parties were always broader in their concerns than the image of a single-issue environmental party suggested, and were keen to demonstrate this to the public, this kind of development is not necessarily surprising. Yet, if green parties are to remain radical they must continue to stand apart from other parties on their interpretation of the scale of structural change required by the ecological crisis, and their view that such changes can lead to a better quality of life.

5 Mass environmental movement organisations

This chapter will examine the almost universal trend towards institutionalisation of environmental movement organisations (EMOs) in western industrial countries by addressing the following questions:

- What form does institutionalisation take?
- To what extent is it inevitable and irreversible?
- What is the relationship between institutionalisation and the pattern of activity of the environmental movement as a whole? Is there a decline in protest? To what extent does the pattern vary across countries?
- Perhaps most crucial, to what extent does institutionalisation mean de-radicalisation? Some critics have argued that by compromising too much the environmental movement is no longer able to offer an ideological challenge to the dominant order.

The main focus will be on EMOs that are organised mainly at the national level, and have a formal bureaucratic structure, as distinct from the grassroots groups examined in Chapter Seven or the NVDA groups in Chapter Six. While the latter have 'organisations' they rarely have a bureaucracy, with paid staff and formal systems of accountability. The use of environment rather than green is intended to signal the concentration of groups dealt with in this chapter on issues of environmental policy. The EMOs are both the most specifically environmental of the groups dealt with in this book and arguably the least radical. Yet, due to their large memberships, many see EMOs as *the* environmental movement, and apart from the voters for green parties, they are its main claim to mass support. Whether the EMOs can be considered part of the green social movement as distinct from the more diffuse environmental movement will be discussed in the final section of this chapter.

The major western EMOs such as Greenpeace, World Wide Fund for Nature (WWF), Friends of the Earth and the Sierra Club have become important political actors. Often consulted by government, courted by business and seen as authoritative and trustworthy representatives of the environment by the media, the EMOs have achieved this status at an astonishing speed. Conservation groups such as the Sierra Club (USA, 1892), Natuurmonmenten (Netherlands 1905) and the Royal

Society for the Protection of Birds (UK, 1899) were part of the first wave of environmental groups in the late nineteenth century. Other groups emerged before the second wave of environmental activism in the late 1960s. WWF was formed in 1961 in order to support the work of the International Union for Nature Conservation, which itself had been formed in 1948 to assist UNESCO. These groups broadened their interests as new environmental issues emerged in the 1980s. Groups concerned with animal welfare found that the protection of animals depended on protecting their habitats and this required campaigning on a wide range of issues from pollution to human poverty.[1] By the 1990s there was a greater convergence between the organisational styles, and agendas of conservation groups and those of the more recently established and more confrontational environmental groups, such as Greenpeace and the various national affiliates of Friends of the Earth[2] (FOE) International. Membership in both old and new EMOs grew very fast in the early 1970s from a low base and again at the end of the1980s, providing them with new legitimacy and new resources (see Table 5.1). As public concern about the environment mounted in response to new evidence about global environmental problems, governments sought to assuage this concern by developing new environmental policies. One result of this was a greater effort by governments to include EMOs in consultation about policy and, in some countries, in its implementation (Hajer 1995: 29).

What form does institutionalisation take?

The institutionalisation of the environmental movement refers to the development of organisations from a more radical and confrontational stance towards a more conventional pressure-group role in which the main aim is to influence government policy or business practices rather than to challenge the social and political order through disruption and the creation of uncertainty for elites. This entails, what Meyer and Tarrow (1998: 21) refer to as a routinisation of conflict in which social movements increasingly adhere to the established rules of engagement; groups that play by these rules are included in policy-consultation and co-option, meaning that challengers neither pursue nor use anti-systemic tactics, pursuing only those claims and tactics 'that can be pursued without disrupting the normal practice of politics' (ibid.).

As we have seen, not all EMOs began as social movement organisations. Therefore this process does not affect groups such as WWF or the National Trust, which have never been radical or pursued disruption and therefore have never been social movement organisations. Of more interest is the process of institutionalisation within groups such as Friends of the Earth and Greenpeace and other new environmental organisations. These newer environmental groups have altered hugely since their formation in the 1970s. At the centre of this is the development of a larger, more formal and bureaucratic organisational structure. This is formalised by van der Heijden (1997) as having three aspects: (1) growth in membership, producing the resources to invest in organisation; (2) internal institutionalisation producing a growth in bureaucracy, professionalisation and specialisation, and; (3) external institutionalisation, involving more co-operation with government and business and less confrontational action. This is a useful distinction because not all EMOs progress in the same direction on all three dimensions. It is the third criterion that is most crucial in determining whether an EMO is no longer a social movement group. Bureaucratic EMOs can still be part of the green movement, despite the constraints that the need to defend the interests of their organisation imposes on their action, if they avoid co-option. Further, we need to see single social movement organisations (SMOs) in relation to the wider green network in which they increasingly play a specialised role, without necessarily isolating themselves from other more radical greens.

Membership growth

The largest and most rapid growth in membership has been for campaigning groups like BUND in Germany, Greenpeace and Friends of the Earth. In Britain, membership of FOE rose by 730% between 1985 and 1993. (Doherty and Rawcliffe, 1995; 243). In the Netherlands the number of Greenpeace contributors rose from 70,000 in 1985 to 830,000 in 1989 (van der Heijden, Koopmans and Giugni 1992: 18). Conservation groups and animal welfare groups such as WWF also experienced significant membership growth, but at a less rapid rate.[3] For both these types of group much of this growth occurred in the late 1980s and was followed by a period of decline in the recession of the early 1990s and then relative stability. In the USA the ten largest organisations grew from 1 million members in 1979 to nearly 7 million in 1990 but slipped to 6.5 million by the mid-1990s (van der Heijden 1999: 201). The largest falls in the early 1990s in most countries were for groups like Greenpeace and Friends of the Earth. In Britain and the USA membership of most conservation groups continued to increase. In Germany, there was less volatility, with a steady growth in membership of all types of EMO in the ten years to 1997 (Rucht and Roose 2000).

Table 5.1 below provides recent figures for membership of EMOs in Germany, the UK, USA and the Netherlands. These countries, along with Australia, Sweden and Denmark[4], have high membership of EMOs in comparison with southern Europe.

For instance, in France, Spain and Greece, membership of environmental groups remains low in comparison with northern Europe. Small regionally based nature associations linked through the federation known as France, Nature, Environment, make up the bulk of the membership of environmental groups in France (Fillieule and Ferrier 2000; van der Heijden, Koopmans and Giugni 1992, Rucht 1989). Only Greenpeace, WWF and FOE operate at the national level (Fillieule and Ferrier: 2000: 7) but membership of these groups is small, for instance, the number of paying supporters of Greenpeace fell from 45,000 in 1995 to 23,000 in 1999 (Gallet 1999) and Les Amis de la Terre (LADT) had only 1,000 members in 2000 (LADT 2000) In Spain (Jimenez 1999a) and Greece (Botezagias 2000), the weak traditions of association membership have been explained as a legacy of the experience of authoritarian rule. Italy has seen a

Table 5.1 Membership of selected environmental movement organisations

Germany (1997, unless otherwise stated)	Netherlands (2000)[2]
Greenpeace 520,000 supporters BUND 365,000 members (2000) WWF 185,000 supporters BBU 100,000 individuals (approx), 220 groups (1994) DNR[1] 5.2 million individuals, in 108 associations (not all environmental) NABU 200,000 members (1994)	Greenpeace 618,000 Natuurmonumenten 950,000 WWF 730,000 Milieudefensie (FOE) 32,500 Dutch Society for Bird Protection 122,600
United Kingdom (1997)	**USA (1998)**
Greenpeace 215,000 supporters Friends of the Earth 114,000 members (excluding some members of local groups) WWF 241,000 members Wildlife Trusts 310,000 members CPRE 45,000 members RSPB 1 million members Woodland Trust 115,000 members National Trust 2.489 million members (England and Wales)	Greenpeace 350,000 supporters Environmental Defense 250,000 WWF 1.2 million members Sierra Club 550,000 members National Audobon Society 575,000 members Wilderness Society 350,000 members Natural Resources Defense Fund 400,000 Nature Conservancy 828,000 members

Sources: Germany (Blühdorn 1995) Rucht and Roose 2000; United Kingdom (Rootes 1999 TEA): USA (Bryner 2001); Netherlands (van der Heijden 2000) and groups' websites.

Notes
1   The DNR (Deutscher Naturschutzring) is a national umbrella group for a very broad spectrum of interests including many users of nature such as hunters. According to Blühdorn (1995: 213), despite its size, it has only a small budget, a low public profile and has little influence on the environmental debate. However, in an analysis of inter-organisational networking between EMOs, Rucht and Roose (2000) the DNR is shown to be widely networked and undertaking joint campaigns with the BUND, BBU and NABU.
2   Total Membership of 96 environmental groups 3.7 million (2000) out of total population of 17 million (van der Heijden 2000)

consistent growth in the membership of environmental groups since the early 1980s, and this reflects the transformation of the Italian political system in which the very polarised politics of the 1970s and early 1980s gave way to a new system in the 1990s in which political parties no longer exerted the same dominance over political activity as before. In this context, environmentalism represented a politics based on issues that were neither part of the tradition of the old left (Communist Party) or the old right (Christian Democrats). Yet Italian environmental associations are not necessarily single-issue groups: leading organisations such as Legambiente have a wide range of concerns, including social justice, nuclear disarmament and human rights (della Porta and Andretta 2000), but they are still nevertheless seen as part of a new pluralist politics in which policy can be made on the basis of political exchange with either left or right governments. While Italy may have the largest memberships of EMOs in southern Europe, they are still relatively small in comparison with most of northern Europe:

Table 5.2 Membership of EMOs in Italy

WWF-Italy 281,000 (1999)
Legambiente 115,000 (1999)
Friends of the Earth 25,680 (1997)
LAV (Anti-vivisection League) 13,500 (1997)

Source: della Porta and Andretta 2000: 15.

Where their membership grew substantially EMOs had greater legitimacy and this was a factor in their improved status with national governments and internationally as interlocutors of governments in UN environmental forums. And, even where membership remained low, as in Spain, the increased salience of environmental issues meant governments were more open to environmental groups (Jimenez 1999). The growth in membership also had important consequences for movement organisations. More members meant more resources. This was particularly the case for the more radical and general environmental groups such as Greenpeace, FOE and Legambiente, who relied overwhelmingly on membership subscriptions for their income. Other groups were less reliant on this source of revenue. In Britain, subscriptions made up only 25% of the income of WWF and the Council for the Protection of Rural England (CPRE), with legacies constituting another 25%. But a much larger number of people were donors to WWF (1.6 million in 1992) than members (Szerszynski: 1995: 28). Merchandising is also an important source of revenue for some conservation groups. The WWF catalogue in Britain reaches 12 million households (Rawcliffe 1998: 82). The Australian Conservation Foundation (ACF) receives around 45% of its income from donations, including legacies, more than twice the amount that comes from membership fees. Government grants made up less than 10% of the ACF's income, with merchandising and product endorsement as the other principal source of revenue.[5] In the US many of the major environmental groups rely more on funding from charitable foundations than on membership income (Gottlieb 1993). Government grants also made up a substantial part of revenue for WWF International, constituting 30% of WWF income in the USA and 13% internationally (Kellow 2000: 18). In Spain and Greece environmental groups relied more on government grants (often from EU funds) for their income, but much of this money was tied to specific projects in education or practical conservation work. In Berlin in the early 1990s there was a remarkable increase in the number of staff employed in EMOs, principally in small local groups from 36 in 1988 to 969 in 1997. This was largely due to a government job creation programme targeted at high unemployment in the eastern Länder (Rucht and Roose 2000).

The concentration of support for EMOs in northern Europe and the USA also means that groups such as Greenpeace cannot really be seen as wholly international at least as regards the source of their funding. While Greenpeace has branches in more than 40 countries, in 1995–7 more than 70% of the income for Greenpeace International came from Greenpeace branches in Germany, USA, UK and the Netherlands. Another 14% came from Switzerland, Austria and Sweden (Kellow 2000: 8). Funding from Germany alone made up nearly half the funds for the Greenpeace International operation in these years. Branches in these countries subsidise the operations of Greenpeace in countries where its support is weaker. As Kellow notes,

> Since voting strength in Greenpeace is determined explicitly according to financial and campaign performance, Greenpeace inevitably represents a means by which an agenda representing northern European values is prosecuted in international politics, even to the extent of being advanced in developing countries and other nations in ways which would not be open to northern European nations themselves.
>
> (ibid.)

The internationalism of Greenpeace is thus at least qualified by the concentration of support and decision-making in northern Europe.[6] Moreover, since Greenpeace, FOE and WWF are all 'very lightly represented in Asia and Africa' (Kellow 200: 18) we cannot really argue that these EMOs represent a global civil society (an issue that will be discussed further in the final chapter).

Those groups most dependent on supporters for their finances suffered most when subscriptions declined, as they did in most countries in the recession of the early 1990s. Friends of the Earth in Britain cut its staff by 20% in 1993, and Greenpeace Australia sacked 30 of its 74 staff and 100 of the 140 casual staff in May 1994. Greenpeace USA suffered a major haemorrhage of 600,000 subscribers during the early 1990s and continued to lose support as the decade went on, which led it to close many regional offices and withdraw from projects (Schlosberg 1999: 144n19). Subscription-dependent groups have to reinvest a considerable part of their income on maintaining their membership base. Friends of the Earth spent 24% of its income in the mid-1990s on recruitment, fund-raising and administration (Jordan and Maloney 1997: 21–2). In 1990 23% of the budget of Greenpeace USA was spent on fund-raising, more than twice the average of the other major EMOs (Shaiko 1993: 96).

Direct mail was first used extensively by US environmental groups in the 1970s and was made cost effective by government subsidies. It only began to be used extensively in Europe in the 1980s, but is now widespread. In Britain WWF had 1.25 million names on its mailing list in the 1990s (Rawcliffe 1998: 82). 'Mobilisation strategies' of this kind, which are intended to reduce the costs for an individual to join an organisation, appear to be effective. For example, of those who joined FOE in Britain, 9% joined through a friend, 5–10% at a FOE meeting – both of which are direct face to face methods of recruitment, which are low in financial costs, but high in time costs for the organisation. In contrast 24% joined through an advertisement; 23% as a result of direct mail and 16% by filling in an application form on a leaflet (Jordan and Maloney 1997: 131). These indirect methods are high in financial cost, but low in time costs.

The internet provides a lower cost means for EMOs to advertise themselves to potential recruits, although at present this is passive, requiring recruits to visit websites. No figures are available to show how many people have joined groups through the net, but the Greenpeace International site shows a steady increase in the number of visitors. By the end of 1999 the site had an average of 370,000 accesses per day.[7] Websites and email lists also help in the diffusion of useful campaign information for supporters. For instance, in 1999 FOE UK provided a detailed map allowing anyone (whether or not they were a member) who visited its website to check the level and type of pollution in their locality. The Greenpeace site concentrates on brief accounts of its main campaigns and seems mainly intended to encourage visitors to become subscribers or to donate money. The British FOE website has much more information that could be used directly by campaigners. If services and information previously only available to members becomes freely available on the web this might reduce the incentive to join. However, for FOE, in contrast to Greenpeace, the priority appears to be to encourage action, even if this poses some threat to its income.

For critics, the use of environmentally costly methods such as direct mail is a sign that environmental groups are becoming too much like business, giving priority to maintaining their own organisation rather than achieving environmental ends (Scarce 1990). Yet since the rate of non-renewal of subscribers is between 30 and 40% for FOE in Britain (Jordan and Maloney 1997: 16) and for EMOs in general in the USA (Dowie 1995: 40) it is understandable that so much attention is paid to subscriptions. Another point made by critics is that actions are chosen because they appeal to potential subscribers not because they are necessarily the most important environmental issue. Greenpeace timed a major direct mail campaign in 60 countries in 1995 to coincide with the expected boarding of one of its ships by the French navy (Rawcliffe 1998: 41). Even if the direct action in this case, which was against the resumption of nuclear testing in the South Pacific, was not simply to expand the base of support, Greenpeace might avoid campaigns that market researchers say will not strike a chord with its supporters. Mark Dowie (1995: 47) reports that Greenpeace USA dropped a plan for a combative advertising campaign based upon the theme 'confrontation works' when market research suggested that it would be perceived negatively by potential supporters. And a Greenpeace campaign in support of a local environmental justice group opposed to a toxic waste incinerator in Liverpool Ohio was abandoned in 1993 when consultants said that it would not appeal to its supporters.

The volatility of income presents a major challenge to groups such as FOE that are seeking to maintain campaigns that may last several years. EMOs have therefore sought other ways of securing their income. Greenpeace suffered a substantial decline in subscriptions in Britain in the mid-1990s but moderated the drop in income by drawing on a group of 4,500 'Frontline members' who were able to raise one third of Greenpeace's annual income of £6 million (Rawcliffe 1998: 43). Although the number of Greenpeace UK supporters continued to fall (from 215,000 in 1997 to 169,000 in 2000) its income increased. The level of Greenpeace UK subscriptions reflects its fund-raising efforts during that year (Rootes and Miller 2000: 12) and as Greenpeace has been able to expand its income it has less need to be concerned about the decline in subscribers. However, the decline in individual subscriptions to Greenpeace branches in most countries appeared to have ended as the number of supporters of Greenpeace globally increased from 2.4 million in 1998 to 2.5 million in 1999 (Greenpeace Annual Report 1999).

Mass membership

One possible consequence of the growth of mass membership is that the central staff of EMOs might lose touch with their supporters and treat them only as a resource. Comments by some staff from EMOs seem to support this. One UK staffer suggested that groups should only tell their supporters what they need to know in order to keep them paying their subscriptions. (Szerszynski 1995). An organiser for FOE (UK) said, 'Members have to decide to back us or not. We make policy and if they don't like it they can join some other group' (Jordan and Maloney 1997: 189). To Szerszynski (1995: 11) it seemed that supporters of FOE were 'little more than a database whose relationship with the organisation was mediated by direct mail and standing orders'.

Given the low levels of participation in EMOs, much of the optimism about such groups as signalling new models of civil society is exaggerated. Only tens of thousands rather than hundreds of thousands take part in any activism that extends beyond the passive cheque book activism of the majority of members of EMOs.[8]. The activity of local Greenpeace groups is largely confined to fund-raising and leafleting. Members of groups such as BUND in Germany and Amici della Terra (AdT) in Italy are mainly encouraged to do practical conservation work and involve themselves in local planning procedures, participation in protest is rare (Blühdorn 1995; della Porta and Andretta 2000). Moreover, in some cases ordinary members have little control over their organisation. For instance, national branches of Greenpeace have very few voting members: Greenpeace Australia had 75,000 supporters (in 2000[9]) but only around 50 voting members and new members had to be nominated by an existing member. Those who pay dues are strictly speaking supporters rather than members

This picture undermines the link between the mass environmental movement and democratic forms of organisation. Nevertheless it does not always mean that there has been a move from democracy towards oligarchy. Many of the largest EMOs were not formed as democracies. Greenpeace and Friends of the Earth (UK) were founded as groups of professional activists. Local Greenpeace support groups were only established when it became clear that groups were better at fund-raising than individuals. Greenpeace has therefore always had a unique status as a group that specialises in protest, but remains hierarchical, excluding its millions of supporters from the right to take protest using its name. In the past this has provoked small breakaway groups such as Robin Wood in Germany, founded in 1982 and the (much smaller) London Greenpeace[10] anarchist group, both of which favoured non-hierarchical internal structures and were committed to encouraging non-violent direct action. In the 1990s Greenpeace in several countries has tried to provide mechanisms for its supporters to take part in more direct action. This was in response to a recognition that there was a body of Greenpeace supporters anxious to take more direct action. Yet these efforts have not really succeeded because of fears that local groups may act in ways that damage the image of the organisation. As the organiser of Greenpeace Italy with responsibility for local groups said in an interview:

> Since Greenpeace is based upon the image, it is important to internalise a series of things before doing an action, because when you are there every movement of your face is enlarged and you must not seem aggressive … communication and the management of the image through gestures that are not coherent, responsible and serious would be suicide for us.
>
> (quoted in della Porta and Andretta 2000: 24)

FOE in Britain established local groups in 1971 when it realised the demand to take part in environmental action following the widespread media coverage of a protest against Schweppes for its failure to encourage recycling of bottles (Lamb 1996). The 250 local FOE groups have considerable autonomy, some run their own businesses, and have some indirect control over the national organisation because they elect 10 of the 17 members of the governing body. This came about after a revolution by local groups in the early 1980s because they had been excluded from consultation over moves to cut costs and professionalise the staff. Nevertheless, in practice the balance of power lies with the national organisation. One of the local group workers said that 'the Local Groups Conference is more of an opportunity for FOE Ltd to educate local groups than it is for them to influence the direction of the national office' (Rootes, Seel and Adams, 2000). Although the national office has a permanent staff that works to support local groups, the latter are supposed to be self-financing and initiate their own campaigns. Its national staff sets the strategy of FOE and it is only a small number of the most experienced local members that are consulted by the national office. Interestingly, the internet has allowed some devolution from the national office in recent years and increased consultation with some regional officers, because a shared intranet allows the latter to read internal documents (Pickerill 2000), but this affects only a fairly small number of people. As Rootes, Seel and Adams (2000) note the growth of local groups in FOE UK and the encouragement of participation in local groups has occurred alongside internal institutionalisation in the central office. Thus this is one case where internal institutionalisation has produced some decentralisation.

National branches of FOE International, vary greatly in forms of organisation and strategy. The original branch of FOE was established in San Francisco in 1970 by Dave Brower, who had been Chief Executive of the Sierra Club but left after disillusion with what he saw as its lack of militancy and its failure to address the new environmental agenda emerging around issues such as pollution. However, Brower came into conflict with a FOE staff that he thought was too focused on administrative and legal matters and not oriented sufficiently to campaigning. Some of the latter left in 1972 to form a new group based in Washington DC, which became known as the Environmental Policy Institute. As FOE membership and resources declined in the early 1980s, a bitter internal dispute led to Brower's departure and the merger of FOE and the EPI in 1984. While FOE still exists in the USA it is primarily focused on lobbying in Washington and has no grassroots campaigning presence (Gottlieb 1993: 148).

One legacy of Brower's initiative is that he set no constraints on how national branches of FOE were to be organised. Thus in some countries, such as Australia, FOE remains very decentralised and green-alternative in character with decision-making by consensus rather than votes at its annual meetings and a strong commitment to grassroots campaigning on social justice. In France LADT is also very decentralised and closely linked to the radical ecology movement, but remains tiny in comparison with other countries with only just over 1,000 members in 2000[11]. In Italy AdT is also relatively decentralised, with only one paid full-time staff member. It had its origins in the campaigns against nuclear energy in the 1970s and like many other parts of the Italian environmental movement, its leaders had experience in New Left and peace groups (ties with the Radical Party were particularly strong). Now, however, it carries out very little protest activity and its main roles are lobbying local and national political institutions on policy questions and supporting practical voluntary work and in contrast to other branches of FOE its principal source of income is from research carried out for public institutions and other bodies (della Porta and Andretta 2000: 29).

NOAH, which became the Danish branch of FOE in 1988, had its roots in New Left student protest in the late 1960s (Jamison, Eyerman and Cramer 1990: 82). This determined its oppositional stance in criticism of public authorities, its open and non-hierarchical structure and criticism of the misuse of power by science and technology. Similar to the later EDA groups, NOAH combined its lack of bureaucracy with a lack of emphasis on doctrine or formal membership. If you were active you were a member, and it contrasted its acceptance of the autonomy and right to difference of its local groups with the emphasis of Marxist groups on having a 'complete' position (ibid.: 89). However, while it retains its radical character[12] by the 1990s it was dwarfed in membership size by groups such as Danmarks Naturfredningsforening (DN) which had a membership of 15% of the adult population in 1988 (ibid.: 113) and Greenpeace. Neither of these groups encouraged grassroots activism.

Other FOE branches differ both in organisation and strategy, with some more professional and lobbying-oriented and others more grassroots and remaining protest-oriented. Thus there is no organisational model associated with FOE as a whole.

In Germany the BUND (German branch of FOE International) retains a delegate council and has seen internal conflicts in the 1990s between grassroots members and office staff. Local chapters of other EMOs have also challenged compromises made by the national staff.[13] In 1991 grassroots members of the Sierra Club took part in a self-styled 'fundamentalist insurrection' to press for more aggressive pro-wilderness policies. This led to some policy changes, though they were insufficient to satisfy the activists, by now working as the John Muir Society, who forced a referendum of members in 1993. Although they lost, Dowie (1995: 218) suggests that the phrasing of the questions by the Sierra Club leaders disadvantaged the dissidents. The Australian Conservation Foundation (ACF) is governed by councillors elected by its members, who have helped to change the character of the organisation when they felt it had not worked well enough. In the 1970s Jack Mundey a communist official in the Building Labourers Federation was elected to the ACF Council because of his success in initiating 'Green Bans' in which builders refused to work on environmentally damaging development projects. More recently, there have been conflicts between the elected councillors and staff members over professionalisation, leading to the departure of two Chief Executives in the 1990s (Doyle 2000: 90). In many older conservation groups such as the National Trust, RSPB and CPRE in Britain members have formal power to elect governing bodies but few members take part in elections. For instance, only approximately 1 in 14 of the members of the National Trust vote in elections (Ward 1999).

There is therefore considerable variation in the rights of members within EMOs and in the extent of participation. While EMOs are not necessarily grassroots controlled, neither are they all controlled by a central bureaucracy. The role of the membership varies according to the traditions of the organisation.

Internal institutionalisation

The growth of membership provided the newer and more radical EMOs with the means to invest in their organisations and to recruit specialists. This resulted in a growth in the number of professional employees, a more formal working environment and greater specialisation and division of labour within the organisation. Whereas in the mid-1980s most of those who worked for EMOs were environmentalists whose work grew out of their commitment to the movement, by the 1990s many groups recruited specialists who had less need to work for an EMO.[14] For instance, in FOE (UK) teachers were recruited to work in education units, journalists in media units, product managers in business units, and sales staff to work in subscription and other departments. While the campaign staff still had a special status as those who defined the values and purpose of the organisation's activity, they also had to take account of the interests and expertise of other groups within the organisation.

One result of professionalisation has been that it has become possible to develop careers within EMOs (Dalton 1994: 119). The risks of working for EMOs appear to have been reduced compared to the 1970s. Pete Wilkinson said that working for Greenpeace used to mean having little chance of getting a 'straight' job, but was now more respectable and a 'test-bed' for other careers (Rawcliffe: 1998: 106). In Britain, the EMOs such as Greenpeace and FOE have provided jobs for some involved in the early waves of Earth First! and many of the leading personnel in British and US EMOs have worked for several different EMOs. For instance, all the senior staff at Greenpeace USA in 2000 had previously worked for other environmental or social movement-related organisations. For some working for an EMO can be a route into the civil service or public office. Jimenez (1999: 155) and Botezagias (2000) note how the expansion of environmental administration in Spain and Greece provided jobs for many that had worked for conservation groups.

Long-term members of staff of EMOs in the USA and Britain have noted the change in their working environment as a result of internal institutionalisation. The growth in the number of staff meant a loss of informality. It is also likely that the professionalisation of work in EMOs has favoured the progress of male staff. They are more likely to have had the time to develop professional skills and a formal CV. This was one reason for the establishment of the Women's Environmental Network in Britain in 1988 (Szerszynski: 1995: 43).[15]

It could be argued that the recruitment of professionals from outside the movement makes EMOs less radical. Campaigners will no longer be able to assume that all their colleagues share their politics and are willing to take the same risks. There is likely to be internal pressure to seek a minimal common ground which takes into account the interests of all sections of the organisation, including those seeking to build links with media, business and civil servants. Yet against this, it should be remembered that although many of the new personnel may be recruited because of their specialist skills, they usually accept lower rates of pay and have a more vulnerable employment status than in other fields. Therefore, they are likely to share the values of the organisation. It should also be stated that indirect evidence does not necessarily support the hypothesis that recruitment of non-movement staff leads to de-radicalisation. This is because, as will be argued below, there is no conclusive evidence showing de-radicalisation in EMOs.

It is not only for maintaining the organisation that new professionals have been needed. EMOs have had to undertake new tasks as the nature of the environmental debate has altered. As risk uncertainties have became an increased focus of debate, the scientific credibility of EMOs has become increasingly important. Most therefore invest considerable resources in research and in sponsoring scientists to carry out research on their behalf. Exemplary in this regard is the German Öko-Institut founded in Freiburg in 1977. This is an independent EMO in its own right, which is funded by individuals, donations and by fees for its research. Legal specialists are also important on the staff of EMOs. Having more resources has allowed major EMOs to pursue court cases more vigorously and often with success, but it has also exposed them to new risks. Companies and governments have been able to secure injunctions against Greenpeace and FOE to prevent them undertaking direct action protests and, if broken, they face a substantial fine or loss of assets endangering their continuity. This certainly limited the kinds of protest carried out by both organisations in Britain in the 1990s. Friends of the Earth withdrew from non-violent direct action intended to prevent the construction of the M3 at Twyford Down in Hampshire in 1992 because it feared legal action. And in 1997 British Petroleum took out an injunction against Greenpeace to prevent it taking action to prevent test drilling for oil in the North Atlantic.

EMOs need to be able to use the media effectively if they are to be successful in framing issues and communicating their position. Effective use of the media is particularly important for the greener EMOs who have weaker ties with government and business. EMOs have developed sophisticated media strategies as a result. Szerzsynski (1995: 79) describes the replacement of scatter-shot press releases with increased targeting of specific journalists and specialist media. Stories are 'versioned' for different audiences. For instance, a special report with 'in house' video footage can be prepared for regional and local TV news or to coincide with a specific event. EMOs seek to develop a close working relationship with specialist environment correspondents. For many of the latter, EMOs are a major source of news. For instance, the Environment Correspondent of *The Guardian* in Britain commented 'that an average of three press releases a day land on his desk from Greenpeace alone' (Rootes 1999: 21). EMOs such as Greenpeace plan action carefully to maximise its resonance in the media. The campaign against Shell's plan to dump the Brent Spar oil platform in the North Sea in 1995 was greatly enhanced by the use of satellite images broadcast from the platform directly to news organisations that had no camera footage of their own (Andersen 1997).

Organisational growth and the new tasks carried out have therefore produced a new division of labour within EMOs. In 1987 the 27 staff of FOE (UK) all reported directly to the Director Jonathon Porritt, who was also simultaneously a leading figure in the Green Party. Following a report by management consultants this structure was replaced by one with several new sub-units, each with their own managers. By 1995 the staff had grown to 110 (Rawcliffe 1998: 80) and it was impossible practically for the Director to be involved in other political activity. This is a pattern now found in most major EMOs. Units within organisations are often 'substantially independent of each other' (Diani and Donati: 1999: 19). A major campaign requires considerable planning and co-ordination of research, press, finances, and information for supporters. Once these resources have been invested in an issue or campaigns it can be difficult for the EMO to back out if circumstances change. For these reasons, the price paid for higher professional standards is a slower and more cumbersome campaign, and a more cautious organisation, which is likely to take the threat of legal action into account in deciding on the form of protest. Thus the need to protect the organisation constrains the kind of action taken by the better-resourced EMOs.

External institutionalisation

Most EMOs now have stronger ties with government and business than they had in the 1980s. Typical activities include testifying before government and parliamentary committees, meeting with officials and partnerships with business to sponsor environmentally conscious policy and consumer choices. One reason for the increase in such ties is the increased influence of EMOs themselves. Neither government nor business wants to appear anti-environmental and both seek the approval of EMOs where possible. EMOs in their turn are under pressure to appear professional, responsible and be prepared to compromise. This requires that they provide credible technical arguments on policy positions and that they deal with issues without continually challenging the principles of business itself, or the main planks of a government's programme. Many of those working in EMOs note how they have come increasingly to speak the same language as government and business. Informal contact between specialists in the civil service and those with similar expertise in EMOs is likely to be based on shared technical and professional norms.

Ties with government

Russell Dalton (1994: 189) reported that in the mid-1980s EMOs tended to evaluate the Ministry of the Environment[16] more positively than other potential allies such as parties, unions, business and the media. In Britain, Tom Burke, the Director of the Green Alliance, a parliamentary lobby group supported by the major EMOs, was a special adviser to successive Conservative Environment Ministers between 1991 and 1997 and provided EMOs with an unprecedented inside track for access to policy-makers. German EMOs were praised by the Ministry of the Environment and given funds by government with the explicit aim of creating a counter-lobby (Brand 1999: 52). But Dalton's comment that even when they gain access they do not feel this access is routinely translated into influence (1994: 190) applies as much now as in the mid-1980s.

In Australia during the 1980s two of the major EMOs became particularly close to the Labor Party. This followed the backing of the Hawke Labor government for important wilderness protection measures in the 1980s and the environmental vote was widely credited with swinging the 1990 federal election towards Labor. The ACF and TWS called on their members to vote Labor, Green or Democrat in 1996 but when Labor was replaced by the Coalition government led by John Howard in 1996 the ACF and TWS found that they were out in the cold (Hutton and Connors 1999; Doyle 2000). The new government preferred to deal with the state-based conservation councils and the national environmental organisations found they were moved from insiders to outsiders in the policy-process. The Labor Party, worried that it would seem anti-development, also made it clear that it was not prepared to make concessions to court the environmental vote.

Relations with the government, and particularly the Environment Ministry might be expected to be stronger where a representative of the Green Party holds the post of Environment Minister. In 1999 this was the case in Italy, Finland, France, Germany and Belgium. Green Party Environment Ministers are under pressure to show that they can deliver solid gains for the environment. Yet the weak position of green parties within coalitions, combined with the lack of power of environment ministries, means that major gains are unlikely to result simply because a green becomes the Environment Minister. This was made clear in Germany when Jürgen Trittin the Green Environment Minister was forced by the Social Democratic Prime Minister Gerhard Schröder to accept a major compromise delaying until the long-term the phasing out of nuclear power in Germany. Dominique Voynet achieved the cancellation of the Rhône–Rhine canal but was forced to accept only a partial closure of the Superphénix Fast Breeder Reactor at Malville (the focus of major opposition from greens in the 1970s), and found herself in opposition to the rest of the cabinet on many other issues (see Chapter Four). In October 1999 the Green Party member Edo Ronchi, who was then the Italian Environment Minister, expressed his support for campaigners opposing the construction of electricity pylons in remote parts of Tuscany but was powerless to intervene because on this issue central government could not over-ride the decisions of regional authorities.

Thus, while EMOs are likely to have good relations with the environment ministries, particularly when the minister is from a green party, the latter tend to be relatively weak in determining government policy (Hanf and Jansen 1998: 304). In fact, Environment Ministers are likely to draw on EMOs and wider environmental policy networks to enhance their bargaining power within the government (Smith, M. in Jimenez 1999a: 150). They have used evidence of environmental activism and support in conflicts with other parts of the state such as those that have occurred in Britain in the 1990s over transport policy (Robinson 2000). EMOs also need to establish stronger ties with other parts of government if they are to be able to achieve greater impact. In the Netherlands the Stichting Natuur en Milieu was represented on over 70 government boards and Dutch environmental groups had much more access to government than those in France or Germany (van der Heijden 1997).

Many in EMOs are cautious about the value of ties with government, even when that government has a pro-environmental reputation (Dalton 1994: 189–95). Roger Higman, a campaigner for FOE (UK), said in an interview[17] that he had resigned from his position as Chair of a government committee on environmental transport because he felt that while he held that responsibility he could not campaign openly. He argued that he could achieve more from outside. In 1992 24 staff from EMOs joined the Clinton Administration. However, EMOs felt misused after vice-president Al Gore approved a new incinerator in the face of criticism from environmental campaigners and Clinton used the dependence of EMOs on the Democrats to secure their backing in the negotiations for the North American Free Trade Agreement (Dowie 1995: 187). In Australia the government committee on sustainable development included representatives from EMOs, but they were unable to prevent the government undermining the value of the Kyoto climate change accords by negotiating an 8% increase in the country's emissions targets (Hutton and Connors 1999: 242). And when FOE (Aus) became more critical of the government it was threatened with expulsion (Doyle 2000: 86). In general government and business tend to be closely agreed on aims since they share a common commitment to the imperative of economic growth. As Doyle says of these consultation processes, 'There can only be advantages in participation if groups recognise that their usefulness is limited, and if this process is pursued along with a range of other options. Unfortunately the Australian environmental movement devotes vast amounts of its limited time and resources to these forums' (ibid: 206).

In practice there remains a difference between green and non-green EMOs regarding their ties with government. Groups such as FOE, while not generally refusing invititations to participate in consultation with government, do not do so at the expense of their right to be critical.[18] This is because for the latter groups specific policy achievements are measured against a long-term and wide-ranging agenda, which entails significant social and political change. For more traditional conservation groups, the utility of building new policy networks within mainstream political and business networks seems clearer because they are uncritical of the principles of power on which these networks depend.[19] Moreover, they are in some sense part of this structure. As Rootes, Seel and Adams comment on the RSPB's non-radical ethos:

> This is achieved not only as a direct result of its approach to science, but also through a non-confontational style of operation. RSPB constantly seeks to be 'positive and constructive', attempting to provide realistic and well-researched solutions through 'problem solving partnerships' among key players and the forging of alliances with many different organisations. This networking ability is linked also to the social status of the key individuals within the organisation, and their ability to access important decision-makers.
>
> (Rootes, Seel and Adams 2000: 5)

A similar strategy is pursued by groups such as Environmental Defense in the USA, which has been particularly criticised by other green groups for its emphasis on policies based on market incentives that involve no critique of business. Governments often establish working parties that include environmental groups but in which they are greatly outnumbered by representatives of business (Jimenez 1999; Dowie, 1995). Moreover, it is usually only EMOs that can be trusted to accept the rules of the policy-making game and speak the language of the 'political court' that are invited. For instance, the US Presidential Council on Sustainable Development established in June 1993 included members of the older established EMOs but no one from the Citizens' Clearinghouse on Hazardous Wastes (CCHW) or Greenpeace.

The same factors also apply at international level. Finger's analysis of the role of Environmental non-governmental organisations (ENGOs) in the UNCED process makes a distinction between the ideology and practice of FOE, Greenpeace and other more political campaigning groups and that of mainstream ENGOs 'such as WWF, IUCN, the World Resources Institute and the "Big Six" in the USA ... these are the ones which struck mutually agreeable bargains with major actors in UNCED. As the process unfolded, they increased their access to secretariat and delegations of northern countries. They were often consulted and sometimes even had representatives at crucial stages of the UNCED process ... Although their credibility grew in the eyes of many, they have at the same time, become somewhat co-opted and have isolated themselves from the rest of the environmental movement' (Finger 1994: 210). In contrast, the more radical groups, although engaging in dialogue with government, also remained confrontational and critical and developed stronger ties with NGOs from the south.

Ties with Business

More controversial even than ties with government are relationships with business. Doyle (2000: 201) says that as the role of the national state has weakened environmental groups have directed greater attention to business, whether as a target of protest or lobbying or as a potential partner. At minimum, the 1990s saw the development of a range of new relationships with business groups whereas previously EMOs dealt with business via the state, concentrating on policy and legislation to constrain business. In 1994 the BUND expanded its core aims to include consumer protection (Blühdorn 1995: 189). Greenpeace Germany invested in an East German company to support the development of non-CFC refrigerators and sponsored the development of a non-fossil fuel car. In the UK, Greenpeace joined a partnership of energy companies to invest in windpower. Other environmental groups such as WWF have supported ecolabelling by allowing their logos to be used to endorse environmentally friendly products in return for a fee. More radical greens regard such actions as encouraging consumerism, and undermining the green argument against productivism. Yet as consumers begin to use their buying power to punish companies that are seen as unjust, the threat of a consumer boycott provides EMOs such as Greenpeace with a powerful weapon. After the boycott of Shell products in much of northern Europe during the Brent Spar controversy in 1995 Shell invested heavily in environmental projects in an effort to repair the damage. However, for greens the failure of Shell to intercede with the Nigerian government to prevent the execution of Ogoni leader Ken Saro-Wiwa meant that it remained a pariah company. In 2001 FOE International and Greenpeace supported a boycott of Exxon oil (owners of Esso) because of its opposition to actions to curb greenhouse gas emissions.

Newer environmental organisations such as Greenpeace and FOE have been careful to avoid ties with companies that are responsible for environmental damage or social injustice. Older conservation groups, however, have welcomed the sponsorship provided by major industry. In the USA business foundations provided 7% of the funds of the main EMOs in the early 1990s (Dowie 1995: 49). Although not decisive, this is an important part of the income of most major conservation groups. Employees from Exxon and Chevron sit on the governing boards of conservation groups such as WWF in the USA. In Britain, the RSPB also gained around 7% of its income from corporate sponsorship deals (Rootes, Seel and Adams 2000: 5). Forum for the Future, a UK environmental lobby group set up by Sara Parkin and Jonathon Porrit (previously leading activists in the Green Party, see Chapter Four) takes money from BP, ICI, Tesco and Blue Circle, all companies that have been the targets of attacks by other greens.[20] In the USA and Australia a company called Earthshare raises money for EMOs through payroll deduction schemes which have been taken up by 'numerous questionable banks and corporations' (Doyle 2000: 207). Whether or not this leads some environmental groups to censor themselves, the most important question is what are the companies buying by supporting EMOs?

By investing relatively small amounts in EMOs major companies are able to argue that they support the environmental movement. In advertising, environmental management measures, usually required by legislation, are presented as evidence of the reality of the company's environmental commitment. The fact that they are accepted by some EMOs means that they can present themselves as part of the movement itself. A WWF US brochure for business said: 'Your company can use a World Wildlife Tie-in to achieve virtually every effort in your market plan ... new product launches; corporate awareness; new business contacts; brand loyalty' (quoted in Dowie 1995: 54). The same brochure reported how the Jaguar car company was funding a jaguar reserve in Belize. When the full balance sheet of the activities of energy and car companies is assessed in relation to environmentalism, the results of their actions contradict this image, but by accepting them in the name of co-operation with business, EMOs make it easier for such companies to defend themselves against environmental criticism. That this was the intention in one case was made clear in a leaked document from the oil company Chevron which, along with BP, made a deal with WWF over a project in Papua New Guinea, promising $1 million dollars for a conservation project. This said that, 'WWF will act as a buffer for the joint venture against ... international environmental criticism' (quoted in Monbiot 2001). In 2000 WWF 'held back from publication of a damning report on tropical rainforest destruction, for fear of upsetting the companies that it named' (ibid.).[21]

Although it is often criticised by radicals Greenpeace stands with them and against the non-radical EMOs such as WWF or Environmental Defense in its attitude to business. While it makes efforts to influence business, through expensive and glossy magazines such as *Greenpeace Business*, it has no corporate sponsorship and is in general critical of dominant business practices. For instance, in his 1999 annual report, entitled 'Public Risk, Private Profit', Thilo Bode of Greenpeace said:

> Today many politicians treat entrepreneurs as heroes, and business people seem to expect a free ride when creating risks. GE crop trials are typically uninsured. A sausage manufacturer whose factory poisons a river may expect a punitive fine but the GE venture capitalist who creates a continent-wide ecological disaster will probably get off free ... For as long as the 'economic argument' remains paramount, truly responsible decision-making will remain out of reach ... We need objective and open political discussions of the role and acceptability of new technologies, without anyone trying to pretend that these are just technical matters that the laboratory and free market can determine. The principles of eco-system sustainability and the rights of future generations are not negotiable. Moreover, their implementation is not merely a question for experts and certainly cannot be left to business.
>
> (Greenpeace International, Annual Report 1999)

While this is not an ideological attack on capitalism *per se*, and would not satisfy more radical greens, it is an attack on the current relationship between principles of economic profit, new technology, and environmental damage and one that would be inconceivable for groups such as WWF.[22] Since government and business have begun to act in ways that can be presented as environmental, the environmental movement has lost its monopoly of the discourse of environmentalism, and the limits of this elite environmentalism can only be exposed by groups that remain critical. Having more radical and critical EMOs with strong public support and prepared to act outside the policy process can help make the bargaining of those who accept the rules of the policy-making process by making them seem more moderate. This is why, for all its failings in the eyes of radical greens, Greenpeace remains a major asset for the green movement.

Ties within the environmental movement

Greenpeace seems to network less with other parts of the environmental movement than many other environmental groups. In a survey of inter-organisational networks between British EMOs (Rootes and Miller 2000: 17), Greenpeace presented itself as only rarely collaborating with FOE. In contrast FOE, though not the best resourced of the EMOs, was the one with most ties and regular collaboration with other EMOs, followed by WWF. The National Trust, the largest organisation in the survey, was marginal in this network, lending support to the view that it is relatively peripheral within the movement and given its limited political character, does not qualify as part of the green movement at all. For some groups, such as FOE, networking within the movement and seeking to build its collective capacity could be regarded as an alternative to prioritising ties with business and government.[23]

Although Greenpeace has a reputation for acting autonomously, it often culti-vates links with other groups involved in non-violent direct action.[24] For instance, Greenpeace has provided assistance and funds to anti-roads direct action groups in Britain, and Greenpeace joined the Women's Environmental Network, Reclaim the Streets and Earth First! in establishing the Genetic Engineering Network in 1997. Friends of the Earth (UK) also held strategy discussions with the latter two groups and they worked together on some anti-roads campaigns. FOE (UK) also worked with the Green Party[25] on drafting legislation in the late 1990s on Road Traffic Reduction and Energy Conservation.

Inter-organisational co-operation within the green movement is common (Dalton 1994: 170)[26], and does not raise the same ideological dilemmas as ties with business and government. Attention to the formal ties between organisa-tions is only part of the story, however. As Purdue's analysis of the genesis of the anti-GM movement in Britain shows (2000), the development of an antago-nistic position on an issue such as GM by greens depended greatly on the networks between individuals in different organisations.[27] The development of a common discourse and trust between these activists produced a counter exper-tise that they used collectively to raise the wider environmental and social dangers of GM. This counter expertise then fed into the wave of anti-GM protests in Britain in the late 1990s, and also into international networks between anti-GM activists. The GM issue also encouraged some within WWF to seek to change its strategy. These activists, who had played a role in the national and international networks of GM counter experts, sought to move WWF's anti-GM position beyond the implications of GM for biodiversity to also include social issues such as the rights of communities to control their own crop seeds. In doing so, they exposed a conflict in WWF between older conser-vationists and those more committed to the relationship between ecology and social issues. The new socially-oriented wing lost, and several were sacked (Purdue 2000: 88).

Summary

In assessing the form of institutionalisation, what is clear is that despite trends towards institutionalisation, considerable diversity remains. In some countries there has been significant membership growth, but in others relatively little. Also, while there has been fairly steady growth in membership of less radical conser-vation groups, support for more radical groups, having risen sharply at the end of the 1980s, has been more uneven since. While there has been a clear trend towards internal institutionalisation in many groups such as Greenpeace, in some cases this is also resisted in organisations with a strong tradition of grassroots democracy, such as NOAH or FOE (Australia). Also, while there has been a trend towards external institutionalisation, there remain important differences between those groups with a commitment to a broader agenda in which political and social changes are seen as necessary before environmental goals can be met (such as various branches of FOE International), and other EMOs, whether old (the RSPB) or new (Environmental Defense in the USA). The question of whether the level of institutionalisation differs between different countries and can be explained by their political context will be dealt with further below, but first it is necessary to examine the implications of this evidence for resource mobilisation theory (RMT), the most influential theory explaining how social movement organisations develop.

Is institutionalisation inevitable and irreversible?

The evidence above suggests that EMOs have some choices about how they organise and are not subject to inevitable institutionalisation, but one argument against this is that the groups that fail to institutionalise are likely to fail as organisations and become weaker. The dramatic fall in supporters for Greenpeace USA in the 1990s might seem to confirm this. Support fell from 2.3 million in 1990 to 350,000 in 1998 and might be attributed to its failure to pursue the kind of external institutionalisation that would produce concrete results. If effective use of resources is a crucial challenge for social movement organisations does the use of resources govern the development of EMOs?

In explaining how movements develop RMT gives priority to the need to mobilise resources (Oberschall 1973; Gamson 1975; McCarthy and Zald 1977). Initially RMT advocates favoured a model in which institutionalisation was part of the normal cycle of successful social movements and a sign of maturity. Hierarchical organisation was more efficient than participatory organisation and moderate groups were likely to have wider support and be able to amass more resources and achieve more political influence than more radical groups. Critics argued that this general model failed to pay sufficient attention to the ideology of groups (Dalton 1994: 8). Those with more radical ideas were not necessarily less effective than more moderate groups, but they made choices about how to act that were consistent with their ideology. For radical environmental groups hierarchies were part of the problem and challenging hierarchies was part of the solution to environmental crisis. Thus for radicals working within existing structures in order to deliver ameliorative policy changes was not as effective as attempting to change consciousness from outside. RMT theorists have tended to accept that a group's judgement of the appropriate action will depend on its ideology (Zald, 2000) and the kind of resources required will vary according to the nature of the group's action. For instance, Oliver and Marwell (1992) argue resources could be conceived of as either 'time' (activists') or 'money' (which may or may not come from activists) and the 'technologies' required differ in each case. Where money is the main resource EMOs have to develop a bureaucracy capable of providing efficient long-term administration. This is consistent with the internal institutionalisation described above. It is also consistent with Weber's model (Gerth and Mills 1991; Beetham 1987) in which a rational form of bureaucratic administration based on principles such as specialisation of tasks, objective rather than personal criteria in decision-making and clear lines of authority is necessary in order to achieve certain tasks, such as supporting the long-term development of policy, media and other expertise. Where time is the principal resource participatory structures are better than hierarchical ones at making members feel valued and necessary.

RMT views social movement organisations (SMOs) such as FOE and Greenpeace as part of a 'social movement industry' (McCarthy and Zald,1977) in which different SMOs co-operate and compete. According to this logic, organisations seek niches in the market and develop strategies that best exploit this specialism. This certainly applies well to the relationship that evolved between the major EMOs within the Washington beltway. After a period of tension and conflict over who owned particular issues, in 1980 groups began to co-operate more through a framework that was known as the Group of Ten (Gottleib, 1993, ch. 4) but also developed an informal division of labour, for instance the NRDC became adept at technical aspects of drafting legislation. According to RMT, SMOs seek to develop their own brand identities to remain distinctive. Thus, the diversity that exists within the environmental movement might be based on appeals to different markets. For instance, if Greenpeace gave up on high visibility protest it would be likely to undermine its subscription base. Support for what Jordan and Maloney (1997) define as a 'protest business' – characterised by minimal participation by the grassroots – is dependent upon the success of such organisations in marketing themselves to potential supporters.

This picture of organisation emphasises the importance of calculation by movement leaders (or entrepreneurs) who assess the best way of achieving specific goals by garnering resources and exploiting political opportunities. But maximising resources is not always the principal aim of EMOs. A 'brand positioning and optimisation' report prepared for the ACF by a market research company 'sought to assess its position in the marketplace relative to its competitors'. It included the recommendation that 'Primary qualities of the ideal [environmental] organisation include: professionalism and working with industry and business' (quoted in Doyle 2000: 207). There was no stress on representing green ideology. If the ACF and other groups had allowed themselves to be led by such considerations they would be behaving as economic actors and viewing their organisation principally as a business. However, it is not simply the nature of competition within the social movement industry that shapes the nature of EMOs. The ACF retains radical commitments and continues to support protests such as those against the Jabiluka uranium mine.[28] As Rawcliffe's study of British EMOs shows (1998: 108–9) the values of an organisation and its informal patterns of work need to be assessed alongside political opportunities and organisational pressures in explaining the practice of EMOs and institutionalisation does not necessarily destroy these values.

Nor, contrary to the expectations of RMT (McCarthy and Zald 1977), are environmental protests usually mobilised from above by EMOs. Critical events seem to generate most protest activity and most protest action is initiated by local groups or NVDA networks or by personal networks of campaigners who already know and trust each other. In this, the green movement is similar to most recent social movements (Meyer and Tarrow 1998: 16). Even, when activists from

EMOs have created a campaign, the example of the anti-GM campaign in Britain shows that this 'had a deceptive double quality'. Purdue charts in great detail the network of counter-experts that developed across a variety of large (such as Greenpeace) and small EMOs, and also included activists from the countries of the south and direct activists from the UK. Although the coalition used the numbers of members in EMOs as a means of establishing its legitimacy with government, the network was built on the trust between individual activists rather than bureaucratic alliances of EMOs. As Kriesi *et al.* (1995: 156) show, the membership of formal organisations tends to rise after protests die down. So protests are not a result of organisations garnering more resources, but rather membership of EMOs might be seen as a more limited and distant kind of commitment, which becomes more attractive when protest is in abeyance, or after the protest draws the organisation more to the attention of a potential supporter.

These criticisms of RMT should not be exaggerated. Resources do matter to all kinds of social movements, but there is no universal model of which resources matter most and or of how best to maximise them. Although there are pressures which push the more formal organisations towards some common forms of internal institutionalisation or bureaucratisation, these will not produce a single homogeneous environmental movement, because, bureaucratisation does not entail external institutionalisation or co-option. Greenpeace and FOE and other green EMOs will remain distinct from WWF and other conservation EMOs. And, although, as will be discussed below, bureaucratisation does limit the range of actions that SMOs can carry out, within the wider green movement the grass-roots space vacated by 'protest businesses' has been filled in many countries by new NVDA protest networks and local protest campaigns. Most importantly, the claim implicit in earlier versions of RMT that the development of a social movement organisation could be understood without reference to its ideas has been rejected. Russell Dalton (1994) argues that RMT can be made stronger by recognising that action is 'ideologically structured'. It is ideology that gives meaning to political actions, and makes what appear as opportunities to some, an unacceptable betrayal to others.

Movements can step back as a result of experience. Greenpeace altered its strategy in the mid-1990s to give renewed emphasis to NVDA when it was criticised by some, including many of its own staff, for having given too much emphasis to trying to influence business. If the ideology of the founders of EMOs explains the diversity of types of environmental groups it is also important to remember that this ideology can change as a result of experience, and not always in a more conservative direction. In explaining the evolution of environmental organisations we need to see ideology, learning experience and the resource-driven pressures towards institutionalisation, as independent factors. The overview of EMOs in western Europe presented by Diani and Donati (1999) shows the complexity of the empirical reality resulting from these multiple pressures.

Resources can be garnered either through appealing to people's values, or by providing them with selective incentives,[29] resulting in a choice between participatory or professionalised models of participation. Political efficacy can be based either on disruption, which has the risk of marginalisation, or compliance with the rules of the game, which has the risk of depoliticisation. This produces four 'ideal types' of EMO: the public interest lobby WWF (most countries), Environmental Defense (USA), NABU Germany; RSPB (Britain); participatory protest organisation (NOAH, Denmark, or Robin Wood Germany or Earth First! in Britain and the USA); professional protest organisation (Greenpeace) and participatory pressure groups (BUND, Germany). In line with the survey above, Diani and Donati say that although there has been a general shift within the formal EMOs away from the participatory model and towards the professional model, the picture is very complex. Many groups have elements of different models. For instance, Friends of the Earth (UK) plays a role as both a participatory protest group and a public interest lobby on occasions. Even WWF has supported moderate local protest actions in Italy on recent occasions (Diani and Donati 1999: 23). Furthermore, as the evidence below shows, there is no clear link between institutionalisation of EMOs and a decline in environmental protest.

Does institutionalisation affect the activity of the wider environmental movement?

Two questions follow from this:

- First, how far does the pattern of institutionalisation vary cross-nationally and, can a comparative model explain this variation?
- Second, is there a relationship between institutionalisation and cycles of environmental protest?

The answer to these two questions seems to be that while there are variations in the form and pattern of institutionalisation they are not as significant as the variation in the pattern of protests cross-nationally, and there is no comparative model to explain either patterns of institutionalisation or levels of protest.

Political opportunity structures (POS) are intended to identify the factors in the political context that explain the actions of social movements. When used as a comparative tool POS has been used to explain cross-national differences between movements with similar aims (Kriesi *et al.* 1995). When the actions and strategy of movements with similar aims differ between countries it is argued that it is likely that the difference may be explained by the nature of their political environment.

In a study of EMOs in France, Germany, Switzerland and the Netherlands van der Heijden (1997) used a POS model to explain different levels of institutionalisation. Consistent with the model, the lower rate of membership in EMOs in France compared with other western European countries can be attributed to the dominance of older political cleavages such as class conflict between workers and bosses and urban/rural divisions. The form of EMOs' organisations varied according to the degree of centralisation in the state. The states with the most decentralised institutions also had the most decentralised EMOs. However, other results were not as the model would have predicted. The model suggested that the more centralised states would have larger numbers of members of national EMOs. This was true for the Netherlands, but not for France, where low membership of EMOs seemed more likely to be due to the low salience of environmental issues. Moreover, levels of professionalisation were not explained so much by the degree of centralisation in a state as by the levels of EMOs' income. These were highest in Germany, which was less centralised than the Netherlands.

Nevertheless applying the model to other countries seems to bear out van der Heijden's general argument. The level of membership of EMOs is relatively high in Denmark, Britain, Australia and the USA where older political cleavages appear to have been more pacified and still relatively low in southern European states, where older cleavages still dominate politics (Eder and Kousis 2001) Where the state made an early attempt to manage environmental issues as in Norway (Bortne 2000; Dryzek and Hunold 2001) and Sweden (Jamison and Ring 1999), membership of EMOs tends to be lower. Centralised states such as Britain and Sweden have centralised EMOs, although in Britain after new national assemblies were created in the late 1990s in Scotland, Wales and northern Ireland, FOE groups operated more independently in these areas. Yet, although decentralised states such as Spain and Australia have relatively decentralised EMOs (consistent with the model), in the USA, despite the federal structure, the formal EMOs are very centralised. In the USA, according to Dowie (1995: 92–3), the major EMOs continued to concentrate their efforts and resources on Washington because they had experience of legislative success in the 1970s. In the 1980s, however, with the rise of anti-environmental 'Sagebrush' and 'wise use' groups with effective grassroots networks the major political battles were being fought in the state legislatures and courts and the regional environmental agencies and a Washington-centred strategy was inappropriate (1995: 64). It seems that strategic errors, plus the difficulty of restructuring organisations explain this anomaly. As van der Heijden acknowledges, the model of POS does not always provide a full explanation.

It is not only political structures in the form of institutions, but also political traditions and ideological choices that shape the likelihood of institutionalisation. In his survey of EMOs in the mid-1980s Dalton (1994) demonstrated that the values of EMOs played a crucial role in determining acceptable allies. Conservation groups viewed government and business more positively than did the newer environmentalist groups. This is consistent with the argument made above about external institutionalisation. However, the situation differed in different countries. For instance, Diani's (1995) analysis of networks among environmental groups in Milan found that while there was still a difference in values between environmentalist groups and conservation groups there was relatively little difference in their political behaviour. Allies were evaluated according to their relevance to the issue and their organisational resources rather than their ideology. This pattern was confirmed by recent analysis of the national level activity of Italian EMOs by della Porta and Andretta (2000). Diani saw the convergence between conservation and environmentalist groups as affected by the Italian political context. As the deep polarisation between left and right of the 1970s diminished in the 1980s, there was more space for new political issues to develop independently of the left/right cleavage. The fact that different sections of the environmental movement concentrated on different issues helped to reduce inter-organisational competition.

Dryzek and Hunold (2001) have argued that the potential for green movements to make progress varies according to both the nature of state structures and the political traditions in different countries. States that are 'actively inclusive' of groups from civil society, such as Norway, tend to demobilise radical environmentalism, by offering opportunities to influence policy and funding, in exchange for a co-operative relationship with the environmental movement. This inclusion, however, is at the expense of the critical and reflexive role of environmental groups, which at best achieve incremental forms of weak ecological modernisation, and have no active external radical environmental constituency pushing for more radical change in civil society. Rephrased in the terms used in this book, this would mean that Norway has a very weak green movement, even if very moderate environmentalism has had some influence on the state. States that are 'passively inclusive' such as the USA do not seek to manage and incorporate interest groups as intensively. However, when it suits the interests of particular governments, groups will be incorporated, as long as they do not threaten the state's core imperatives. Thus in the USA there have been periods of partial incorporation of the more moderate and professionalised interest groups, particularly in the early 1970s, when the adoption of moderate environmentalism made Nixon look less intransigent in the face of system-threatening action from the New Left. Incorporation was less far reaching than in Norway and sufficiently limited in what was offered that significant parts of the environmental movement (both in the green networks and the grassroots environmental justice movements) resisted opportunities for incorporation. One consequence of this was that the movement was able to preserve its critical capacity, though without achieving any major advances in the direction of ecological modernisation, either through the state or in civil society. This raises the question of the relationship between institutionalisation and de-radicalisation.

Does institutionalisation entail de-radicalisation?

For radicals institutionalisation is in itself a form of de-radicalisation (van der Heijden 1997; Jamison, Eyerman and Cramer 1990: 110–17). The argument that EMOs have de-radicalised can be divided into three. First, for Jamison *et al.* the major difference between the environmental movement of the 1970s and now is that the movement is no longer capable of creating a new cosmology. Institutionalisation has changed the 'cognitive praxis' of the movement so that it concentrates on technical and scientific issues and no longer offers an alternative set of values or a challenge to the existing order. This challenge was reflected in the non-hierarchical forms of organisation and the critical attitude to knowledge developed by environmentalists in the 1970s. Now environmental groups provide information rather than new knowledge.

Second, as we have seen, EMOs cannot be seen as the sources of a new participatory society. At best they provide a mechanism for what Kriesi *et al.* (1995) have called 'vicarious activism' in which new middle-class professionals with little free time but some spare money buy a service from campaign groups. Given the limited form of passive membership in WWF and Greenpeace, van der Heijden argues that:

> This leads to the paradoxical conclusion that the environmental movement of the 1970s and early 1980s with its much smaller (but also more active and more concerned) constituency, in a sense could be conceived of as stronger than the movement of the past decade. Many environmental organisations have lost their unique movement character and therefore an important part of their strength. It is doubtful whether their stronger position at some negotiating table will compensate for this.
>
> (1997: 46)

Third, the priority given to the development of the organisation and the monetary resources necessary to sustain it has become an end in itself that is an obstacle to achieving more radical change. Groups reliant on subscriptions from mass support in order to maintain their organisation are likely to avoid risky protests that might lead to sequestration of assets and also may be too influenced by marketing surveys of how their strategies play with key groups of subscribers.

Mark Dowie's trenchant critique of the major American EMOs suggests that they have betrayed the heritage of earlier environmentalists by accepting too many compromises and by giving priority to the maintenance of their own organisations. He shows that the major American EMOs were ill prepared to respond to worsening political circumstances in the 1980s. They had become too reliant on the 'good faith and authority of Federal Government' (Dowie 1995: xi) and failed to broaden their focus from wilderness and animals in response to the growth of environmental justice activism. Worse, the actions of the EMOs actually weakened the movement. This occurred when groups accepted executives from major corporations onto their boards, thus bolstering the credibility of the claim that business supported the environmental movement.

Against this pessimistic view optimists argue that institutionalisation has not changed the values of EMOs. In Australia, argue Hutton and Connors, there has never been any major disagreement over philosophy within the movement (1999: 14) and the movement remains committed to a radical project. The shift in strategy towards more ties with government and business can be defended as necessary when the political climate is less promising, as in the recession of the 1990s. Rawcliffe (1998) also argues that EMOs may retain radical values while taking advantage of the greater influence that institutionalisation provides. Professionalisation has improved the campaigning ability of EMOs and increased their impact on public opinion. It is unlikely that this would have occurred without institutionalisation. The main question at stake in this debate then is whether the EMOs still count as radical, and whether therefore we can regard them as part of the green movement.

Are the EMOS still radical?

Some EMOs were never radical in ideology and were therefore not properly part of the green movement as a social movement. This applies particularly to conservation groups such as the WWF and the National Trust.[30] However, it is primarily the national affiliates of FOE International and Greenpeace as well as smaller national environmental groups that are seen as having become less radical.

While some groups such as NOAH in Denmark and FOE in Australia are still unquestionably green in their combination of commitments to social justice, ecology and democracy, others are more difficult to judge. In the Netherlands for instance, Milieudefensie has shifted from being a participatory protest organisation to a public interest-type lobby and van der Heijden remarks that most discussion of environmental issues in the Netherlands is now based on practical ameliorative schemes, such as energy saving and recycling. 'On practical solutions ... a high level of sophistication has been reached, but most environmental groups have stopped talking about topics that lie outside this "discursive space": reduction in meat consumption, car driving and air traffic; living in smaller houses, etc.' (2000: 22). In Italy Amici della Terra and Legamabiente are also now much more pragmatic in their campaigns than in the 1980s (della Porta and Andretta 2000). Other groups, however, retain a broader radical discourse, as we will see below.

Greenpeace represents a special case. From the outset, Greenpeace deliberately avoided ideological statements that would place it on the left or the right. It concentrated on single issues that could be framed as moral causes in their own right. Nevertheless, part of its self-cultivated image was as a critic of big power, whether in the form of government or big business, and in that sense it had a radical image. Its criticism of the nuclear industry and nuclear weapons as well as of business meant that despite its moral rather than ideological framing of issues, it was widely associated with the anti-authoritarian position and its support for issues of social justice places it implicitly on the left. But, this consistency with green ideology is so *sotto voce* that if the green movement had to rely on Greenpeace its ideological arguments would be largely invisible.

EMOs are also charged with failing to mobilise their supporters to take action. The action repertoires of EMOs are less radical in many cases than previously. In Italy, while EMOs still take part in protest, in contrast to the 1980s these protests are rarely disruptive. Yet, as Carmin (1999) shows for the USA, and as was discussed above, it was probably rarely the case that national groups initiated mass environmental protests. Some groups such as FOE (UK) did not pursue disruptive protests. Leafletting, public information stalls, press releases, petitions and non-confrontational demonstrations were and remain the predominant repertoire of many EMOs. When Dalton surveyed EMOs in the mid-1980s the frequency of protest was relatively low and they viewed their protest activities primarily as a means to mobilise public opinion through symbolic action and not as a challenge to power (1994: 196–7). The advertising of Greenpeace can be seen as encouraging a kind of passivity, suggesting that environmental problems can be solved by its committed professionals. And many members of EMOs are willing to allow the organisation to act on their behalf: for 70% of members of FOE UK 'playing a part in politics' as individuals was not important (Jordan and Maloney 1997: 192), so there is little basis for thinking that the mass membership of EMOs is being thwarted from taking more radical action by an oligarchic elite in the EMOs. Email has provided a new low-cost form of protest activity for groups that seek to expand participation by their members without demanding too much of them. The websites of most EMOs now include email protest letters, and both Friends of the Earth International and Greenpeace have recently supported disruptive forms of cyberprotest. Friends of the Earth closed down the White House website several times early in 2001 as over 100,000 people protested against President Bush's stance on climate change.[31]

Yet, the lack of radical action by EMOs does not entail the absence of protest. In fact, there is no relationship between the institutionalisation of EMOs and the levels of environmental protest. The Transformation of Environmental Activism (TEA) project surveyed levels of environmental protest in ten European Union countries and found no consistent cross-national pattern between 1988 and 1997. The most remarkable rise in confrontational and disruptive protest was in Britain, (Rootes 2000) and this was despite the fact that this was a period when EMOs had unprecedented access to policy-makers. In Germany, contrary to general impressions that the movement had become de-radicalised, the research showed that there had been no decline in protest. Although the largest mass membership national groups carried out mainly 'conventional' protest action such as petitions, referenda, leafleting, lobbying and legal actions, there was no decline in more confrontational actions such as vigils, boycotts, demonstrations blockades and occupations, which were carried out mainly by the medium sized and smaller EMOs (Rucht and Roose 2000: 17–19). Germany was also notable for the continued vitality of protests against nuclear energy, which elsewhere had declined (Rucht and Roose 1999). In Spain there was also no clear decline in protest during the 1990s, but the main characteristic was the predominance of local protest and the diversity of issues motivating protest (Jimenez 1999b). In contrast in France, Italy and Sweden[32] the level of protest appeared to have declined substantially since the late 1980s. And, as Fillieule and Ferrier (1999) note, the form of environmental protest contradicts the more general pattern of high levels of confrontational protest in France. Actions taken by the environmental movement are overwhelmingly non-violent and less confrontational than is the norm for protest in France. Thus the confrontational pattern of action by the French environmental movement noted by van der Heijden (1997) for 1975–89 no longer holds and appears to have applied largely to the period of anti-nuclear protest in the late 1970s.[33]

The TEA surveys and evidence from the USA (Carmin 1999) and Australia (Doyle 2000: ch.12) show a general pattern in which the institutionalisation of EMOs was paralleled by a growth (though variable and more difficult to measure) in grassroots protests outside the formal EMOs. This suggests that even if part of the green movement becomes institutionalised, in some countries new grassroots groups have reacted reflexively, taking more radical action. This is particularly notable in the rise of Earth First! in Britain, whose activists were well aware of the limits of existing EMOs (Wall 1999; Doherty 1999). At local level new grassroots protests often include members of local groups of the national EMOs, as was the case in the anti-roads protests in Britain (McNeish 2000; Cathles 2000), qualifying the picture of institutionalisation further. Moreover, the rise of more confrontational protest and the evidence that it is accepted and tolerated, (on which see Chapter Six) increases the space for EMOs to combine protest with other institutional activity. In Britain FOE, having initially avoided involvement in direct action, came to support the new direct action networks, as long as they remained non-violent. In those countries where grassroots protests have not emerged so strongly, the explanation is more likely to lie with other political factors such as the weakness of the alternative milieu, or its engagement with other issues, than with the institutionalisation of the EMOs.

A major change for environmental groups was the dramatic increase in the salience of environmental issues as a focus for public concern in the late 1980s. When, in response, government, political parties and business began to argue that they too recognised the seriousness of environmental problems EMOs had to adapt and respond. For van der Heijden the turning point came with the acceptance by EMOs of the discourses of ecological modernisation. This 'massive surrender' (1999: 203–4) meant replacing the changing of capitalist industrial structures with an agreement that change could occur within capitalism. However, as noted in Chapter Three, ecological modernisation is not an uncontested concept. Moreover, some EMOs have either refrained from endorsing it, or stepped back from it. In Germany, after a period when EMOs were increasingly consulted and praised by government and business, there was a return to a more critical stance towards government from the early 1990s. Economic recession brought a return to a pro-economic growth discourse from political elites and reinforced a growing scepticism within EMOs about the discourse of ecological modernisation (Brand 1999: 52). Environmental groups complained that despite increased access to government, particularly following the accession of the Greens to government, they still had much less influence than industry. Moreover, this scepticism has produced a return to a stronger critique of business and the unsustainable nature of western affluence. Brand also argues that as environmental problematic has become more obviously global, it has provided the movement with the opportunity to develop sustainable development[34] as an alternative to the hegemony of neo-liberal discourse. A return to the more confrontational positions of the 1970s is not feasible, but EMOs have returned to a more critical position regarding government and business.

A similar more critical shift can be identified in some British EMOs. Under Charles Secrett FOE has been more explicit in its focus on the structural causes of environmental degradation and its link with western affluence. FOE and other EMOs such as Wildlife Trusts, WWF, Transport 2000 and Pesticide Action Network have participated in a coalition with church, development and peace groups called The Real World Coalition. In two reports (Jacobs 1996; Christie and Warburton 2001) this coalition has argued that the central political issues are 'pushed to the margins of public debate and political manifestos'. These include 'environmental degradation; gross inequalities between the developing and the developed countries; endemic poverty and conflicts in the developing world; growing inequality in the UK, and the persistence of major gaps in social and economic opportunity between different groups according to race, gender, age and disability; widespread disaffection towards the political system; towards political parties and for many, towards society as a whole; a pervasive sense that community life and families are being damaged by changes in the economy and the operation of organisation; the gap between standard measures of economic success and the realities of quality of life' (Christie and Warburton 2001: ix). While the agenda is one of 'radical reform' rather than system transformation, it is most notable that it reads sustainability as requiring a considerable expansion of democracy in a more participatory form, action against a broad range of social inequalities, and the rejection of neo-liberal economic policies. In its critique of car use, of quantitative forms of economic growth and emphasis on redistribution it gives priority to the kinds of questions absent from debates in the Netherlands.

While sustainable development has many critics and is by no means an unproblematic concept, the most important conclusion that can be drawn from the return to a more radical interpretation of it by EMOs in Britain and Germany is that internal institutionalisation does not entail permanent de-radicalisation. On the other hand, it is not possible for EMOs to return to the same discourse as before. The attitudes of the public, business and government are different from the 1970s and EMOs have adapted in response. It is no longer enough to provide information about environmental problems. There is sufficient awareness of these that the principal role of the green EMOs has become more policy-focused, and based upon detailed research, combined with the legitimacy provided by their mass membership. So, although the values of many EMOs remain radical, EMOs may no longer be in a position to initiate or co-ordinate contentious opposition. By investing so much in professionalisation and by concentrating on the mobilisation of material rather than human resources they have developed a position as authoritative actors within the political system. Organisationally, as we have seen, it is difficult to take a more oppositional stance, because campaigns based on professional expertise have to be planned so long in advance. Discursively, the EMOs are constrained by their shift to constructive maturity. In these respects, the development of organisation has shaped the strategy of EMOs.

Yet, although protests may have become routinised, and there has been a partial inclusion of EMOs, it is less clear that co-option has occurred. The willingness of groups to co-operate with elites is still very much shaped by ideology. And even groups that have moved to a less radical position have sometimes moved back again to a more radical critique. Moreover, SMOs cannot be analysed in isolation from the rest of the green movement. In some countries institutionalisation of EMOs, has not prevented a rise in grassroots protests, and this shows that institutionalisation does not determine the actions of the whole movement. Moreover, since the more radical EMOs remain well networked with other parts of the movement, and, at local levels often overlap in social networks and mutual membership (Doherty, Plows and Wall 2001), it is plausible to view this as a mutual division of labour.

Conclusion

EMOs are very diverse and only some are green, and even then, when tested against the ideal type outlined in Chapter One, the green EMOs are only partly green. Those that retain a collective identity compatible with the three core green commitments and with green practices have the strongest status as greens. Here we might include Friends of the Earth International as the principal example. In some respects these groups, along with some conservation groups such as ACF, have developed a stronger commitment to issues of social equality and democratisation in recent years. Like other parts of the green movement their character has changed over the last three decades and in this respect, continued diversity is as much the story of EMOs as a move towards institutionalisation. Organisational institutionalisation, has been much more general than external institutionalisation.

The EMOs are sometimes seen as the core of the green movement, because other groups such as direct action groups, green parties and local environmental campaigns have much smaller memberships. Some have commented that the model of the green movement as radical is misplaced because it is based on the characteristics of only a small part of the movement, which is far outweighed by the mass membership of EMOs (Sutton 2000; Jordan and Maloney 1997). Against this argument, two points can be made. First, it is important to make a distinction within EMOs between groups that have some social movement characteristics, including evidence of commitment to a green ideological framework (FOE, Greenpeace and other smaller national groups), and those that do not (WWF, and many other conservation and animal welfare groups). Second, when we compare levels of activism and categories of support, it might be argued that the EMOs do not dwarf the other parts of the green movement. For instance, if the great majority of members of EMOs do no more than send in an annual subscription, the number of activists attending meetings, taking part in protests and doing organisational work in EMOs may not be greater than in green parties or direct action groups. Furthermore, voting green might be a stronger sign of ideological commitment than joining an EMO. Green voters certainly provide another source of evidence of support for green policy, if not necessarily all for green ideology. Moreover, since poll evidence and the actions of juries in the UK, who have regularly acquitted protesters against GM food and nuclear missiles, seem to indicate rising tolerance of direct action and a degree of support for their aims, it is not even clear that direct action groups lack the support of mass publics. The fact that this support is not institutionalised in the form of a mass membership makes it less tangible, but it may be more character-istic of contemporary forms of social movement action in which the actions of small groups can have a much wider ideological impact. Since we know that there are some overlaps in membership between these different parts of the green movement (McNeish 2000; Rüdig, Bennie and Franklin 1991; Jordan and Maloney 1997) it is likely that there is also a green public, who while not neces-sarily engaged in green protest action will support the direct action of more radical green groups, perhaps vote green, and provide the financial resources for Greenpeace or FOE to continue with their more policy-focused campaigning. Since the evidence to prove this is not available and the meaning of 'support' is very broad, this makes more plausible the model of the green movement based on the view that it is green activism which creates new discourses that renew and extend the meaning of green ideology by challenging the boundaries of current societies. In the next chapter the most radical form of green challenge, that of direct action will be examined.

6 Ecological direct action groups

After a week of rumours the raid came with a few hours warning. Our roving security team spotted a two-mile long column of cars, trucks and buses entering a hangar at Minneapolis-St Paul International Airport, one mile south of the site. Our whistles blew as Ryder trucks pulled up in front of each house, disgorging masked, black-clad police wielding rifles and battering rams. It was immediately apparent the authorities knew where each and every barricaded site and lock-down was, despite some having been built only two weeks prior. Police filled entire houses with tear gas, choking and nauseating the protesters whose arms were locked into barrels. Many unlocked voluntarily, only to be pepper-sprayed while trying to make their way out. Police also tipped over a tripod while a sitter was still on top and maced a man who was sitting on a roof.

(Minnehaha Free State in Minnesota in December
1998 on Earth First! (US) website[1])

This account describes the eviction of seven houses squatted by Earth First! activists in the USA protesting against the re-routing of a highway through Native American land. The scenario would be familiar to radical environmental protesters in other countries, many of whom also act under the banner of Earth First! This is not a formal organisation. You cannot become a member and not all radical environmental protesters regard themselves as Earth First!ers. It is by taking autonomous direct action in defence of environmental goals, including being prepared to break the law, that one becomes an Earth First!er. Direct action is, therefore, the defining characteristic and purpose of Earth First! and similar radical environmental groups. The spread of the name Earth First! is due to the inspiration provided by the US activists who formed the first Earth First! group in 1980. Other groups in Europe and Australia share their rejection of the compromises made by established environmental pressure groups and their strategy of militant confrontation with those seen as responsible for the worst environmental abuses.

This chapter focuses on ecological direct action (EDA) as a type of environmental group as well as a form of action. Other environmental groups such as Greenpeace[2] that use ecological direct action do not have the counter-cultural ethos, commitment to non-hierarchical forms of organisation or the radical ideology of EDA groups such as Earth First! (Rucht 1995). There is no single organisation, no national office or administrative centre, or any other formal national co-ordinating structure for these groups in the UK or Australia and only a relatively weak central base in the USA. Instead, it is a network of groups based in local areas, some calling themselves Earth First! groups, others with different names, but sharing a common identity, form of action and linked by network ties. There is also a national network of those involved in environmental direct action – in the UK activists meet at Winter and Summer Earth First! gatherings and sometimes to plan national campaigns or particular protest actions. In the USA there are annual gatherings to discuss strategy. In Britain there is a regular (approximately monthly) national newsletter called *Action Update*, the editorship of which is rotated between local groups annually, and other national-level publications such as the more hard core and discussion-based journal *Do or Die!* In the USA the monthly *Earth First! Journal* has long been the main source of movement news. In all three countries in the last few years email news lists have also become more important in networking actions. Personal ties and solidarity based on common participation in protest actions over the past nine years have created strong bonds between activists from different areas. The separation between local groups is also blurred by the frequent movement of individuals between local areas. Thus while there is a national EDA network, it is paralleled by autonomous local groups which often have a strong sense of their own identity.

The protest action of these groups is often in support of specific local environmental campaigns, and many EDA activists are also involved in non-counter-cultural local campaign groups (McNeish 2000) but despite the fluid boundaries between them, ecological direct action groups are distinct analytically from local environmental groups in important ways.

First, they are much more likely to justify their protest as a strategy for bringing about comprehensive social and political change towards a radical green society.

Second, in contrast to both local environmental groups and environmental movement organisations (EMOs) the protest actions of EDA groups are rooted in an alternative culture. This is manifested in their clothes, their ecological lifestyle, the rejection of dominant values and institutions and their non-hierarchical form of organisation. The groups involved in ecological direct action are at the radical end of the green spectrum.

Third, they reject the existing EMOs as inadequate or, at best, insufficient to deal with the environmental crisis. Those who initiated EDA groups in the USA, Britain and Australia did so as a result of disillusion with the compromises made by EMOs. They argue that when EMOs accept incremental gains, they help to legitimise policies and a decision-making process that favours development over the environment. Fears over the loss of assets from fines or legal actions taken against them or the loss of support from cheque-book subscribers, make it difficult for EMOs to take radical action. But for radical environmentalists, unless there is action to achieve radical change, the future looks bleak. The style of organisation adopted by EDA groups is intended to offset the limitations of the EMOs. Because they have a decentralised and informal organisation EDA groups have no resources which could be targeted by their opponents and which might limit their ability to act. Moreover, since action is the responsibility of the individual or of small groups it is the individual rather than the organisation who has to take responsibility for the consequences.

Fourth, individual moral obligations to do what is right count for everything, political obligation to the state counts for little or nothing. Although the heritage of Gandhi and Martin Luther King is acknowledged, the ecological direct action of radical environmental groups is more confrontational than earlier forms of civil disobedience. King's emphasis on upholding the law and American and Christian values is not echoed by contemporary environmental EDAers. Radical environmentalists often see themselves as acting consciously in opposition to dominant values that help to sustain environmental destruction. Illegal action is justified both because the political, social and economic system itself is illegitimate and because it is seen as the most effective means to achieve change.

Ecological direct action by these groups therefore includes the kind of civil disobedience where arrests are anticipated and even desirable, and in which protesters are prepared to account for their actions publicly, but it is also broader. It can include actions where protesters seek to evade arrest, and covert actions such as 'monkey-wrenching ... unlawful sabotage of industrial extraction/development equipment and infrastructure, as a means of striking at the Earth's destroyers at the point where they commit their crimes'[3]. Ecological direct action refers to 'action where protesters engage in forms of action designed not only or necessarily to change government policy or to shift the climate of public opinion through the media, but to change environmental conditions around them directly' (Doherty, Paterson and Seel 2000: 1).

Damage to property is usually reported as violence in the media and such actions are condemned by the major EMOs. For EDA groups, however, property damage is non-violent because it is 'aimed only at inanimate objects'. As the website of US *Earth First! Journal* states: 'It is the final step in the defense of the wild, the deliberate action taken by the Earth defender when all other measures have failed, the process whereby the wilderness defender becomes the wilderness acting in self-defense'. More controversial, however, is the question of whether violence against people is justifiable. For some activists non-violence is a principle that justifies their action, but for others this is a dogma that needs to be challenged. This debate will be examined further below.

Attempts to generalise about EDA groups are made difficult by their rapid evolution and change, and their informal character. Groups of this kind can be found throughout northern and eastern Europe, but are strongest and have greatest impact in English-speaking countries, notably the USA, Australia and Britain. However, as the analysis of groups in those countries shows, even where there has been most mutual influence, there are still significant differences. In the USA and Australia the defence of wilderness has been the primary *raison d'être* for Earth First! and similar groups, although in recent years this has been complemented by their involvement in protest against capitalist institutions such as the World Trade Organisation (WTO). In Britain and the rest of Europe anti-capitalist urban protest has been more important for longer, and direct action in rural areas has more often been alongside existing local environmental campaigns. Another obstacle to universal definitions is that, more so even than in the green parties, radical environmentalists tend to be anti-doctrinal, and there is remarkably little interest in developing a common political theory within countries or between them. Only in the USA has there been a significant theoretical dispute – between those who gave absolute priority to the defence of wilderness and those who saw social and ecological concerns as inter-related. In the USA deep ecology has had a more obvious influence, elsewhere activists tend to be linked by a shared identity and ethos that is reflected more in common practices and ideas implicit in the nature of their protest than by a definable philosophy. While in recent years more activists have adopted an anarchist identity there is no sense that particular writers such as Kropotkin or even the principal modern theorist of eco-anarchism Murray Bookchin (1982), are widely read or discussed. This is certainly a part of the green movement that has no required reading.

The first part of this chapter will analyse the characteristics of EDA groups, drawing mainly on evidence from Britain, the USA and Australia. This will be followed by an examination of two related questions often asked about these groups. The first is the debate over the use of violence; the second is their role within what has been called the global anti-capitalist movement.

Ecological direct action in the USA, Australia and Britain

The USA

In the USA ecological direct action is more synonymous with Earth First! than in Australia and Britain. Earth First! was founded in the USA in April 1980 by five environmentalists disillusioned by their experiences with the major environmental organisations. They viewed the compromises of the pressure groups as inadequate in the face of the erosion of wilderness and threats to biodiversity. This was brought to a head after the Forest Service's second Roadless Area Review and Evaluation (RARE II) in which less than a quarter of the area under consideration had been designated as protected wilderness (Scarce 1990). According to movement legend, it was after a hike in the Picante desert in Mexico that the five decided to found a more militant group. In doing so, they drew on the guidance provided by Edward Abbey's *The Monkey Wrench Gang* (1975). This novel was a thinly disguised call to arms for wilderness defenders. It heroised the actions of four eco-saboteurs prepared to 'undermine road building, bridge building, dam building – anything that threatens the wild places of Utah and the surrounding states' (dust jacket). Like the monkey wrench gang, Earth First! was prepared to support the use of ecological sabotage (or ecotage) wherever it was strategically appropriate. It did not directly advocate ecotage or any illegal actions, but in publishing *Ecodefense: a field guide to monkeywrenching* (Foreman 1985) and in calling those who took part 'heroes', the position of the organisation was clear.

The driving force in the formation of Earth First! was Dave Foreman, who had been senior campaigner with the Wilderness Society in Washington and came from a right-wing and military background. The initial base of the movement was in the south-west, but effective publicity and Earth First! roadshows produced new recruits nation-wide. Since there was no formal membership of Earth First! after 1982 there are no reliable estimates of the numbers of activists, but there were probably no more than a few thousand at any one time.[4] In the 1980s more than half the journal's subscribers were in California (Zakin 1993: 358) and the West Coast remains the strongest base of activity. Most activity was based in local networks with a national gathering of the most committed activists at the annual Round River Rendezvous. In the intervening period the main linkage between local groups was through the movement's journal. While the movement was still very small in the early 1980s the strategic direction provided by Foreman was accepted. However, the journal's editorial policy became a matter of dispute in the second half of the 1980s when strategic divisions began to emerge within Earth First! On one side were those like Foreman who saw Earth First!'s purpose as to take practical actions that produced tangible results in saving wilderness. Carefully targeted sabotage of development projects or logging in wilderness areas would increase the costs for developers and make development uneconomical. Public protests were also useful, as when tree sitting was used to increase the costs of logging in the forests of the Pacific north-west. But Foreman, and others such as Chris Manes, rejected the broadening of Earth First!'s goals to include alliances with other organisations campaigning on issues of social justice. In their view, human nature was irredeemable and humanity needed to 'return to the Neanderthal'. They expected an ecological meltdown and speculated that only an ecological elite with the correct 'wilderness gene' would be able to survive it (Lee 1995).

The second 'holistic' group accepted the priority of defending wilderness and the likelihood of ecological collapse but thought that progress to a better society was still possible. The best means to defend the wilderness was through direct action, but based upon the inter-relationship between issues of social justice and ecology. Thus, in tackling the logging industry, activists such as Judi Bari tried to build an alliance with logging communities by showing how clear-cutting undermined the long-term future of the timber industry and only served the interests of the largest companies. The 'holies' were influenced by new social movement ideas and wanted Earth First! to develop as a political force. Bari made links with the remnants of the International Workers of the World, or the Wobblies as they are better known, whose agitprop and direct action tradition had earlier inspired Foreman (Zakin 1993: 361). While actions that directly preserved wilderness were seen as justifiable, they needed to be complemented by actions that built the base of the movement. As Bari put it: 'Dave Foreman wants Earth First! to remain small, pure and radical. I want it to be big, impure and radical' (quoted in Lee 1995: 128).

The split deepened following the furore created by two infamous articles written by Chris Manes in the *Earth First! Journal* in 1987, touching on an issue that divided the two groups, namely poverty and the causes of over-population. Manes wrote that Aids and famine could have good effects in reducing over-population, although he did not disregard the suffering of Aids victims. The holies rejected Manes' argument because it did not tackle the causes of over-population, primary among which was inequality. No letters on Manes's article were published in the *Earth First! Journal*, edited by Foreman. The 'holies' felt that debate was being stifled and the Round River Rendezvous were often bedevilled by disagreements between supporters of the two positions. Among the more important incidents were the expulsion of an eco-anarchist group from the 1987 Round River Rendezvous after a confrontation with Abbey and the burning of an American flag by 'holies' at the 1989 Rendezvous (Zakin 1993: 409). Another bone of contention was the suspicion among holies that Foreman was using the resources of the journal to finance activities that he supported to the exclusion of other groups. Mike Roselle, a leading 'holie', and like Foreman, one of the five original founders of Earth First! established a 'Nomadic Action Group' (using the name of an Australian EDA group) of Earth First!ers committed to direct action, which was funded separately from the journal. Disputes continued over the failure by the *Earth First! Journal* to cover the public EDA actions favoured by the holies (Maenz 2000).

Two major incidents increased the stakes of the conflict. In 1989 after an undercover operation by an infiltrator, the Federal Bureau of Investigations (FBI) arrested a group of Earth First!ers in Arizona who were trying to cut a power line. Foreman was later arrested and charged with financing the operation. Although the FBI had failed in its aim of entrapping Foreman in carrying out ecotage, it had for the first time been able to successfully prosecute Earth First! activists for serious ecotage. The second incident was the bomb that exploded under the seat of Judi Bari's car in April 1990, causing her serious injury. She had been organising a Redwood Summer of protest against logging of old growth forest, evoking the 1964 Freedom Summer campaign of the civil rights movement. Police charges against Bari and her partner, a musician and poet, Darryl Cherney, were dropped, and it is extremely unlikely that they were, as the police initially professed to believe, carrying a home-made bomb (Littletree 2000). Neither of these traumas was sufficient to reunite the movement. Arguments over the journal and the philosophy of the movement continued until in August 1990 Foreman announced he was leaving Earth First! because it had been 'taken over by West Coast hippies' (Lee 1995: 137). This was followed by the resignation of the journal's staff.

From 1990 onwards the 'holies' were in control. Since then there has been a clearer commitment to the importance of public EDA and to taking social justice seriously. The journal now carries extensive reports on direct action campaigns in the USA and other countries. Most activists have remained within

Earth First! rather than following Foreman and other wildies. Although it remained an uneasy combination of biocentrists and ecological leftists, most of those who remained were more prepared to live with their differences. Activists such as Bari, who had socialist parents and past experience in trade union, feminist and anti-nuclear energy campaigns, succeeded in contradicting Foreman's assertion that Earth First! was not on the left (Lee 1995: 106). Earth First! continued to campaign against logging, with long-term camps at Headwaters, Cascadia and other parts of the Pacific North West and also against other development projects such as new roads through Native American lands at Minnehaha. Earth First!ers also played an important role in the network that co-ordinated the protests against the meeting of the WTO in Seattle in 1999 and subsequent protests such as those in Washington in April 2000 and Quebec in April 2001.

A new dimension of EDA campaigning in the US has developed in the form of the Earth Liberation Front (ELF). Although the ELF emerged from within the EDA network, its membership and activities are clandestine and its actions are not claimed publicly by Earth First! The ELF claims to have been inspired by British EDA[5] in the 1990s but whereas there is no evidence that the ELF has ever really existed as a distinct group in Britain, the ELF in the USA has a press officer and a website and has claimed responsibility for many acts of ecological sabotage (or ecotage). The ELF describes its main aim as 'To inflict economic damage to those who profit from the destruction and exploitation of the natural environment' and it has worked alongside the Animal Liberation Front in major incidents of sabotage since 1998. In October 1998 the ELF claimed responsibility for arson attacks in Vail, Colorado. The attacks were targeted at developers who, it was thought, were threatening the viability of a scheme to reintroduce Lynx to the area (*The Observer*, 25 October 1998). Since then targets for arson have included new houses on Long Island, in Kentucky and Arizona, deemed guilty of adding to urban sprawl and encroaching on nature, warehouses containing genetically modified (GM) cotton and offices of the US Forest Service and logging companies. In March 2001 damage by the ELF was calculated at more than $45 million dollars (*The Guardian*, 6 March 2001).

The ELF is linked to the Animal Liberation Front (ALF), and the latter hosts the ELF website[6], which states that environmental and animal rights groups have to 'learn from each other what needs to happen to make other movements a real threat to the death industries'. In some areas the ELF and ALF overlap, but the ELF is also the inheritor of a long-standing tradition of ecotage in the US which has been revived by activists who have taken heart from the renewed militancy of direct action in Britain in the 1990s, Seattle and other protests. ELF groups can function alongside public campaigning of groups such as Earth First! without activists from the latter knowing who, if any, of their friends is carrying out ecotage. Small groups of this kind, or even individuals can work covertly and autonomously simply reporting action to the press or through the internet without needing to develop any public organisation. The analysis of Britain below shows that a similar repertoire of ecotage exists in Britain, but perhaps because there have been fewer large-scale cases of ecotage it has been less widely reported.[7]

Australia

Although it has had less attention than Earth First! in the USA. EDA on environmental issues began at least as early and possibly earlier in Australia. From 1979 to 1983 direct action helped the environmental movement to achieve success in three important conflicts over forest issues. The first at Terania Creek in northern New South Wales covered only a small area, but provided a model for the two more significant campaigns that followed, at Daintree in Queensland and the Franklin Dam in Tasmania. Importantly, the Terania Creek campaign had its roots in existing alternative-left groups. The area was popular with alternative communities and many of those involved in this protest had previous campaigning experience in the anti-Vietnam movement and other radical campaigns in the 1970s. When logging began in August 1979, protesters had already established a camp and after their initial attempts to block the road with cars failed they 'climbed trees and attempted, tree by tree, to slow the process down' (Hutton and Connors 1999: 153). The novelty of such tactics attracted television coverage and national attention. The most controversial action was taken by two protesters who acted in contravention of a decision by the protest group, by spiking trees with nails and cutting up logs on the ground to make the wood unusable commercially. Their action was condemned by all, including the protesters, but as Hutton and Connors note, it stopped the logging.[8] Tree spiking has always caused controversy as a tactic because of the dangers that it poses to timber workers cutting the trees. The nails can break the saws used to cut the timber and cause injury to the operator of the saw. Its use as a tactic has been most widespread in the USA where much EDA protest has concentrated on the timber industry, but even there it has been criticised by those within Earth First! trying to build an alliance with those logging workers against the timber companies.[9]

The activists at Terania Creek used the organisational principles of new social movement campaigns, based on consensus decision-making and informal organisation and established these as part of the culture of radical environmental groups in Australia. As also in other parts of the green movement, the culture of EDA was shaped by a process of rethinking that occurred within the broader alternative left. Epstein's comments on what she calls the direct action movement in the USA apply to EDA in Australia and Britain as well: Most activists 'have seen the specific objectives of the movement as inseparable from a vision of an ecologically balanced, non-violent, egalitarian society' (1991: 16). And, although Terania was the first use of EDA on forest issues, the militancy seen at Terania and Daintree had precedents in the broad-based movement against uranium mining, which had been able to mobilise large numbers for demonstrations in the second half of the 1970s. Thus, while recognising the novelty of using such tactics on forest issues, we also need to see the forms of action and organisation adopted as a product of existing social movement networks on the left.

Another important feature of the Australian environmental movement was importance of alliances with groups outside the mainstream movement. Environmentalists worked with aboriginal communities in opposition to the expansion of uranium mines, and trade unions played an important role in urban conservation conflicts in the 1970s. The 'green bans' declared by the Building Labourers' Federation (Burgmann and Burgmann 1998) succeeded in preventing new building in areas where local residents were campaigning against developers. A similar alliance with sympathetic unions, known as the Earthworker–Green Alliance, was established in 1997 by Friends of the Earth and other groups. In contrast to the left in Britain,[10] environmental themes were taken seriously by important parts of the mainstream Australian left,[11] particularly in the 1970s and early 1980s, however the defeat of the Mundey forces within the unions and the decline of the anti-uranium movement, led most environmental groups to focus more on wilderness protection (Hutton and Connors 1999: 197) and turn away from a wider ideological engagement with issues of ecological justice (Doyle 2000: 215).

The environmental movement succeeded in making forest conservation an important political issue in the early 1980s particularly as a result of the successful campaign against the plan to build a hydro-electric power dam on the Franklin River in south-west Tasmania. This would have threatened a major wilderness area, which had been nominated for World Heritage status. The Tasmanian Wilderness Society (TWS), led by the charismatic figure of Bob Brown, who later represented the Greens in the Australian Senate, organised blockades at the site from December 1982 until the Federal Election of 1983, which made the dam a major election issue. Over 6,000 people took part in the campaign and over 1500 were arrested. The new national Labor government imposed the decision to accord the region World Heritage status on the Liberal state government of Tasmania and forced the cancellation of the dam. The TWS campaign had been based upon the principle that wilderness should be defended for its own sake not because of its use to humans (Hutton and Connors 1999: 163). However, the leaders of the campaign were criticised for failing to develop a strategy based on using direct action as a basis for a movement committed to longer-term change. For instance, Martin saw the TWS as reducing the issue to an 'appeal to elites' and thus failing to develop a radical movement (Doyle 2000: 130). Moreover, by relying increasingly on lobbying decision-makers Doyle suggest that this campaign 'ultimately led to the more corporatist style of environmental politics' (ibid.: 132) dominated by the EMOs.

The third campaign was against the building of a new road through a rainforest in a Queensland national park. The Daintree protests of 1983 were more militant than the previous conflicts and protesters used more confrontational tactics, diving in front of bulldozers, burying themselves up to the neck, and climbing high into the trees (Doyle 1994). Many had been at Terania Creek, the Franklin Dam and the protest against uranium mining at Roxby Downs. In this case some of the early confrontations were especially dangerous as building workers faced protesters without any police or media presence. Australians seem to have been the first to seek to develop tactics that used technical devices to manufacture situations in which they became especially vulnerable. One example was the tripod, (then) made of wooden poles from which a protester was attached in such a way that moving one pole would cause the others to collapse and cause injury. Many of the tactics and techniques developed in the Australian rainforest protests were exported to Britain and the USA and used in different contexts. For instance, the tripod later appeared on sites as varied as major London streets and isolated logging roads in the US Pacific north-west (Restless 2000).

Australian radical environmentalists have concentrated mainly on forest and wilderness issues, particularly opposition to uranium mining.[12] In this they are more similar to US Earth First! Yet, in their links with a broad range of left movements and the lack of any significant philosophical divisions (Hutton and Connors 1999) they are more like European radical environmental movements. Whereas earlier the example of the rainforest campaigner John Seed who wrote a column from Australia for the US *Earth First! Journal*, showed the importance of US–Australian links, now there are more signs of Australian–European contact. British activists who established Earth First! in the UK were influenced by the experience of George Marshall and Shelley Braithwaite in the Australian Rainforest Action Group (Wall 1999: 48). More recently, there may have been more influence from the other direction. One of the founders of Earth First! in the UK, Jason Torrance, spent several years in Australia in the mid-1990s and spoke of the importance of the interchange between British and Australian activism, one sign of which was the first Australian Reclaim the Streets party in Sydney in 1997. Doyle (2000: 48) notes that in Australia there has been a shift from more traditional forms of EDA in which arrests were accepted, to what he defines as militant direct action (MDA). This is still non-violent towards living things, but poses a more direct challenge to authority. Those pursuing MDA are likely to do their best to avoid arrest and seek to disguise their identity, in ways similar to British EDA campaigners. In recent years Australian EDA activists have participated in global anti-capitalist protests paralleling the similar pattern in the UK and the USA. Attempts to blockade the meeting of the World Economic Forum in Sydney in September 2000 involved a broad coalition of anti-corporate activists from urban anarchists to Green Party politicians as well as EDA activists. As with similar protests in the USA and Britain, while the experience of EDA activists was important in the organisation and tactics used, they were part of a broader coalition of green and other groups and part of the anti-capitalist 'movement' rather than the whole movement.

Perhaps more than in the USA and Britain there have been conflicts between EDA groups and other parts of the green movement in Australia. The use of EDA by some of the mainstream EMOs in the 1980s and 1990s such as the Wilderness Society[13] meant that there was less distance between EDA groups and EMOs than in Britain and the USA. However, when EMOs gained more influence, continued EDA by radicals posed a threat to this status. For instance, an occupation of the office of the Forestry Commission in Sydney in 1992 resulted in the loss of consultation for the North East Forest Alliance (Doyle 2000: 57), some of whose activists were part of the occupying group. In the 1980s many wilderness groups became more involved in consultation with state governments and with the Labor government of Bob Hawke. Wilderness activists have been criticised for their lack of attention to social justice issues and many younger activists are uncomfortable with the concept of wilderness because of the history of the dispossession of the indigenous Aboriginal peoples. In the 1990s anti-nuclear networks have re-emerged as a more oppositional environmental voice, campaigning against uranium mining, but within a social justice framework (Doyle 2000: 198). In Britain, in contrast, EDA has not been divided into wilderness and social justice-oriented networks and has complemented the activity of more mainstream EMOs such as Friends of the Earth (FOE) who avoided confrontational and illegal direct action.

Britain

Before 1991 there had been very little EDA in Britain, in contrast with the USA and Australia. There had been small and brief protest actions against nuclear power in the 1970s and 1980s and FOE and Greenpeace had also carried out media-oriented illegal protests, such as the mass return of glass bottles to the doorsteps of the drinks company Schweppes which launched FOE (Lamb 1996: 38), and Greenpeace actions against the Orkney Seal Cull and the sea dumping of British nuclear waste but there was no sustained grassroots EDA network. Yet by 1999 the British EDA network appeared to be stronger and more militant than those in the USA and Australia. What, then, led to the growth of EDA in Britain?

The first part of the answer is that the EDA already had roots in British social movements. Although they were much smaller in scale than the mass protests in many other countries the direct action protests against nuclear power in Britain between 1979 and 1981 established many of the features of EDA in Britain. The emergence of a non-violent direct action network against nuclear power followed the failure of the public inquiry to decide against the Thermal Oxide Reprocessing Plant at Windscale (later Sellafield) in 1978. As Welsh shows (2000: chapter 6) the network that emerged came from readers of *Peace News*,[14] *The Ecologist* and anti-nuclear groups such as the Scottish Campaign to Resist the Atomic Menace (SCRAM). These groups split with FOE over strategy, arguing that the disregard of all the evidence against nuclear energy in the public inquiry meant that a more confrontational approach was now justified. Direct action involved a number of site occupations, most notably at Torness in south-east Scotland in 1979 and a cross-over between direct action radicals and local campaigners which spread the challenging ideas associated with this radical culture. Welsh shows that practices such as consensus decision-making, debates about the politics of gender in protest movements, and the disputes over what forms of action can be regarded as non-violent, as well as broader strategic understandings about the importance of personal transformation as a means of achieving political change, were part of the culture of non-voilent direct action (NVDA) in the late 1970s. A similar culture was also shared with anti-nuclear and other direct action groups in the USA in the 1970s and 1980s (Epstein 1991) and US activism had a clear influence on how British activists organised. As Welsh argues, these practices continued and re-emerged in movements focused on different issues, such as the peace movement in the 1980s, and personal links between anti-nuclear activists, peace movement activists in the 1980s and green activists in the 1990s supporting the view that there is significant continuity in the direct action milieu in Britain.[15]

A second and often overlapping strand of British oppositional culture was the counter-cultural wing of the green movement. Green communes, Tipi villages and permaculture projects had spread during the 1980s and were bolstered by other institutional features of the wider alternative milieu such as alternative health and healing networks, food co-ops, radical cafes and bookshops. As Jowers *et al.* note (1999), this milieu provides a means for many to practise politics in ways that are not measured by conventional political analysis. Alternative lifestyles represent a challenge to 'normal' ways of life. Even when legal they involve a cultural challenge to the mainstream.[16] By the 1990s there were more reasons for the diverse urban alternative communities to take protest action. Alternative lifestyles were under attack, beginning with the 1986 'Battle of the Beanfield' in Wiltshire, when New Age Travellers were evicted in a violent confrontation with the police. After the urban riots of the early 1980s police methods of dealing with protest and disorder (viewed as the same thing in operations manuals) had become more aggressive. The miner's strike (1984–5); the eviction of Molesworth Peace Camp (1984); police attacks on New Age Travellers; conflicts with ravers over squatted warehouses and the Poll Tax Riot in London in 1990 all convinced radicals that the policing of protest had become tougher and less tolerant. The opposition to the poll tax was particularly important as a lesson to radicals. Opposition to this tax, which replaced a tax based on property with one based on individuals, was primarily outside the main political parties (with the exception of Scotland where the Scottish National Party led opposition). The fact that the Conservative government abandoned the tax following a major riot in London convinced many radicals that more confrontational actions worked.[17] Thus not only was there an existing tradition of non-violent direct action, but by 1990 it had become more open to militancy.

Another causal factor was the sense among the alternative milieu of political closure. Until the late 1980s the Labour Party had supported the main demands of the peace movement and many Labour activists, local councils and some of its leading figures were committed to other new social movement demands. While environmental commitments were never significant for Labour, an important section of the party was committed to a politics of the social movements, particularly feminism, gay and lesbian and anti-racist movements. Most radicals were probably not Labour voters, but the possibility that a radicalised Labour

Party could change politics was taken seriously by some in the alternative networks, particularly in the 1980s. However, after failure in the 1987 election, Labour began to move away from commitments perceived as too left wing. Thus, while the long reign of the Conservative Party helped to cement alienation, the Labour Party, particularly as it moved to the right under Blair, has also been viewed with increasing hostility.

The Green Party seemed to represent an alternative avenue of change for a brief period following its success in the European elections in 1989 but its rapid return to electoral marginality, and the absence of environmental commitments by any of the major parties, also reinforced the sense among radicals that the political system was closed to greens.

Earth First! groups were first established in Britain in 1991. Derek Wall's (1999) research on the founders of UK Earth First! shows that the early activists all had some prior experience in green or peace movement activism. The founders were influenced by the example set by Earth First! USA and also by contacts with Australian anti-rainforest activists, and tactics were borrowed from both. The Green Student Network provided a means for recruitment and the first Earth First! groups were based in cities and towns with substantial universities and already strong alternative left milieux. The initial targets of actions were importers of tropical hardwoods; and those seen as causing direct environmental damage in Britain, notably supermarket chains such as Tesco and the Tarmac construction company. Protest against these targets drew some media coverage, but police intelligence and numbers countered most attempts at mass site invasion. It was only with protest against road building that the movement really grew and was able to make a significant political impact.

In 1989 the Conservative Government had announced a mass road-building project costing an estimated £26 billion. The new schemes met strong opposition from numerous groups of local residents who formed a new national network ALARM UK in 1992 (see Chapter Seven on this group). The major environmental movement organisations underestimated the potential of the mobilisation against roads and did not believe that they had the resources to mount a major campaign, although FOE, in particular, supported local groups in fighting their case through public inquiries and the courts. However, FOE was not prepared to risk sequestration of its assets and therefore its other operations by taking action that could be illegal. The new Earth First! groups and others had no resources to lose and were committed to a strategy of EDA and threw themselves into the battle against new roads. Roads represented 'progress culture' (Plows 1998) at its most destructive. In the USA and Australia resistance to roads concentrated on the effects of roads in opening wilderness areas to humans. In Britain, roads were opposed because of the direct damage to countryside and homes and also because of their role in a transport policy that encouraged unsustainable use of cars and lorries. Even when there was no wilderness, roads destroyed areas of special scientific significance and wildlife habitats. Government research (Robinson 2000) showed that building roads increased traffic overall and brought infill development.

Beginning in 1992 with the resistance to the destruction of Twyford Down in Hampshire, a hill that was cut through to save a few minutes for drivers on the M3 motorway, 'the roads wars' were the principal focus of action for the EDA movement for more than five years. The novelty of environmental EDA protest by counter-cultural groups captured the attention of the media, reaching its apogee in the cultivation of Swampy and a few other protesters as symbolic referents for idealistic youth in 1997.[18] At this stage it was difficult for the EDA movement to present itself as other than a single-issue struggle. Yet, roads had never been its only target, and the movement was able to adapt to the decline in road building after 1997. Despite the lack of leading personalities and minimal theoretical debate, it continued to produce innovative actions and to maintain a high level of activity based upon only a loose consensus on goals and strategy.

There was no split of the kind that divided the movement in the USA. This was not only because of the absence of wilderness in Britain, but also because of the roots of the new EDA movement in an existing counter-culture which already had its own social and political traditions. This is not to say that there was no diversity within the movement. The lack of awareness of feminism and even patriarchal practices of macho men at protest camps was the subject of criticism in both Britain (Anon. 1998a) and the USA (Anon. 2001a). There was criticism of the commitment to non-violence as dogmatic (Aufheben 1998) and there was also an informal distinction between more spiritually influenced campaigns based at rural sites and those in urban areas which had a stronger social and anarchist influence. But perhaps the main reason why there was no split between ecocentrism and social concerns was the simultaneous growth of other related protests with which the EDA network identified. The Criminal Justice Act 1994 was an attempt to deal with marginal social groups seen as posing a threat to public order. By restricting the rights to protest and assembly and movement of other groups such as ravers, the Act helped to establish the basis for a common cause between large parts of youth culture. The relatively small numbers of road protesters were therefore bolstered on occasion by larger groups of supporters, especially in the shorter urban actions such as Reclaim the Streets (RTS) 'street parties' which centred around rave dance music. As Jowers *et al.* argue (1999), one response of radicals to the closure of British politics had been the development of alternative lifestyles outside formal politics. Once the Conservative government began to intervene to outlaw these, the strength of the latent British counter-culture became more obvious.

Where EDAers occupied a site earmarked for new development the fact that they often had the support of local residents provided them with legitimacy and some of the practical help needed to be able to maintain their occupation. For this reason, whereas previous EDA protest had been generally treated in the media with hostility, the reporting of anti-roads EDA was often supportive. When campaigning to defend countryside, EDAers were often supported by relatively conservative local groups which encouraged even conservative tabloid newspapers to seek to 'normalise' radical protesters as less different from the mainstream than had been assumed (Paterson 2000).

However, the British EDA networks, like those in other countries do not orient their activity to gain media coverage. While there was considerable media support[19] for actions targeted against genetically modified food, the confrontational protests against global capitalism received only condemnation in the media. Protests organised by RTS against the G8 Summit in Birmingham in 1998 and the 'Carnival Against Capitalism' in the City of London in June 1999, led to violent confrontations with the police and attacks on the property of hated multinationals such as Macdonald's.[20] While few were hurt and most were not involved in the violence, few in the EDA movement were willing to condemn those who attacked the police, and the contrast between the peace movement's more strictly non-violent protest in the 1980s and the more confrontational style of the late 1990s was clear.

The British EDA network has been greatly shaped by its interaction with other often non-green groups, notably those opposing the Criminal Justice Act, local campaigns against new roads, housing or airports and also by its roots in an existing green alternative milieu. EDA in all three countries examined above can only be understood by examining its relationship to a wider green alternative milieu. The latter has shaped it in all three countries, but alone in the USA, the alternative and green character of EDA emerged only after a battle with non-green wilderness defenders opposed to the grassroots democratic practice of new social movements (NSMs) and to the linkage of ecology with social justice.

Repertoires and logics of action

One of the main features that links EDA groups in these countries is their shared repertoire. Because tactics can spread so easily, by reading or hearing about them from activists, by downloading a manual from a website, or by seeing a tactic on television, it is difficult to trace the pattern of diffusion with any confidence. What is clear, however, is that once a tactic is used successfully by one of these movements it is likely to become part of what was defined in Chapter One as the 'modular' repertoire for EDA groups in other countries too. For example, climbing trees to prevent logging probably spread from Australia to the USA in the mid-1980s and subsequently to Britain. Australian activists buried themselves up to the neck in the early 1980s and this provided the inspiration for British activists to expand upon this idea in the 1990s by digging tunnels: when protesters are inside a tunnel evictors moving heavy machinery close to trees run the risk that it might collapse. Activists have remained in tunnels for up to three weeks. A further common tactic was the use of bicycle D-locks to attach activists by the neck to machinery. Starting the machine would lead to injury or death for the activist. This was later developed in all three countries in the 1990s in the form of static lock-ons. These allowed activists defending a site against eviction to clip their arm onto a bar at the end of a plastic tube, which had been embedded in concrete within a wall, in the ground or in a barrel placed in a tree. The activist could not be evicted without digging out the concrete around them. A further innovation was the use of plastic tubing around various forms of locks by which activists were attached to each other on sit-down blockades. This meant that activists could maintain the blockade for much longer than was the case in the 1980s and earlier when they simply linked arms.

These tactics require considerable technical ingenuity. In all cases tools of some kind are used to create an artificial and extreme form of vulnerability in which activists deliberately place themselves at the mercy of their opponents. Since all EDA involves some acceptance of vulnerability by protesters, the novelty of this form of action does not lie in vulnerability *per se*, but its manufactured quality (Doherty 1999b). This use of technical tools is significant, however, in taking the display of bodily vulnerability to a new extreme. In situations where visible violence from the police is rare, manufactured vulnerability allows protesters to juxtapose their own exposure to danger with the destruction being caused by their opponents. In practical terms the political costs of using violence against defenceless protesters means that the authorities are usually reluctant to use it to speed up the eviction. When violence has been used by the police, as in the evictions at Minnehaha described above, or in indiscriminate attacks on those blockading the World Economic Forum in Sydney[21], and most notably in the attack on non-violent protesters in Genoa in July 2001 and the beatings of those arrested, police have been heavily criticised.

The logic of this form of action has been defined by della Porta and Diani as the 'logic of bearing witness'. In contrast to other forms of protest:

> Such action is not designed to convince the public or decision-makers that the protesters constitute a majority or a threat. Rather, it seeks to demonstrate a strong commitment to an objective deemed vital for humanity's future ... The right to influence decision-making processes comes from neither formal investiture nor intrinsic power but from force of commitment. Actions of this kind tend to reinforce the moral message being conveyed by a movement because activists are willing to run personal risks to demonstrate their convictions.
>
> (della Porta and Diani 1999: 178)

The second logic underlying EDA actions is the logic of material damage. Action based on this logic seeks to impose costs on the movement's opponents, which may be achieved through damaging property, but could also be achieved through causing financial damage through lost production or sales. Strikes, boycotts and disruptive direct action, all depend upon the logic of material damage. Many radical environmentalists justify their action according to this logic. They do not view politicians or transnational corporations as persuadable, and argue that it is only by making development uneconomic that the environment can be defended. This is the logic underlying the actions of the ELF in the USA and those who carry out covert ecotage in Britain:

> economic sabotage is far more effective against corporate enterprises to cost them money and the most effective way to do that is to physically damage

their property … It's the only thing they recognise and take notice of …
That's my strategy.

(Anon. in interview with author 2001)

The same logic is found in the animal liberation movement in justifying arson
attacks on stores that sell furs, or breaking the windows in butchers' shops. In
Britain and the USA, activists have been influenced by the example set by animal
rights actions. At local level in Britain there is little overlap in membership in
animal rights and EDA groups, and the networks are distinct. Studies of local
communities of activists in Britain showed that while a few activists were involved
in both networks, most were not (Doherty, Plows and Wall 2001) and each
provided support for the other only on occasional big events. There are also differences between ecotage in Britian and the USA. In Britain, in contrast to the USA,
there is no separate ELF network. Although there were efforts to establish one by
militants in the early stages of Earth First! (UK), these did not produce any
sustained actions claimed by the ELF (Plows, Wall and Doherty 2001). Also in
contrast to the USA, much covert action in the UK is not claimed publicly, and it
is only through occasional reports in activist newsletters and through interviews
that it becomes clear that ecotage is a widespread part of the repertoire of British
activists. The campaign against test sites for GM crops since 1998 means that
ecotage is now more reported (because making the test programme unworkable is
part of the aim of this campaign) but as well as this there are regular if small-scale
acts of unreported and unclaimed sabotage against construction sites and pariah
companies. Typical actions include putting sugar in the petrol tanks of machinery
of a targeted construction company, sometimes at a site miles from the one being
defended, or supergluing the locks of branches of MacDonald's. However, it is
impossible to measure the extent of these kinds of actions and this also means that
attempts to measure the extent of direct action have to be qualified since they can
only be measurements of action that is reported (Plows, Wall and Doherty 1991).

Differences in the nature of the opposition and repression faced by EDA
networks also shape the use of tactics. In the USA sentences for offences such
as arson have been much more severe than those experienced by UK EDA
activists tactics (Warcry 2001). While EDA networks in all three countries
analysed above have been the target of police surveillance and new legislation
designed to repress their activity, it is only in the US that activists have had to
face regular counter-violence, particularly from 'wise use' groups representing
anti-environmental right-wing groups (Rowell, 1996: 157–67; Taylor 1995: 27).
Related groups such as MOVE a small African-American network based on the
teachings of John Africa combining ecological, religious and social justice
commitments, have been the target of severe repression, most notably the
attack by the FBI on a MOVE household in Philadelphia in 1985 in which 15
people were killed.[22]

Both the logic of bearing witness and the logic of material damage are
present in the justifications offered by many EDA activists. The logic of material
damage is used to defend the effectiveness of protest that delays construction and

causes extra costs as well as to justify covert ecotage. The two can also be linked, as for instance, when protesters create a Temporary Autonomous Zone (TAZ) (Bey 1996) by blockading urban streets and organising a street party. The TAZ is seen as tangible and physical because it is the result of protesters taking action themselves to create free space and to push back the dominance of cars and roads. It also becomes a site for a theatre of the possible in which activists can affirm their commitment to other ways of life in which community, public life and pleasure are valued more than paid work in formal employment and consumerism.

RTS actions in particular have had a deliberate theatrical quality which owes much to pre-modern traditions of protest.[23] Modern repertoires of protest, which developed in the nineteenth century (Tilly 1995), centred on making claims as citizens within the framework of the national state. The repertoire that evolved – demonstrations, public meetings, petitions, and strikes – was based on the logic of numbers showing protesters were a large force. Seriousness of purpose was important. In EDA protests the carnivalesque features of direct action are very obvious and music and humour are central. Pre-modern carnivals used the grotesque display of bodily excess as a form of ordered and tolerated reversal of social norms. The festivals of Reclaim the Streets and the pink and silver EDA groups within international anti-neo-liberal demonstrations are intended to reproduce this provocative dimension of carnival, claiming a space in order to display and express an alternative identity. Many earlier forms of protest were based on the claim that those in authority had transgressed the moral code. Protesters then felt entitled to transgress norms in order to express their outrage and seek redress. Contemporary eco-activists also use transgression as a challenge to authority, refusing to negotiate or be contained by the police. Like pre-modern protesters they use effigies of their opponents and mock trials. Cross-dressing was common among pre-modern protesters: EDA protesters develop their own tribal form of clothing, combining in 'monstrous' form combat clothes, dreadlocks, ethnic styles and, more recently, pink and silver fairy outfits in a display of tribal style, which marks out their difference.

At the same time there are other features of their culture and action which reflect the emergence of global politics. Not only do EDA protesters carry out actions in solidarity with those in other countries, and target multinationals for their actions abroad, but they also sometimes define themselves in tribal terms, in ways intended to express their empathy with oppressed groups in other countries. Nodal ideas such as enclosure are used to make a link between contemporary politics, such as the enclosure of urban space by roads and cars, and past struggles against enclosure and for radical politics such as the Diggers (in the English revolution of the seventeenth century). Similarly the attempt by government to prevent nomadic alternative lifestyles in Britain is linked with contemporary environmental struggles in the third world, which in rural areas mostly include land rights.

A characteristic of EDA protest has been the hybrid character of organising networks, which draw upon overlapping networks of groupings with little formal organisation. It would be impossible to understand how these groups have mobilised by focusing only on the history of organisations, whether formal or informal. Research on activists' networks in the south-west of England (Jowers *et al.* 1999) shows that the gestation of EDA-type protest in the 1990s is best understood through examining the wider alternative milieu and its associated networks. This is consistent with the point made above about the character of EDA groups in the USA, Australia and Britain. In the south-west of England networks connected to music-making, organic food production and distribution, festivals, and local exchange and trading schemes provided 'a seedbed for a range of transformed practices of everyday life. "Political activity" is but one of a range of practices which question the rationality of the state, markets and subjectivity in diverse ways' (1999: 100). Jowers *et al.* emphasise that such networks should not be reduced to 'proto-political' forms whose real purpose is to produce overt protest. Many of those involved in this milieu never take part in political protest. 'Expressive practices are not a backdrop from which a more essential politics necessarily emerges' (ibid.: 102). Jowers *et al.* want to challenge the dominant conceptions of political participation which influence the understanding of social movements. This assumes that political activity is aimed at influencing national governments. But, for EDAers, who are oriented towards global concerns the meaning of the political cannot be contained at the level of the nation-state. The justification of their actions is really an appeal to common shared global interests. Their politics is better defined as 'striving to affect the collective understandings by and through which we live together. Moreover, they are also part of a transnational movement which does not only share a political commitment, but also styles of clothes, expressions and exchanges of information' (ibid.)

A transnational movement?

The development of the internet has been important in the rapid globalisation of these networks. The net speeds up the distribution of information that was expensive to distribute for groups with minimal time (Anon. 1999) and money and allows activists to be part of the same events in real time. The 18 June 1999 protests against the effects of economic globalisation were the first major example of this. Protests were organised in 43 countries and many countries had protests in several places. The form and level of protest and the groups involved varied, but all were linked through a single website. Anyone accessing this site could switch between live web-casts of the end of the day's protests in Sydney and the beginnings of actions in London and later read reports of actions in Toronto, New York and Los Angeles.

The internet compresses time and it provides the environmental movement with a means of accessing what Castells (1997) calls 'the space of flows' in which global media and instantaneous financial transactions create a kind of 'timeless time' that is normally dominated by capital. Thus although the compression of time and space is an impetus for the intensification of global capitalism, it also aids the mobilisation of opposition. The web makes it possible to mobilise activists, for virtual meetings to take place more or less continuously (see Chapter Two) and for activists to provide reports on their own actions independently of the mass media. Activists are able to read instant accounts of events reported from their own side, something that was hitherto not possible and this means they are less likely to be undermined by hostile media responses. Conventional journalists also have access to a much fuller perspective from inside the activist scene (Carey 1998). However, while the internet is important in reducing the costs of exchanging information, not all activists have regular internet access and few activists are directly recruited by visiting websites (Pickerill 2000).

The principal change in EDA protest in the late 1990s was the growth of international anti-capitalist protest. EDA activists played an important though variable role in initiating this protest, beginning with the involvement of British activists in establishing the coalition known as People's Global Action (PGA) in Geneva in 1998. The PGA was a diverse coalition of EDA activists, Indian farmers, the Brazilian Landless movement and other grassroots groups which had formed from meetings of supporters of the Mexican Zapatista uprising. It co-ordinated protests against the meetings of the WTO in Geneva in 1998 and protests in several countries against the G8 summit in Cologne in June 1999. However, from the beginning the principal feature of the new anti-capitalist protests was the diversity of constituencies and groups involved. As one US/UK leaflet discussing these 'communities' said: 'we are talking about the anti-corporate movement, greens, anti-racist movements, the women's movement, the queer movement, as well as the traditional left and trade union movements' (York Earth First! 2000). In Britain EDA groups, particularly RTS, were the main initiators of anti-capitalist protests, however, they did not control them. In the 18 June Carnival Against Capitalism in 1999 EDA groups and others from the wider green and alternative milieux such as Campaign Against the Arms Trade took part in over 30 separate actions in London alone involving around 10,000 protesters and more than 30 in other UK towns and cities. There was chaos in the city of London and considerable property damage and some attacks on the police alongside strictly non-violent action. As one protester commented:

> The June 18th demo against the G8 Summit in the City of London was amazing! It was possibly the best riot in London since the Poll Tax one. The cops totally lost control of the situation and got a good beating, and various businesses like MacDonald's, car showrooms, banks and the Futures Exchange were trashed. According to the press four cops were hospitalised. The City was also covered in anarchist graffiti. Every time you were with a large mob thinking this is great all these people, you'd turn the corner and there's be an even larger crowd there creating mayhem.
>
> (London: a personal account, 18 June Website:
> <http://bak.spc.org/j18/site/uk.html>)

In the past, large-scale demonstrations were mostly marches from A to B ending in a rally with speeches by movement leaders. The London 1999 and subsequent protests in Seattle, Prague, Genoa and elsewhere were much more anarchic, confrontational and difficult to police.[24] The arguments about violence which have arisen as a result will be examined further below. In the USA anti-corporate protest has been targeted at inter-governmental meetings to make treaties that would reduce environmental and social regulations on business in the name of free trade and involved a broad range of left groups, trade unions as well as counter-cultural EDA groups. The Seattle protests of November 1999 which succeeded in preventing some sessions of the WTO from taking place were a turning point for this movement. Seattle drew international attention to the changes in global governance occurring in the name of free trade and the diffuse network of anti-neo-liberal protesters became identified as the principal opposition. As a result this diverse coalition was attributed a unity which it did not have and then held responsible for its lack of a clear alternative. This view, widespread in the media, was based on a misunderstanding of the nature of the groups involved or the potential for common agenda.

Transnational protests are not new but in the past they have been targeted at national governments, as with the mass peace protests of 1981–3 in western Europe and the USA. The novelty of the anti-capitalist protests is that they target international institutions themselves. The policies of the WTO, International Monetary Fund (IMF), World Bank and other institutions are dominated by the richest western countries and are largely decided in secret. Literature by EDA activists provides well-researched critiques of the effects of these institutions in maintaining the wealth of the rich countries, while imposing trade agreements, debt repayment programmes and structural adjustment programmes on poorer countries that lack either the bargaining power to resist or are dominated by corrupt elites. Democracy, ecological rationality and social justice are interrelated in the attack on these institutions.

The free-trade agenda is opposed by a wide variety of groups, including church-based campaigners, development activists, the Marxist left, many trade unions, and anarchists, as well as the EDA groups and other greens.. Some within this coalition want to see reforms to transnational institutions, others to do away with them. Some have very little concern for ecological rationality, and some are primarily single-issue campaigners, such as those involved in Jubilee 2000, Drop the Debt campaigns or groups such as ATTAC[25] (Aguitton 2001: 28–9).

The diversity of the anti-capitalist coalition is a consequence of the new strategic dilemmas that left movements have faced since the early 1990s. The principal change since then is not so much the loss of the example of socialist societies in eastern Europe, since the latter had little support on the western left. Instead it is the 'new world order' proclaimed by George Bush in 1991 after the Gulf War, which reflected the dominance of western strategic interests in global security issues and the turn to the right of mainstream left parties that provided the conditions for this coalescence. As mainstream left parties such as the British and Australian Labor Parties, the German SPD and the Italian PDS and also the

Democrats in the USA have accepted the dictates of international competitive-ness and reducing barriers to free trade there seems to be no voice within the political institutions questioning this. Yet, to a diverse range of largely marginal groups the new world order is seen as unaccountable democratically, militarily brutal, increasing inequality and suffering and doing nothing to avert ecological catastrophe. While there are many alternatives, such as the arguments for reducing material consumption and increasing local self-reliance popular among EDA communities, the finer points of alternatives are less pressing than the importance of creating a new opposition.

Second, the lack of agreement on alternatives is deliberate. EDA groups along with others – including possibly some of the Marxist groups such as the British Socialist Workers Party (SWP) – do not want to repeat the mistakes of past left sectarianism in which groups failed to take common action because of disagree-ments over questions of doctrine. One of the conditions of success of EDA groups has been their willingness to work with groups with whom they may have major differences. For instance, British EDA activists worked with farmers following the fuel protests to discuss common ground in opposing an industrialised agricultural system. They also took action with striking dockers in Liverpool who had been abandoned by their trade union and with many conservative local communities opposing new developments or GM crops. And this willingness to look beyond the traditional sources of alliances for the left also reflects their views on power. While opposing class inequality is part of the EDA worldview, activists believe that there are multiple forms of power that require multiple forms of resistance. Protest actions may always be to some extent contradictory – for instance, travelling in cars to demonstrations – or allying with livestock farmers – but they are justifiable pragmatically because they can be related to the overall goals of green ideology, namely expanding democracy; opposing inequality; and advancing ecological rationality. However, one negative consequence of this loose strategy is that the consensus on non-violence is difficult to sustain.

The violence/non-violence debate

The debate over non-violence is a perennial source of disagreement in EDA groups and widely misunderstood by observers, particularly in the media. The central areas of disagreement are over harm to people; there is relatively little disagreement about damaging property or sabotage. This distinguishes EDA groups from Greenpeace and many in the peace movement, who, like Gandhi, regard sabotage as violent and unacceptable (Scarce 1990).

The early adoption of EDA by Australian greens and the overlap between peace and environmental struggles in Australian campaigns against uranium mining led to clashes in the 1980s between anarchistic EDA and those in the peace movement committed to non-violence as a philosophy:

> There was bitter criticism of having to buy the whole non-violent package
> as well as the concept, particularly from those who were driven by a deep

anger directed against the state and all its institutions. One such movement, known originally as the Nomadic Action Group, travelled around different actions in Australia, charging into ongoing campaigns with little if any respect for those already involved, and usually causing havoc, before moving on as self-appointed saviours of the Earth.

(Jones, Pestorius and Law 1995)

There were also major debates among anti-nuclear activists in Britain and the USA in the late 1970s about whether damage to property counted as non-violent (Welsh 2000, ch. 6: Gottleib 1993: 180; Epstein 1991) and this issue remained controversial among peace activists in the 1980s, although this did not prevent criminal damage at Greenham Common[26] and other military bases. There is much less debate now among EDA activists about how to define violence. Violence is defined as harming living things,[27] including people, animals and (in some cases) plants.[28] Damage to property, as when activists break machinery or 'trash' the offices of a despoiling corporation, is not defined as violence and so ecotage of the kind discussed above, is not an issue in the violence/non-violence debate within EDA. As this British activist put it in interview:

> I don't count violence to property as violence. I don't count the destruction of property … the destruction of machinery I think is perfectly valid, something that I enjoy immensely and is something that is important to do.
>
> (Anon. in interview with author 1999)

Such a definition allows activists to view a considerable range of illegal confrontational action as non-violent. Most are not committed to pacifism as a philosophic principle. There is no controversy about the right of indigenous peoples such as the Zapatistas or tribes in Indonesia, Papua New Guinea or the Amazon region, to use weapons in what are seen as struggles of self-defence. Nor would it be illegitimate in the view of many activists for individuals to use violence in self-defence when attacked. It is principally planning to use violence that is rejected.

Tensions emerge, however, over how to react to the violence of allied groups. This has become a more difficult issue as the number of anti-capitalist demonstrations has increased. On protest camps and other small-scale actions EDA activists could establish a consensus on non-violence because they knew those they were with, but this is not possible in the open and non-directed protests such as those in London, Seattle, Prague and Genoa and many are reluctant to condemn the violence of the black bloc. EDA activists know what it is like to be on the receiving end of police violence. In Genoa, violence by the police was indiscriminate during the protests and after they ended, it was the headquarters of the non-violent Genoa Social Forum that was attacked by police seeking revenge, rather than the black bloc's camp. The dilemmas created by these large 'open' (to all) actions are the subject of much discussion by activists. For instance in a paper analysing the post-Seattle situation York

EF! activists (adapting a paper by US activist Michael Albert) commented that 'it would be wrong for a relatively few activists to hijack the demo without the many having a say in what is affecting them. But second, it would also be wrong for even a large majority to imagine that it owns "all demo space"'. The solution proposed is that demonstrations should be organised to permit groups to act separately. This was the principle followed at Prague in September 2000, where there were three separate marches, one for 'Black Bloc' anarchists – prepared to use violence, one for international socialists, in support of the Italian group Ya Basta, who sought to confront the police, but to avoid violence, and another, Pink and Silver group, for EDA activists, who aimed to reduce confrontation with the police with samba music, carnival costumes and by evasion of police lines. None of these groups escaped attacks by the police, but only the EDA group managed to get close to the conference centre, not least because of the skills and experience of the EDA activists in evading arrest. Yet it was primarily the pitched battle fought by the black bloc that was covered in the media.

In contrast to traditional civil disobedience it is usually accepted by EDA activists that it is legitimate to avoid arrest. Only a few campaigns have followed the strategy of openly courting arrest to bring an issue to public attention. The Ploughshares campaign in which women successfully defended themselves in court after damaging Hawk fighter planes and nuclear submarines, provoked a brief debate among British activists about whether it might be better to carry out 'accountable' actions of this kind.[29] The Genetix Snowball campaign against genetically modified food, which came from part of the EDA network in Britain, later followed this open method. Most activists seem to feel that they should seek to get away without arrest whenever they can, not least because they feel no sense of obligation to laws which are intended to prevent their actions.[30] Thus the practice of non-violence is not civil disobedience in the tradition of Thoreau, Gandhi and Martin Luther King and is better understood as a form of resistance politics. A critique of Genetix Snowball by activists from Leeds Earth First! makes this clear:

> It is our view that resistance does not need to legitimise itself according to the terms of the system and its ideology. It is legitimate precisely because it resists these things. Genetix Snowball, in contrast to this, does not contextualise itself within capitalism or the resistance to it. It is therefore forced to justify itself by emphasising its ultra non-violence and accountability and the moderation and 'reasonableness' of its demands. A case of fuck the disobedient let's get civil. Genetix Snowball is not resistance but souped up civil disobedience, spectacular lobbying, and therefore shoots itself in the foot.
>
> (<http://www.leedsef.ukf.net/critique.htm>)

Genetix Snowball did not last long, although many of its activists continued to carry out crop 'decontamination', and while its activists retained the support of most in the EDA network, and like the Ploughshares campaigners were admired for their stand in court cases, most crop trashings were covert, and based on the aim of avoiding arrest.

Nevertheless, there is still a division between those who see their protest as an attempt to build public support and those who see it as a power struggle in which the views of the public are secondary. In Britain, even those prepared to use violence, want to build a stronger movement and argue that this is not incompatible with carefully justified use of violence on occasion. In the USA there is stronger support among a minority of activists (principally in the ELF), for more militant action irrespective of public reaction. As 'Snap Dragon' put it in a debate on violence in the US *Earth First! Journal* in 1998:

> There simply is no moral or strategic imperative to adhere to non-violence and engage in civil disobedience. We don't need to convert the public; we need to protect wild places. Without its symbolic underpinnings, CD (civil disobedience) is a terribly inefficient way to stop logging, road building and developing. Every day 137 species become extinct and 176,000 acres of forest are lost forever. We don't have the luxury of civility. We must do whatever is necessary to defend our home and protect our ecological family. Once it is gone, we can only wish we had done more.
>
> (Snap Dragon 1998)

Violence can also be effective in getting an issue noticed, something long accepted by analysts of social movements (Piven and Cloward 1977; Gamson 1990). As one letter writer to *The Guardian* pointed out, following fights with the police at the 18 June 'Carnival Against Capitalism' in the City of London: 'I am glad that "we" caused trouble, trashed buildings and some personal property. I am glad that "we" provoked the police into riotous scenes. I am glad because "you" had to notice us, "you" had to wake up and see us express what "we" passionately feel: capitalism hurts everybody' (Marie Mulley 1999).

Although the property damage and mayhem of the 18 June and Mayday 2000 demonstrations were front-page news in nearly every newspaper the following days, many British activists argued that the demonstrations were successful because a peaceful demonstration would not have attracted much attention. No recent protest against capitalism had ever attracted so much coverage. By expressing their anger violently protesters were demanding a reaction and a response.

These are rational analyses of the strategy of the movement based upon the impossibility of achieving change within a system, which is seen as based on a closed network of political and economic power sustained by cultural norms. Much of this reasoning is reminiscent of that of many Marxist groups.[31] Other, less common, responses to this problem emphasise the importance of emotion in answering the problem of whether the use of violence can ever be justified. As US Earth First!er Derrick Jensen puts it:

> I don't believe the question of whether to use violence is the right one. Instead the question should be: Do you sufficiently feel the loss? So long as

we discuss this in the abstract, we still have too much to lose. If we begin to feel in our bodies the immensity and emptiness of what we lose daily – intact natural communities, hours sold for wages, childhoods lost to violence, women's capacity to walk unafraid – we'll know precisely what to do.

(Jensen 1998)

Jensen's argument is essentially a call to transcend reason and act on natural instinct. He says: 'Discussion presupposes distance, and the fact that we're talking about whether violence is appropriate tells me we don't yet care enough. There's a kind of action that doesn't emerge from discussion, from theory, but instead from our bodies and from the land'. It would be difficult to challenge this argument without also challenging some of the more strongly held deep ecological commitments of many US Earth First!ers. Jensen compares the actions of humans defending the wilderness to grizzlies defending their cubs. Only by reaching this level of instinct and stripping away the false accretions of civilisation will people be able to act naturally.

Although not usually based upon such an ecocentric position, the arguments of the fairly small number of EDAers in Britain who defend the movement's right to use violence (e.g. see Aufheben 1998) share one characteristic with Jensen: both are based on an absolutist view of power. They see the existing form of society as a closed single structure, which is undifferentiated and cannot be opposed from within. Only a struggle for power in which tough means are justified can achieve the end of destroying the old system and creating something new.

Those who defend non-violence also do so from a variety of positions. Some emphasise the importance of the relationship between means and ends, others defend non-violence in more strategic terms. As Mike Roselle now a Greenpeace employee and associated with the social justice wing of Earth First!, puts it: 'Losing the message is a real risk in non-violence. It is a bigger risk in monkey-wrenching, a greater risk still with violence' (Roselle 1998). Mikal Jakubal says that turning to violence would be the quickest way to alienate potential support: 'Our enemies understand this only too well, which is why we regularly have to fend off attempts to link us to the Unabomber. Here on the north coast of California, the timber industry concocts a steady stream of fake sabotage and fake threats of violence because they know the best way to discredit and undermine support for non-violent activists is to attribute violence and sabotage to us'. (Jakubal 1998). Advocates of non-violence have greater confidence in being able to mobilise public support and play a role, if a confrontational one, in achieving change through protest.

A further reason for the commitment to non-violence is its relationship to the form of organisation favoured by activists. A loose, non-hierarchical and open form of organisation cannot carry out an effective campaign of violence. Even the small scale sabotage carried out by some US Earth First! activists and more recently the ELF brought the attention of FBI undercover investigators upon the movement. In Britain, the strength of the direct action groups has led to increased attention from the police. This has taken the form of specialist units, disingenuous warnings of the dangers of environmental terrorism, (related) demands for new resources and legislation and revisions to tactical manuals. The legal definition of terrorism as including damage to property even where there has been no harm to people, and the use by the British Prime Minister of rhetoric such as 'mindless thugs' to describe Mayday protesters seem to EDAers to be paving the way for further repression. A shift by them to violence would provide justification for such a move, end both the style of open debate which characterises the movement, and prevent many of the activities that depend upon long-term links with local communities. The network would become secretive and even smaller than at present.

NVDA increased in importance as a form of protest in the twentieth century. The use of violence in protests has declined and public hostility to violence in internal politics has increased in step with the establishment of democratic rights (Meyer and Tarrow 1998; Rucht 1990). This may not be a permanent feature of modern politics,[32] nevertheless two factors make a return to violence unlikely. First, the repertoire of non-violent protest methods has expanded substantially during the twentieth century, particularly in the use of symbols to attempt to change the collective understanding of a problem. The expansion of access to the mass media has also substantially increased the ability of protesters to diffuse such symbols.[33] Thus the protester at the beginning of the twenty-first century has a more substantial and tested tool-kit of tactics. Moreover, there is evidence that public opinion is becoming more tolerant of illegal protest, as long as it is non-violent.[34] In situations where violence by protesters in democracies has been common, as in Italy up to the late 1970s, the violence tended to decline when policing became more sensitive (della Porta and Reiter 1998). There is no case within an established democracy where violence by protesters had strong popular support. But this makes it all the more remarkable that illegal but non-violent protest seems to have become more acceptable. And this is important for EDA, because government violence against non-violent protest, such as that at Seattle in 1999 and Genoa in 2001, invariably increases support for the protesters (Gamson 1990: 81).

Moreover, the fact that the EDA movement is so small and based on specific social groups and a distinct sub-culture, means that it is vital that it does not undermine the basis of its alliances with larger groups of people. A shift to violence would have this result. At present, by refusing a reductionist interpretation of what green means and focusing on how environmental issues raise broader issues of power and democracy which are no longer contained by the formal political system, or national in scope, EDA groups strengthen civil society and democracy.

Conclusion

The rise of ecological direct action appears to be linked to the perception of governments' weakness in the face of globalisation. The adoption of EDA by groups of young people is also an effort to compensate for weaknesses in other parts of the green movement. The EMOs are unwilling to risk their relationships with government and business or their mass support by pursuing a more challenging agenda. Green parties are at best minority parties often pushed into decisions that seem to promise only minor change in the distant future. Local environmental campaigns provide useful allies but even when most radical they do not challenge power on a global level. EDA activists see themselves as pursuing the radical agenda that is necessary if progress towards the goals of green ideology is to be made. However, they are all too aware of their weakness. As one British activist put it:

> The way I look at it is, political movements grow in two directions, vertically and horizontally. And vertically, that means, generally your politics becomes more radical and oppositional and more coherent outside the mainstream, but equally your methods of taking action become more effective, you learn what works so your basic tactical and strategic armoury gets more effective … but the thing is as you progress vertically, as your politics becomes more radical and more distant from the mainstream … you run a real risk of being marginalised from, you know, everyone else. And you become this tiny bunch of wackos that no one wants to deal with. Political movements I think are doomed to failure if they only grow vertically and if they don't combine that with a simultaneous horizontal growth, and horizontal growth means that you firstly get bigger in terms of numbers of bodies but you also draw from more different backgrounds, you basically embed yourself in a greater chunk of civil society. And Earth First!, as I say, has done the vertical growth very effectively and the horizontal growth extremely ineffectively.
>
> (Anon. in interview with author 2000)

Most activists are well aware that EDA cannot become a mass movement. Direct action is demanding and it is noticeable that activists rarely have children and most are under 35. Second, one of the strengths of EDA is the alternative culture which creates emotional bonds and solidarity between activists. However, this also poses barriers to 'straight' people, who feel that they cannot become involved and maintain their existing responsibilities to work, home and family. Thus, it seems likely that the movement will remain a tiny, if creative and powerful, minority, which may go through cycles in different countries. However, just as it is unlikely to become a mass movement, it is also unlikely to disappear.

The common source of continuity in each of the three countries examined here, is a wider green and alternative milieu, in which the repertoires of EDA have now become well established. Moreover, these milieux seem more extensive and stronger than before, not least because of the range of groups with which EDA has been able to form alliances. The growth of a diverse international movement of protest against neo-liberalism has provided an opportunity for the small number of EDAers to link with other groups and a space in which their specialised direct action and anarchism can play an avant-garde role. Like other parts of the green movement, EDA cannot achieve a transformation alone. But the diffusion of protest and the new space for opposition to neo-liberalism has opened up new, if uncertain, possibilities for further diffusion of protest and green ideology. The local action of grassroots environmental campaigns, exam-ined in the next chapter does not generally have the international or counter-cultural character of EDA, but its roots in non-green communities make it important for green activists of all kinds.

7 Local environmental protest groups

Local environmental groups as social movement actors?

Greens place a high value on the local sphere. The slogan 'Act Locally, Think Globally' is used by greens in many countries because it evokes their commitment to political decentralisation balanced by global responsibility. Action at the local level is valued because it is more likely to provide opportunities for meaningful participation for individual citizens. Local groups are less beset by the problems with large-scale decision-making that reduce the role of the individual. The ecological arguments for decentralisation usually depend upon individuals taking responsibility for their local environment and learning to live within its limits. One of the promised gains from this is a stronger affective bond with particular places. Thus the green sense of community combines the natural world and the social world in a way that makes the local particularly valuable. Other reasons for valuing local action include the fact that greens do not believe that central government or transnational corporations are capable of protecting particular local environments. Their bureaucratic character excludes consideration of factors such as the subjective importance of local aesthetic concerns, and their commitment to expanded production means that they treat risks to particular localities as acceptable for the greater good. Moreover, by challenging these bodies local communities are empowering themselves and strengthening their community bonds.

But, if judged by their ability to fulfil these aims most local environmental protest groups[1] appear to be weak from a green point of view. These groups are often represented as motivated by essentially selfish concerns, which have led to them being labelled as NIMBY (Not In My Back Yard), implying that as long as their's is not, they would not care if someone else's backyard was despoiled. At best, local protest groups are motivated by an immediate threat to their health, or to the amenity value of their immediate environment, at worst they are merely concerned to preserve the value of their property, which might be undermined by new development.

Local groups are often presented by their opponents as standing in the way of developments which would be of benefit to other groups through providing jobs or by contributing to greater prosperity. But in fact the evidence suggests that most local protest campaigns do not deserve the NIMBY label. The label itself has become a way of undermining the legitimacy of local groups, and its use suggests that any action in defence of local material interests is somehow unjustified. Viewed from the green perspective, however, local interests should be defended, particularly when they are in opposition to environmentally damaging or socially regressive business or industry. Even groups that begin based on essentially NIMBY concerns, once faced with the accusation of being NIMBY, have to respond. To do so effectively they must be able to show that their arguments are not wholly selfish and that their opponents' project is not in the wider interests of society. For instance, rather than simply argue that a new development should be in somebody else's backyard many local groups argue that the causes of toxic waste or increased traffic need to be dealt with rather than simply accepted as a necessary cost of progress.

Even if the NIMBY label is often unjustified, greens (defined in this book as radical environmentalists sharing a green ideology), still see the limited political horizon of local environmental protest as a weakness. Because they tend to be concerned with particular new developments such as a waste incinerator or a road, local groups look unexciting and unpromising when compared to the more radical green agenda. Local groups are also much more likely to work through the institutions provided by the political system, using licensing mechanisms, the courts and pressure on local authorities in order to pursue their case. Direct action for them is likely to be a last resort and many would reject it if it meant breaking the law. Thus direct action groups often regard local protest campaigns as hampered by strategic naïveté.

Local groups are often outflanked because they lack equivalent resources to their opponents. Transnational corporations can threaten to relocate their business leading to a loss of jobs locally if threatened with opposition on environmental grounds. Local environmental protest groups are also often handicapped by the secrecy of decision-making and the formal and informal relationships between officials and business. Local activists also need to be scientifically competent and able to question experts on their own terms. All of these obstacles suggest to the more radical greens that local groups cannot succeed within the existing system. A more radical and confrontational strategy, which challenged the assumptions of policy-makers in business and government, would contribute more in the long run than the piecemeal approach of local groups.

Local environmental groups therefore appear politically weak, due to the imbalance of power between them and their opponents, and also relatively marginal within the green movement, because of their moderate strategy and their focus on specific projects. Given this picture it is hard to find a place for local environmental groups that fits the definition of the green movement (as a social movement) in this book. As we saw in Chapter One, there are criticisms of the use of the term social movement as a way of classifying environmental movements (Jordan and Maloney 1997; Jordan 1998; Baggott 1998). Many of the points made by those critical of using social movement concepts to explain environmental groups are relevant in the case of local environmental groups, which do not usually have the characteristics expected of social movements: most are short-lived, many never undertake protest action and may not see participation as an end in itself of their activity, and most do not seek far-reaching changes in society and politics.

Yet, it also seems that there are often aspects of the activities of local environmental groups that are better understood using social movement analysis than by the alternatives as used in political science such as local pressure groups or campaign groups, in which the focus tends to be on success or failure in achieving changes in policy. The strength of social movement analysis is in the focus on the motivation of activists and the structural constraints and opportunities that condition their action. As the evidence below suggests, many political novices speak of the radicalising effects of participation in campaigning, including their loss of trust in political representatives and the democratic process. For some, as has been charted in the case of US anti-toxics and environmental justice groups, this leads to participation in other political action after the demise of their local groups[2] (Epstein 1997; Szasz 1994; and Schlosberg 1999). In Europe activists from local environmental groups in the 1970s and 1980s were important in many countries in the development of green parties. In Germany the BBU (network of local citizens' initiatives) had a political platform that prefigured much of that of the Green Party and leaders of the BBU such as Petra Kelly played an important role in the formation of Die Grünen.

Local groups are often willing to undertake protest action if it is likely to be effective instrumentally, but can also be cautious for tactical reasons, believing that it will alienate other locals (Lichterman 1996). Yet, given the evidence of loss of trust in political authorities the absence of protest does not necessarily mean that local groups are anti-radical. In fact, local groups often seem to welcome and work with outsiders who are prepared to undertake non-violent direct action (Stewart, Bray and Must 1995; Cathles 2000). And while initial participation in local groups may not have been motivated by a wider analysis of democracy, studies show that activists speak of a strong sense of obligation to take action once they appreciated the nature of the issue (Berglund 1998; Lichterman 1996).

Furthermore, in the course of their campaigns local groups often develop broader arguments based on environmental principles alongside their technical objections to development projects. The best example of this is the emergence of the category of environmental justice and the related theme of environmental racism in the USA. The concept of environmental racism was based on evidence that emerged in the late 1980s that environmental hazards such as toxic wastes were much greater in areas where the population was mostly African American or Hispanic (Commission on Racial Justice, 1987). It was also a deliberate challenge to the overwhelmingly white mainstream environmental movement organisations (EMOs) that had focussed too much on conservation and preservation of the non-human natural world and disregarded the environmental hazards faced by people. The call for environmental justice was intended not only to achieve remedial action on local issues, but also to challenge the structural causes of environmental hazards, linking race, poverty, the production system and the weakness of the democratic system.[3]

An account that suggested that all local environmental groups shared commitments to broad and radical political goals and radical democratic practices would, however, be incorrect. Many case studies also exist of groups that show no signs of the radicalisation necessary for social movement analysis to become useful (Hayes 2001; Jimenez 2001; Botezagias 2001; Rootes 2001; Fillieule and Ferrier 2001). But, the fact that not all environmental groups share all the characteristics of an 'ideal type' of social movement does not make the use of social movement concepts invalid. It is not the case, as has been claimed (Jordan and Maloney 1997), that the interest group term can capture all the dimensions of what is meant by social movements and do so with more analytical precision. The grey areas that are criticised in the use of the term social movement are such because they deal with questions of culture and identity, which are not clear-cut. Thus the more radical groups in the green movement might best approximate the ideal type of a social movement. However, if we reject its use for more ambiguous cases, such as local environmental groups, then we will lose useful tools for trying to understand and explain them.

The analysis of local environmental protest in this chapter is not therefore divided neatly into social movement and interest group sections. Instead, there is a focus in the first part of the chapter on cultural questions such as the status of science in local group arguments and the meaning of community for different kinds of local environmental groups. The second part of the chapter examines comparative explanations of factors that affect the political impact of local environmental campaigns. First, however, I will identify the most prominent of the common features of local environmental protest groups.

The nature of local environmental groups

It is difficult to make reliable, empirically based generalisations about local environmental protest groups. Since so much protest is short-lived and is usually not reported by national media, it is very difficult to assess the numbers involved. Estimates have been made of the numbers of 'grassroots' groups in the USA based on the numbers linked with national co-ordination networks such as the Citizens' Clearinghouse on Hazardous Waste (CCHW) and the (now defunct) Anti-Toxics Network varying between 7–8,000 (Szasz 1994: 72) and 12,000 (Gould, Schnaiberg and Weinberg 1996: 21). In Germany there were over 1,000 groups affiliated to the BBU in 1982, when local environmental activism appeared to be at its peak (Burns and van der Will 1988: 183) with between 300,000 and 500,000 members. In Britain in 1980 there were approximately 350 local groups opposed to nuclear energy (Welsh 2000: 185) and in 1996 around 300 groups were affiliated to ALARM[4] UK, the federation of local anti-roads campaigns. Other attempts to survey local environmental activism have tended to provide snapshots of regions such as Berlin (Rucht and

Roose 2001), Florence (della Porta and Andretta 2001) or south-east London and parts of Kent (Rootes, Adams and Saunders 2001). In Florence, citizen committees (some NIMBY, some more grassroots) campaigning on issues of local governance have replaced EMOs and the green party as the main source of environmental protest activity, and this rise of independent and local protest appears to be a common feature in many countries since the 1990s.[5] But since there is so little systematic evidence it is difficult to demonstrate conclusively whether participation in local environmental activism is rising or falling overall.[6]

And, because most local environmental groups do not have formal membership this makes surveys of their members difficult.[7] Moreover, many of these groups would not necessarily define themselves as environmental groups. Some, such as many environmental justice groups in the USA, began from pre-existing civil rights groups which took up particular environmental issues (Sandweiss 1998). Others, including some groups within the BBU in Germany, were concerned with civic amenities or housing issues. Most of the evidence about local environmental groups is therefore based upon case studies of particular groups or campaigns. These suggest five features that are common, though not necessarily universal, for local environmental protest groups in First World countries.[8]

- They are usually the result of autonomous local initiatives and not created from above.
- Local mobilisation is most likely to be provoked by a new development and based upon local and immediate risks or environmental costs.
- Local opposition is most likely to become confrontational when political opportunities are restricted and also when officials or opponents are seen to be acting arrogantly, or unfairly.
- Local environmental groups require some technical expertise and must define their attitude to science and technology.
- The arguments of local environmental groups depend on a socially constructed conception of the interests of the local community.

They are usually the result of autonomous local initiatives and not created from above

The decision to form a group and take protest action is usually made by concerned local people, who may sometimes include activists in national environmental movement organisations.[9] Political parties are notable by their absence, even in southern Europe where parties have tended to be associated with other kinds of community activism (Kousis 1999; della Porta and Andretta 2001; Botetzagias 2001). Nationally based EMOs may offer training and resources to support local networks as in the case of the 'Wetland watchers' initiative by US national environmental organisations (Gould, Schnaiberg and Weinberg 1996). However, such groups appear to be exceptions. Most local groups appear to be initiated independently of EMOs.

Despite the expansion in resources of most major EMOs since the 1980s they are usually too thinly stretched to be able to provide much support to local groups. Most EMOs can only sustain a few campaigns at any one time (Dowie 1995; Rootes 1999). Groups such as Greenpeace do not encourage their local support groups to engage in their own campaigns. The primary role of local groups is to raise funds. Other groups such as Friends of the Earth (FOE) in Britain or the Sierra Club in the USA encourage their own local branches to engage in campaigns, but often based upon themes set by the national office. In the USA in the 1980s, for instance, as the weakness of regulatory agencies became increasingly apparent, local branches of environmental groups were encouraged to become citizen-regulators, doing the job of enforcing regulations that the EPA (Environmental Protection Agency) had failed to do. While local branches of groups such as FOE often do play an important role in local campaigns, it is generally a new environmental threat that mobilises local protest (see next section), not an on-going and more general environmental campaign.

A characteristic of the environmental justice groups (Schlosberg 1999) other grassroots local groups such as the earlier wave of local, mainly white, anti-toxics campaigners in the USA, (on which more will be said below) and the Bürgerinitiativen (Citizen's Initiatives: linked in a network as the BBU) in Germany in the 1970s, was their commitment to network-type organisation. In each case, groups were linked through a decentralised form of organisation that protected the autonomy of local groups. The role of the national offices was to disseminate information between groups rather than to develop a national unified strategy. The other important dimension of this network-type organisation was that it encouraged exchanges of information, advice and resources between local groups rather than through the national office. As a result of their grassroots orientation these network organisations are often impermanent. If the grassroots groups begin to decline or are no longer needed, the groups may decide it is time to disband, as was the case with ALARM UK in 1998.

In the USA the adoption of decentralised forms of organisation was associated with a particularly strong critique of the mainstream environmental organisations:

> There has been anger at the lackluster and ineffective campaigns of the larger groups, disappointment at the lack of attention given to the diversity of the grassroots, distrust of the professional atmosphere of organizations, frustration with control by the major funding organisations rather than memberships and criticism of the centralized, professionalized organizations that are not accountable to memberships or local communities.
>
> (Schlosberg 1999: 107)

Schlosberg calls the form of organisation developed by the environmental justice movement 'rhizomatic'. A rhizome is a root system that spreads in numerous directions and sprouts in numerous places. The metaphor of rhizome captures the complex trans-local relations between different environmental justice groups.

For instance, alliances against a particular company or industry might link groups from different classes or different regions, women's groups, church groups and trade union groups, and groups within this alliance might have other ties to neighbourhood groups or groups working on other issues. These networks are constructed around a sense of solidarity and shared interest which transcends the local. Alliances such as the Citizens' Clearinghouse for Hazardous Wastes (CCHW) in the USA and ALARM UK stipulate as one of their few require-ments of members that local groups avoid NIMBY strategies: 'Our biggest problem in the early days when we started nationally and in London as well was NIMBYism ... to combat this ALARM in London had only one rule – the rule was that the groups, local residents, etc. had to say "no roads anywhere in London" and then they could join ALARM' (John Stewart quoted in McNeish 2000: 188). When local groups begin to form networks of this kind, share resources, develop common critical frames, and debate strategy they are becoming more a social movement and less local campaigns.

Local mobilisation is most likely to be provoked by a new development and based upon local and immediate risks or environmental costs

Wolfgang Rüdig's encyclopaedic analysis of local opposition to nuclear power plants worldwide showed that existing sites were much less likely to attract opposition (1990: 151). Similarly, Gould, Schnaiberg and Weinberg (1996) showed that opposition was much less likely if a hazardous facility was already in place.[10] Instrumental calculations provide the main explanation of this. There are more opportunities provided by planning regulations and licensing procedures for opposing new developments. Also, when a development is already in place, opposing it may mean local job losses, making it harder to mobilise opposition and making counter-mobilisation from within the local community much more likely. Local environmental groups need to be able to claim that they speak for the community and this is much harder if they are opposed by other locals. Other schemes, such as new roads, which damage ecosystems are impossible to reverse once completed. There are also psycholog-ical reasons why existing sites are less likely to provoke opposition. It can be argued that communities learn to suppress fears of risk because the conse-quences of facing up to them are too great (Beck 1992).

In cases where the risks from an existing hazard are severe and where resi-dents do not have the option of flight, strong opposition can also emerge. This was the case at Love Canal, the campaign that was the catalyst for the anti-toxics movement (Szasz 1994; Gibbs 1982; Newman, R., 2001). Love Canal in Niagara had been used as a toxic waste dump and then later a school and housing was built over the dump. Lois Gibbs, a mother of a child at the school began to research the history of the site after her son and other children became ill. After initial denials by government officials some residents were relocated but most were deemed to be not at risk. The Love Canal Homeowner's Association was formed in 1978 by local women, who had no previous experience of campaigning, to pursue the issue. Their campaign, which included kidnapping a public official, eventually succeeded in forcing the closure of the school and the relocation of all residents. The publicity given to the campaign encouraged other local campaigns and Gibbs was contacted by many of these for advice, leading to the formation of the Citizens' Clearinghouse for Hazardous Wastes in 1981. The Government sued the company responsible and Congress enacted new legislation in 1981 creating the Superfund and a new agency to deal with the cleaning up of toxic waste sites.

Love Canal activists stressed their sense of being trapped and their shock at the disinterest of officials. Emotions evoked by our attachments to places are important in catalysing the action of local environmental groups. As Jasper says: 'The prospect of sudden and unexpected changes in one's surroundings can produce feelings of dread and anger. The former can paralyze, the latter can be the basis for mobilization' (1997: 106–7). Many case studies of local environmental protests stress the mixture of positive sense of pride or love for a place and the sense of outrage at those who threaten it.

It is perhaps surprising that objective levels of ecological risk do not correlate highly with levels of local community activism (Kousis 1999; 173). Although clearly, there must be some sense of an environmental problem it is not the scale of degradation, but how it is perceived, both in terms of injustice and in terms of the possibility of mobilising action, that explains why some communities mobilise while others facing worse hazards do not. The effects of new information about an existing site or the threat of a new hazard on mobilisation can be explained in part by what has been called relative deprivation. Relative deprivation describes a situation in which expectations are not met by reality (emphasis is on how things are perceived by those affected). When new impositions occur suddenly, grievances increase and deprivation or the sense of threat is heightened, the motivation to act is likely to be greater.[11] However, as those who have used relative deprivation theory in relation to local environmental groups are at pains to point out, this alone is not sufficient to explain why protest groups develop. They also need the resources, and commitment of activists as well as political opportunities to develop (Szasz 1994: 83; Rüdig 1990).

An example is the case of a campaign in 2000 against a proposed 110 km line of electricity pylons which passed through spectacular mountain scenery in Donegal in north-west Ireland.[12] Opposition began among a handful of middle-class activists with no campaigning experience who were mainly outsiders or 'blow-ins'. Although motivated by a suddenly imposed grievance, they were pessimistic about their chances because they regarded the Electricity Supply Board (ESB) as too powerful and local people as too fatalistic to mobilise. Effective work with local MPs and a range of political parties and the involvement of prominent personalities from the local music and literary scene (who knew two of the initiators of the group who were themselves artists), helped to create a broader movement that had more legitimacy among the local community. This allowed them to raise sufficient funds to work with experts able to talk effectively about the health risks of radiation from pylons and to question the ESB's projections regarding electricity supply. The opposition succeeded in forcing a planning inquiry, which reversed the local authority's decision to grant the ESB planning permission. This success was due to a combination of the motivating effect of a suddenly imposed grievance, success in gaining support and resources from other social groups and some (but not all) of the local political leaders and the determination of campaigners to overcome their own defeatist perceptions of the local community.

Local opposition is most likely to become confrontational when political opportunities are restricted and also when officials or opponents are seen to be acting arrogantly, or unfairly

Many of the best-known local environmental campaigns escalated into larger confrontations as local opposition groups began to realise that they were being excluded from decision-making. For instance Celine Krauss (1993) shows how women in the anti-toxics movement went though a process which began with approaching what they assumed to be a receptive government, but which, when they found government agencies and officials to be indifferent, led them to a more critical view of power as shaped by gender and class. From then on their activism was not only about the specific hazards facing their communities but was also intended to make the system more democratic. Although the opportunities for citizens to influence local policy directly vary considerably between countries, there is common ground at least in the shared sense of exclusion expressed by campaigners in countries such as the USA, where there are more legal mechanisms allowing for influence on the legal process and the UK, where there are few such opportunities. The case of the movement against the extension of Frankfurt Airport in the 1970s and early 1980s followed a pattern of gradual exhaustion of legal avenues and a growing radicalisation of local activists who were unused to protest (Burns and van der Will 1988). This also occurred in other cases such as anti-nuclear power campaigns in Gorleben, Whyl and Wackersdorf in West Germany, and at Plogoff in Brittany (Rüdig 1990) and in Britain opposition to a new road at Twyford Down (Doherty 1998) and the extension of Manchester Airport (Griggs, Howarth and Jacobs 1998).

The pressure to expand Frankfurt Airport was partly due to the fact that a third of the flights were by the US military. The expansion would destroy woods used by locals. The site was occupied but this failed to prevent trees being felled. The diversity of the protesters was important in the response of the media. As the alternative-left newspaper *die tageszeitung* put it:

> The hippy worked alongside the housewife, the worker alongside the student, the steel helmet next to the Tirolean hat, the leather jacket next to

the Lödon coat: here was the moment when a coalition was possible between the long-haired and the grey-haired.

(quoted in Burns and van der Will 1988: 191)

Public hearings intended to placate the protest were seen as a sham. A petition of 300,000 people demanding a referendum was refused by the regional government and several large demonstrations of up to 150,000 took place but failed to overturn the decision. Violent clashes between radicals and the police divided the movement and exposed the loose nature of the coalition. Nevertheless, this and other cases in the 1970s and 1980s such as Whyl, Gorleben and Wackersdorf also led to alliances in Germany between diverse radicals and less politicised locals, outraged at their lack of power to influence policy. Often it is not simply the result of decisions, but the sense that the protesters' case has not been taken seriously that activists emphasise.

Jasper's (1997) account of the role of a group called Mothers for Peace in the campaign against a proposed nuclear power plant at Diablo Canyon in California is typical of many cases:

> The mothers were hardly unsophisticated when they decided to oppose Diablo Canyon in 1973. The group had formed in 1969 in opposition to the Vietnam war and it began to study nuclear energy as the war wound down and plans for Diablo Canyon were working their way through official channels. They had activist identities, and some networks derived from that activism. Coming to the nuclear issue with a clear political ideology, the Mothers' real shock came at how they were treated by the Nuclear Regulatory Commission (NRC) at hearings. The Pacific Gas and Electricity Company, owner of Diablo canyon, barely deigned to discredit them. NRC representatives did not even begin to take them seriously; they were a bunch of housewives challenging testimony by an extensive staff of professional engineers. They and their husbands valiantly and angrily represented themselves at hearings for the first three years, but in 1976 they began to hire lawyers.
>
> (Jasper 1997: 105)

A result of such experiences is a growing disenchantment with the lack of opportunity to participate in decision-making. Grau and Egea (2001) note that the right to participate was more important than the content of decisions for local environmental campaigners in Murcia in Spain. Chris Gilham of the Twyford Down Association speaks passionately about the disillusion with formal processes of consultation after failing to prevent the destruction of Twyford Down following a 22 year campaign:

> I came from a background of concerned but respectable and restrained involvement. I spent years in formal committees of preservation groups, not achieving very much. Here is the justification, whenever it is needed, for non-violent direct action. The system allowed us to spend decades in argu-

ment, and huge sums of money, making an intellectually unshakeable case, only for the system to brush it all aside. When you hear the words 'democratic process' and 'rule of law', reply quietly with 'Twyford Down'.

(in Stewart, Bray and Must 1995)

In contrast, Fischer (2000) argues that it is possible to construct decision-making structures that allow for meaningful participation by ordinary citizens alongside experts in conflicts over environmental risk. In some cases these have been successful in achieving compromises. As he shows, 'Hard evidence demonstrates that the ordinary citizen is capable of a great deal more participation than is generally recognized or acknowledged' (2000: xi). However, this means changing the nature of decision-making to allow more space for local contextual knowledge in such a way that the superiority of abstract and generalisable analysis by risk experts is no longer taken for granted.

Local environmental groups require some technical expertise and must define their attitude to science and technology

Scientific expertise is an issue in most of the conflicts engaged in by local environmental groups. The health risks of nuclear power and toxic-waste dumps, electricity pylons or new industrial plants are all important to local environmental campaigns. Cost-benefit analyses based on traffic forecasts have to be placed alongside studies that suggest that new roads lead to increased traffic. Local politics tends to be less ideological than national politics and the prevailing assumption among local policy-makers is that environmental issues can be defined in technical terms and are best left to experts. Unless they can draw on technical expertise local environmental groups find it difficult to gain credibility. Thus, for most local environmental groups self-education is an important aspect of their campaign. Finding out about the scientific research on issues provides them with a means to assess their own concerns rationally, but it also allows them to oppose the arguments of local authorities or businesses on apparently neutral grounds. For groups who tend to be viewed as self-interested and emotional there are also strong tactical pressures to adopt the objective and apparently neutral language of science. Moreover, in doing so they are following the pattern already established by the mainstream EMOs. On the other hand, as the discussion below suggests local environmental groups are often forced to acknowledge the limits of purely technical or scientific arguments and need to draw on other arguments.

How do local environmental groups deal with the question of the status of scientific knowledge? In her study of local environmental groups in a city in northern German Eeva Berglund found a common faith in the privileged status of scientific knowledge across different kinds of environmental groups.

It is not science but its misuse that tends to be the target for local environmental groups. 'Science has features that activists are proud of in themselves' (Berglund 1998: 166) such as objectivity. Science provides a basis for proving the opposition wrong. Since it is often argued that it is the economic interests of business that create new hazards, local campaigners see science as a neutral means of challenging dominant interests (Kousis 1993).

In order to engage in an effective dialogue with opponents local environmental groups must first educate themselves and often this includes seeking specialist advice. A campaigner against a waste landfill at Lamberhurst Farm in Kent said that when the campaign started they: 'knew nothing about the waste industry. So we had to find out really what this application was ... we set out on an investigative mission to find out everything we could about waste. We read all the rules and regulations' (quoted in Rootes 1997b: 13–14). But what is seen as relevant knowledge goes beyond laboratory based expertise. The same campaigner visited other sites where the same developer had waste treatment facilities:

> Then I got in the car and I went to these places and I was knocking on people's doors, round where they live and asking them what sort of things happen to them ... we found that children in local schools ... were getting ill, stomach disorders and all the rest of it. And from that we discovered that seagulls have spread botulism, and in one particular area they were putting fly screens onto the windows of their houses because of the millions of flies in summer. There were people having to wash their cars everyday because of the dust, and all that sort of thing that we would discover.
>
> (quoted in Rootes 1997b: 14)

Evidence which may not be backed by clear scientific research but accords with personal experience can mobilise support because of the increased uncertainty about what kind of scientific information can be trusted. Moreover, when scientists make claims that contradict local knowledge they undermine their own legitimacy. In Breisach in Germany the statements of a government agrometeorologist did not reassure wine growers concerned about the potential effect of cooling towers for a nuclear power plant on their microclimate. 'His statement that potatoes would require more sun than the vines was ridiculed and frequently quoted in oral accounts to demonstrate the wine growers disbelief in the "experts"' (Rüdig 1990: 130). In similar vein, the opposition to a new road in Gloucestershire was greatly assisted by errors made by road engineers in their presentation of the plans:

> the DoT [Department of Transport] held their exhibition in the village hall. There were elegant pictures of embankments and cuttings with the usual trees and small cars to add to the effect. Some villagers were going around saying that it was mad to build a long embankment across the flood plain, because it would make our regular winter flooding worse. ... These were professional engineers and would never make such a basic mistake. Unknown to me, Sydney, who was soon to become a fellow campaigner, had already

asked one of the professionals about the flood plain. He received the immortal reply, "flood plain, what flood plain?"

(David Marcer quoted in Stewart, Bray and Must 1995: 4)

If experts deny the perceived risks and claims made by locals[13] their expertise will be viewed sceptically (Gould, Schnaiberg and Weinberg 1996: 195). McKechnie argues that when presented with complex scientific information, people 'measure trust through familiar categories' (quoted in Welsh 2000: 185). Wynne (1996) showed that sheep farmers in Cumbria responding to risk assessment by experts in radiation after the Chernobyl reactor disaster raised many questions that were outside the terms of the scientists' discourse. They wanted to know whether the scientists had integrated analysis of farming practice and the precise condition of the land into their analysis. They wanted to know about the social interests of the scientists and their record for honesty. And they also wanted to know how to compare the different levels of uncertainty expressed by the scientists. Unconvincing answers led farmers to distrust the scientists.

The case of Love Canal described above is only the best known of many examples of campaigns which began from awareness by local people that there was a hazard which was not being officially recognised or acknowledged. In such cases it is concerned locals themselves who initiate a scientific investigation and often their own research goes hand in hand with the mobilisation of the local community. As Tesh (2000) shows, the view that the conflict is necessarily one between lay person and expert is misplaced. Local activists often become experts themselves and draw heavily on the networks of counter-experts prepared to help such campaigns. This reflects the dual diffusion of professional skills and knowledge as the effects of the expansion of education provide more citizens with the intellectual resources to challenge official experts, and the effects of movement critiques on experts, creating a network of critical counter-experts able to expose the limits or errors of official or business expertise.

Activists often become aware of the difficulty of proving their case using science when the relevant expertise is often spread across many specialisms and evidence often conflicts. The process of education can show activists that there is no clear scientific answer. Moreover, risk calculations by industry or government experts often define risks in terms of predicted fatality or harm, ignoring or defining as irrational the sense of risk that troubles most campaigners, which is the threat to a whole community or a way of life. When new risks are introduced they threaten what Giddens (1984: 375) has called the 'ontological security' necessary for social life to work. By this he means a sense that life is grounded in routines that are largely predictable and controllable. Fear of the possibility of catastrophe clearly undermines this. Given the imperviousness of official calculations of risk to this less calculable perception, in order to challenge a particular decision, local environmental groups often become increasingly inclined to challenge an industry as a whole, or the nature of the decision-making process. If campaigners believe that there is a hazard and this is reinforced by scientific evidence, and by ideas disseminated by national EMOs and others, then efforts to deny this by industry and government are likely to seem economically or politically motivated.

While there are good reasons for local environmental groups to develop their own expertise, the high hopes that result from finding evidence that counters the arguments of their opponents are often dashed. Many activists in campaigns who have immersed themselves in new knowledge about the particular issue at stake in their campaign find that local politicians or MPs are far less well informed than they expect (Jasper 1997: 107). And they are then surprised to discover that despite having manifestly superior knowledge, their discussions with politicians have no effect (Berglund 1998).

In many cases, the expertise provided by consultants who are prepared to challenge the status quo, has helped campaigns.[14] A weighty report gives activists confidence but the neutral discourses in a purely technical argument tend to be to the advantage of those with power (Berglund 1998: 96). Local environmental groups strive hard to appear objective, and base their case on expertise, but usually need more than simply scientific arguments to mobilise support. Yet, when asked to say what motivated their activism, activists in Berglund's study stressed ethical concerns about nature and the rights of future generations.

Local environmental activism provides an arena in which concerns that are not accepted as central by decision-makers can be expressed collectively. Being involved with local environmental activism provides meaning insofar as it provides a means of making actions consistent with beliefs. On the other hand knowledge is not purely an antecedent of action. Groups learn to define their own cognitive interests in ways that reflect their concerns over being represented as NIMBYs and depend on the issue at stake. In general, where possible, they seek to base their arguments on both science and the non-NIMBY and universal grounds of ethics. If campaigners who were first motivated by essentially NIMBY concerns are able to link their arguments to a more universal goal they are more likely to be able to mobilise wider support locally and through alliances with politicians, sympathetic experts and environmental groups. This capacity has been linked to the chances of success for such groups (Walsh, Warland and Clayton Smith 1983) but it is also part of the educational effect of campaigning.[15]

Local environmental protest groups need to use science in order to be credible, but science alone is rarely enough, not least because science rarely provides conclusive evidence in the case of risks, and the question of what risk is acceptable is at root a social one. Relying only on science means reducing the argument to one dimension, and usually means relying on others from outside the local group who may put their own professional and social interests first. Moreover, when government or industry is met with scientifically aware local opposition they can often ignore or outgun it. This is much less the case with appeals to ethical issues, which may also be used by local environmental groups strategically. One basis for the latter is being able to show that they can speak for the local community.

The arguments of local environmental groups depend on a socially constructed conception of the interests of the local community

Local environmental groups must argue that they have the best interests of their community at heart. In doing so, they tend to operate with a taken for granted conception of what that community is. As the evidence below will show the conceptions of community appealed to can vary significantly and this affects the style of commitment and the forms of action and organisation adopted by local groups.

Local environmental groups have a tendency to see their own concerns as universal and believe that they can speak for the interests of the whole community. Underlying this attitude are assumptions about community that are often taken for granted and which shape the nature of activism. In a study of the culture of activism and style of commitment in different local environmental groups Paul Lichterman (1996) illustrates this point. Lichterman makes a distinction between communitarian bases of commitment in which individuals act out of a sense of obligation to traditional social structures and what he calls personalism, in which individuals establish their political identities through reflecting on their own biographies. Personalism is a form of individualism that reflects the focus in contemporary societies on highly individualised forms of expression. In political theory the personalist style of commitment is linked to the rational and self-critical conception of political activism advanced by radical democratic theorists such as Jürgen Habermas and Chantal Mouffe. Consensus decision-making, opposition to internal hierarchy, maximum autonomy for individuals and valuing self-expression and self-criticism, which are all features of green political practice, reflect personalist assumptions. Personalism provides a shared cultural basis for radicals to work together and, in contrast to those who see the individualism of greens and other radicals as a threat to community, Lichterman argues that it does not undermine participation. Rather personalism provides a form of participation which is appropriate for left-libertarians such as the greens and an alternative to traditional communitarian participation.

Two examples from Lichterman's study illustrate how the style of commitment and assumptions about the nature of the local community shaped the action of local environmental protest groups. The first group, Airedale Citizens for Environmental Sanity (ACES), a mainly white and middle-class group protesting against a new incinerator at a local military factory believed that activism would be seen as abnormal by others in their local area. 'Fending off expected derision was a regular part of being involved in ACES.' (Lichterman 1996: 87). Protest would be regarded as 'making a scene.' Activists had a psychological developmental model of activism in which individuals moved through 'denial' to empowerment. As with many of the activists in other studies, being involved with the campaign was seen as having given them a new identity.

A contrast is provided by HAT (Hillviewers Against Toxics) that was a mainly African American group campaigning against toxic waste from a local petrochemical plant and which had a more communitarian style of commitment.

Although, the group was open to all members of the community, most whites that joined were outside experts or representatives of other groups. HAT had grown out of an earlier campaign connected with civil rights activism and it was able to draw on a strong sense of assumed common interest. It was taken for granted that Church leaders would allow HAT members to address congregations and although not all members were Christians, HAT meetings would begin with a prayer. This was in contrast to ACES, where there was an assumption that no individual would press their particular beliefs. For activists in HAT 'Getting involved meant acting more as a member of a wronged community than as an individual with an individual cost-benefit ledger' (Lichterman 1996: 109). And HAT activists did not feel that they had adopted new identities as a result of their activism.

The communitarian character of HAT meant that there was less perceived need to nurture and support individual activists. Political activism was not perceived as unusual or risky. Also, in contrast with personalist styles of organisation where leadership tended to be more diffused, leadership and power were concentrated in formal leaders, with the drawback that when these leaders failed to deliver there was no cultural basis for holding them accountable. To criticise them would have been a betrayal, and yet activists were dissatisfied at times with the nature of decision-making or the mistakes made by leaders. The communitarian culture also made it difficult for HAT to deal with diversity. When members of the community did not support them it was assumed that they were acting out of selfishness.

As Lichterman shows both styles of commitment have strengths and each is more appropriate for different kinds of community. The core constituency of greens is from the young and new middle class, both geographically relatively mobile and unlikely to be integrated into stable community structures. Personalism, reflected in green culture, provides a basis for greens to create a community, which is often both local and transnational in its symbols and content. However, this tends to blind greens to the ways that other forms of community make action possible for different groups. The degree of such differences poses an obstacle to the aim of building multicultural or cross-class alliances. Such cultural obstacles could undermine many of the assumptions made by greens about the potential for linking green goals with those of local environmental groups. For instance, greens, with their strong emphasis on individual autonomy, tend to emphasise individual transformation, which can include critical examination of their own role in perpetuating hierarchy. However, if community traditions provide a source of defence against oppression, criticising aspects of those traditions from the point of view of the ideal green political agenda can be seen as threatening. On the other hand, 'members of groups such as HAT would have little basis for taking seriously someone who acts as an individual political agent' (Lichterman 1996: 143).

While other evidence (Schlosberg 1999) suggests that multicultural alliances exist and can work in rhizomatic form, Lichterman's research has important consequences in that it suggests that greens will have to be aware of the benefits of not seeking to universalise their culture and forms of organisation if they want to be able to work with a wider range of social groups. In truth many greens are aware of this, but not always aware of the strengths and emotions underlying attachment to different, even hierarchical, styles of commitment.

The remainder of this chapter will concentrate on two issues: first, the evidence that activists in local protest groups are often transformed by their experience of even limited political conflict. Second, an assessment of the political impact of local environmental groups and their role within the larger environmental movement.

The experience of transformation

Most often it is the sense of a changed political perspective that is most striking in accounts from those involved in local environmental protests. Ethnographic studies have shown how profound the process of personal change experienced by campaigners can be even in what appear to be relatively conventional local environmental groups. There are many examples from interviews[16] in which participants speak of their sense of transformation and political awakening. The following extracts are from an interview by Welsh in 1981 with the managing director of a medium-sized firm involved in the protests against a proposed nuclear power plant at Druridge Bay in Northumberland:

> I'm absolutely amazed and shattered by just how difficult it is to fight the establishment. I mean I have no doubts now ... I know that they classed us – say as trouble maker, communist, anti-establishment, subversive. And I know, like all people that are branded that way ... that you are not, you are doing something that you believe in very much.
>
> That's another thing that you lose confidence in once you start reading about inquiries. It really shatters you when you think about democracy. You become, I think, anti-establishment, they force you that way.
>
> (Welsh 2000: 191)

John Stewart of ALARM UK the anti-roads federation saw the process of transformation as one with several complementary causes:

> The majority of our groups here, will have started out as fairly NIMBY in their views ... Oh my God there's a road coming my way, it's certainly going to disturb my house. But what does happen is that as groups, a) meet other groups; b) become aware of the issues; and c) meet the DoT (Department of Transport) who are a wonderful recruit to our cause, they become radicalised.
>
> (interview with author, 5 October 1995)

When local residents become angry enough to take direct action spontaneously they pose significant problems for the authorities. Two examples illustrate this point. The occupation of the site of the proposed nuclear power plant an Whyl in south-west Germany in 1975 was the protest that kick-started the anti-nuclear movement in Germany. Local farmers and residents had been worried about the possible effects of the plant and the ancillary industrial development on their vines and on the local microclimate. The *Land* (regional) Government had paid little attention to such concerns. The Land Economic Minister was also on the governing Board of the electricity company developing the plant and the Prime Minister was its Chairman. The developer and the licensing authority appeared to be the same people (Rüdig 1990: 131). When construction work began without warning while the courts where still considering legal challenges to the plant, the local opponents of the schemes were furious:

> Any trust in the fairness of government and the law had been severely undermined ... The sudden start of construction work forced the citizens' groups into action. Representatives of the individual groups met on February 17, and called for a rally and a press conference to be held near the site in the morning of the following day. Apart from the press, about 300 people from neighbouring villages, mainly women, appeared at the site. Work on clearing away trees continued. While the rally was still underway, it was a group of women which first entered the site and tried to stop the machines and building workers. Building equipment, and the site itself, was occupied in the process. Work stopped.
>
> (Rüdig 1990: 133)

Although accounts stressed the role of spontaneous anger, the occupation was also made more likely because the protesters had worked previously with protesters against a lead factory at Marckolsheim across the Rhine, in Alsace. Not only were land occupations by farmers more common in France, but also, the successful campaign against the lead factory had included a planned site occupation led by women because it had been thought that women would be more likely to defuse the anger of construction workers (Rüdig 1990: 133). It is clear that at Whyl there was no clear plan for a permanent site occupation and the local residents drew on the experience and commitment to stay of allies from Alsace and students from Freiburg. The violence of the police eviction and their obvious concentration on arresting 'outsiders' strengthened bonds between locals and outsiders and led to further site occupations until the scheme was eventually dropped.

A similar event, although on a smaller scale, played an important part in strengthening alliances between direct action radicals and local campaigners opposed to the building of the M11 Link Road in East London. The local campaign against the road was more than ten years old and had included disruption of public inquiries. But with all institutional avenues of protest exhausted most local campaigners were resigned to the road being built when the direct action campaigners arrived in September 1993:

Direct action remained a fairly specialist activity for the first couple of months and while some locals may have sympathised with what was happening, few of them felt it to be 'their' campaign. But on November 6, this all changed. A 250-year-old sweet chestnut tree on George Green, Wanstead, had been ominously fenced off. This tree lived on common land, land which had survived the Enclosures Act, land which local residents had been led to believe would be preserved intact, since the road would be in a 'tunnel'. Having gathered for a 'tree dressing' ceremony, schoolchildren, pensioners, respectable (and less respectable) Wanstead locals now found their way barred. What happened next, neither security guards, the police, local residents or even the organisers had expected. Together, everyone pushed down the fences, reclaimed the land and rescued the tree. With 70 year old women now cheerfully committing 'criminal damage', any division between activist and resident dissolved.

(McLeish in Stewart, Bray and Must 1995: 20)

Another important example of direct action, which had a powerful impact on the development of environmental justice activism, took place in Warren County, Northern Carolina in 1982. This rural and poor area had a large African American and Native American population and it was mainly local people from these groups who mobilised against a proposed PCB (polychlorinated biphenyls) incinerator. Although the campaign failed to prevent the incinerator being built the strength of opposition, which included an attempt to block the road to the site in which over 500 were arrested, inspired other campaigns particularly among people of colour. Each of these campaigns was much more powerful because they could genuinely claim to be based on local protest and undermined claims that radical outsiders were behind protests. As Rüdig comments in relation to Whyl: 'West German television audiences were quite used to violent clashes between police and students, but harsh treatment being meted out to what had been law-abiding, ordinary citizens of mainly conservative inclinations was something new' (1990: 135). Second, local residents took action because of their sense of outrage at being ignored or misled by the authorities and this de-legitimation of the authorities motivated them to take confrontational action. And, as one interviewee commented on the M11 case: 'You've got people doing remarkable things that before they'd never have done in a month of Sundays. You know, women of 70 pushing walls down and Wanstead is a very Tory area' (interview with a volunteer at ALARM UK Office, 5 October 1995).

It is notable that in both Whyl and Wanstead, as also in most of the anti-toxics campaigns in the USA and local environmental groups in Australia (Doyle 2000: 33), women were in the majority. The simplest explanation for this was situational. There were more women who could be mobilised during the day to join such protests. Szasz notes (1994: 152) that the main opponents that anti-toxic campaigners met were men, and that this led the many women activists to a wider confrontation with patriarchal attitudes. However, it is also possible that local environmental protest has deeper gendered characteristics. Groups based on pre-existing women's networks in neighbourhoods and Church groups have often taken the initiative in local environmental campaigns, particularly in anti-toxics and environmental justice campaigns in the USA. It is not so much that the issues raised in local protest are specifically gendered, or that they occur in a form which divides women and men, but that they raise issues which expose the tensions between the instrumental logic of business and policy-making – fields in which men are dominant, and areas such as family health, community and neighbourhood, which connect with women's disproportionately high responsibility as carers.[17]

Examples of NVDA in local environmental campaigns are fairly rare.[18] But this does not mean that local environmental groups are less brave than direct action groups such as Earth First! To take political action when such action is seen as abnormal and associated with counter-cultural social groups requires considerable courage. If campaigners perceive the local community as apolitical, taking action publicly can endanger relationships with friends and neighbours. Also, actions undertaken by counter-cultural direct action groups would be much more risky for local activists. In Britain, NVDA radicals have sometimes used aliases making it difficult to trace them for court cases. Local activists with permanent homes, children and jobs cannot use such methods. On the other hand they also have more to lose. Whereas the radicals often have no assets and cannot be made to pay significant compensation or fines, local campaigners could be made to pay (and there were attempts by the developers of Twyford Down to sue local campaigners for compensation, although these were ultimately unsuccessful). When local campaigners opposing Birmingham Northern Relief Road (Cathles 2000) and defending Oxleas Wood in south-east London used their houses as security in order to fund legal challenges in the High Court they were arguably taking a greater risk than campaigners who faced arrest and trial.

Since so many local environmental campaigns are short-lived and many are not radicalised it is difficult to assess the effect of such transformation. One consequence of the spread of experience of protest is that it is no longer so restricted to those groups previously seen as 'typical' protesters, particularly the young and those in the organised left. The involvement of a greater diversity of social groups in protest is part of what Meyer and Tarrow (1998) call the transition to a social movement society, in which protest is no longer unconventional. While it is a feature of this normalisation of protest that it tends to be increasingly routinised and thus institutionalised, as Tarrow and Meyer note (1998: 24), it does not follow that this means that it necessarily leading to inclusion and co-option. The evidence above suggests that the growth of local environmental protest may often be associated with a transformation to a more critical view of authority.

The political impact of local environmental campaigns

Although most analyses of local environmental campaigns are case studies of one or a few specific campaigns, such studies can be related to more general evidence from the literature on political participation and social movements. A key issue is the extent to which structural features of communities such as the level of affluence, the opportunities provided by political institutions and the physical environment, affect mobilisation. Alternatively, more action-oriented approaches stress the skills and strategies of campaigners, the availability of allies and the effects of contingent and exogenous events.

It was noted above that the nature of the issue at stake shapes the potential of environmental groups to mobilise support. For instance, it is much harder to mobilise opposition to an existing plant than to one that is merely planned. It is also harder to campaign for a goal which might threaten local employment. When the object of opposition provides a major source of employment locally opposition is even harder.

Local authorities, regulatory agencies and the courts are all important targets, where they play a significant role in planning approval, licensing or approving the implementation of regulations. Accounts from successful local anti-roads groups in Britain, have in common pressure on a local authority that converted them from opponents or neutrals into allies of the campaign (Stewart, Bray and Must 1995). In Rootes' study (1997b) the successful campaign against the Lamberhurst Farm landfill site won the support of the local authorities and the local MP. Planning guidelines and the technical advice received by the local authority raised some doubts about the need for the scheme but they did not rule it out. It was the campaign by the local environmental groups, which increased the political costs of the scheme, which determined the outcome. The local council was not under the control of any one party and so councillors were anxious to be seen to be responsive to local concerns. Similarly, the local MP had just had the boundaries of his constituency redrawn to include the site and was also anxious to win support in the area. Thus the configuration of power presented objectors with opportunities to put pressure on elected representatives. The political situation provided greater opportunities in this case than it did for the opposition to the second runway at Manchester Airport. Anti-runway campaigners were helped by the financial support provided initially by Cheshire County Council, but after some concessions were made by the developers the County Council changed sides. The developers had the backing of local Conservative and Labour Party MPs, and all the local authorities in Greater Manchester (Griggs, Howarth and Jacobs 1998: 362) and the runway was built despite an effective case for alternatives at the public inquiry and a site occupation by direct action protesters.

While local political opportunities can be contingent, dependent upon the local configuration of power, they can also be structural, insofar as they are shaped by the nature of institutions. For instance, it is notable that in Germany and the USA, campaigners are able to make much greater use of the courts than in Britain. Legal action by protesters at Wackersdorf in Germany eventually succeeded in delaying construction of the proposed nuclear reprocessing plant until the electricity company re-evaluated the economic benefits of the scheme and it was dropped.[19]

Resource mobilisation theory emphasises the importance of influential or well resourced allies in achieving successful outcomes, and the most obvious allies for local groups are national EMOs but the relationship between EMOs and local environmental campaigns is complex. EMOs can provide important knowledge and expertise for local groups. FOE played an important role in strengthening the campaign against the Birmingham Northern Relief Road by putting the disparate local campaigns in touch with one another. An employee seconded from FoE also played an important part in establishing ALARM. In Amsterdam the City's environment centre serves as a resource in which activists from local groups and national EMOs meet and work on campaigns together. In the campaign against the Ijburg housing project, local groups were helped by national EMOs and by an employee of the Milieucentrum who had good contacts with national and local groups (Hoad 1999). And in the USA, despite the origins of the environmental justice movement in criticism of the lack of attention to social issues of the 'Big Ten', there have been many examples of alliances on particular campaigns between major EMOs and environmental justice groups (Schlosberg 1999: 134).

However, despite these examples, EMOs rarely play a major role in local environmental campaigns. They do not have the resources to be able to support every local campaign and they give priority to a limited range of themes at any one time. The institutionalisation of EMOs has therefore made them less responsive and flexible to grassroots environmentalism. And, while local branches of EMOs might fill this gap, they are often also small and find that their activities are taken up with fund-raising and campaigning on issues defined by their national office. Thus as well as being initiated autonomously, local environmental protests tend to campaign autonomously, or develop their own networks of support between related local campaigns.

The social structural characteristics of communities seem to matter less than might be expected in assessing the potential for mobilisation. In a study of two campaigns against waste incinerators in Pennsylvania Walsh, Warland and Clayton Smith (1993) showed that there were no significant social differences between the successful campaign and the one that failed. The prior experience of protest in the area was crucial for the successful campaign, and related to this, good organisation and effective pressure on local authorities. It is hardly surprising that many studies find that good organisation and tactics have contributed to successful campaigns. Like national campaigns, local campaigns are more likely to succeed when they are able to mobilise large petitions, hold successful public meetings, make alliances with important local actors and influential actors outside the locality, and make imaginative use of the media. Baggott's (1998) analysis of the 17 years of campaigning by the Drudrige Bay Campaign (DBC) provides evidence of all of these. This campaign succeeded in delaying the decision on construction of a nuclear power plant and in preventing sand extraction from a local beach. During its campaign the DBC had to reconstruct its organisation after it was beset by internal tensions and in doing so constructed a new coalition of groups with ties with a wider range of groups. There was also greater emphasis on activities that strengthened community feeling, which helped sustain support over the longer term. The campaign successfully exploited the 'David versus Goliath' image in the local media and took advantage of opportunities at national level such as those provided by the privatisation of electricity to reinforce the strength of the campaign through lobbying and local action.

Such cases raise the question of what kind of organisation is good organisation? Can it be defined technically, in terms of efficient use of resources, as it is in resource mobilisation theory, or does it also depend on strong cultural bonds between those involved that sustain the extra and perhaps irrational costs of activism? For instance Jasper (1997) argues that the continuation of protest against the nuclear plant at Diablo Canyon in California long after the plant was built was due to the attachment to the issue and solidarity developed by a strong campaign. Given the diversity of cultures of commitment involved in local environmental campaigning it seems unlikely that there is any clear blueprint for successful organisation. A central factor in explaining strong campaigns appears to be accumulated political experience.

Prior experience of campaigning was important in many of the key examples of successful campaigning. At Whyl in Germany, the local opponents drew on earlier opposition to a nuclear plant in neighbouring Breisach and successful resistance to a lead factory by allies in Alsace. The anti-toxics campaigners in HAT studied by Lichterman had previously worked successfully to gain compensation for the effects of a new freeway in the 1970s. In Berglund's (1998) study relatively neophyte activists in anti-motorway groups were initially schooled in how to protest by experienced environmental activists. In the UK, ALARM began as a group to co-ordinate opposition to plans for major new roads throughout London in the late 1980s. By organising London-wide events, and by networking between groups they were able to make sure that groups recognised the importance of transcending NIMBY perspectives. This meant that they avoided being played off against each other and tackled the assumptions about traffic as a whole. After two years of high profile campaigning the proposals were dropped in 1990. The formation and strategy of ALARM had benefited from the experience of anti-roads campaigns in London in the 1970s. Thus, having succeeded in London, its activists were in a good position to help the myriad, unconnected local groups outside the capital when the British government announced a major national road-building programme in 1989. In essence, ALARM acted as a repository of knowledge for new campaigners so that they could avoid the mistakes of the past. The essence of ALARM's strategy was that campaigners should use every means possible to create pressure on local authorities and MPs prior to the calling of a public inquiry. This was based on the accumulated experience that suggested that public inquiries were in general so expensive, and had such a limited remit, that they provided objectors with very little chance of success (McNeish 2000; Stewart, Bray and Must 1995).

The evidence in these case studies and others (Rootes 1997) stressing previous experience of political protest or campaigning is consistent with more general findings in surveys of political participation. The major study of political participation in Britain (Parry, Moyser and Day 1992) showed that the strongest predictor of political participation was organisational membership. This factor was more important than the level of education, affluence or other resources of those surveyed.

While it is clear that local environmental campaigns have been able to build capacity through accumulated experience and rhizomatic networks, they often fail to achieve their primary aim and might thus appear largely futile.

Are local environmental protests ultimately weak?

This pessimistic view has been put forward by commentators sceptical about the power of local environmental groups. Gould, Schnaiberg and Weinberg (1996) argue that local environmental groups usually fail to cope with the imbalance of power between local environmental groups and their opponents. Since the structures of political power and economic organisation function to support the 'treadmill of production,' it is relatively easy for companies to argue that regulations that threaten their competitiveness should be set aside. Profitability and the defence of jobs matter more to local and national political authorities than do the quality of life, or biodiversity. Moreover, most local environmental groups are comprised of citizen-workers, who want to both protect local jobs and prosperity and to preserve their own quality of life through maintaining their local environment. Since many companies can pressure political authorities and threaten to relocate their business, they have the advantages in such conflicts.

Chris Rootes (1999) also argues that local actions are shaped and their outcomes usually determined by non-local and often transnational forces. Looking at cases of local environmental campaigns in Kent, he shows that it was the action by more powerful actors than environmental campaigners that proved decisive. In one case it was the cash flow forecasts of the developer which led it to reconsider a leisure park development, in another it was only when agricultural land owned by a major insurance company was affected by pollution and the company threatened legal action that the electricity generator decided to close a power station. In the third case it was the intervention of the EU and links to the BSE crisis that pushed an animal rendering plant to national status.

Local environmental campaigns are often reliant on the energy and skills of a handful of individuals with few resources and who can expect little help from national EMOs. This provides local groups with incentives to work within the system, but that usually precludes the kind of radical action which will tackle the root of the problem. 'This suggests that no truly radical environmental action … can remain purely local; in order to be effective, it is compelled to raise itself to the national, or even the transnational level' (Rootes 1999: 299). NGOs such as Greenpeace that operate at the transnational level, have considerable skills and legitimacy and are able to put national governments under pressure. Their success is in part dependent on their elitism. The less well off and less educated have neither the resources nor the means to challenge larger structures. He is therefore pessimistic about the prospects for environmentalism from below.

The realism of Rootes and Gould *et al.* is difficult to gainsay, but it is not the whole story. The cases that Rootes examined were in Britain, which is particularly centralised. As was noted above, in less centralised states the courts and regional and state authorities may have greater power. Moreover, the anti-waste campaigns lacked the network ties that ALARM UK provided for local anti-roads groups, whose impact on changing government policy on roads and heightening public debate about car culture was clear (Doherty 1998). Network ties between local environmental groups greatly increase their capacity to resist the imposition of unwelcome decisions, as the record of the US environmental justice movement shows. Moreover, it is also possible to view the relationship between local environmentalism and professional environmental movement organisations from another angle, in which the legitimacy of professional campaigners is strengthened by the upsurge in grassroots environmentalism among 'ordinary' people. This certainly applied in the case of direct action campaigns in Britain in the 1990s. Even when the direct involvement of green groups in local conflicts is minimal, the ideas associated with environmentalism are taken up by local environmental groups, who as we have seen, tend to situate their case in relation to a broader understanding of environmentalism.

Substantive impacts depend on political battles in which local groups demonstrably alter the decisions taken by authorities or companies by creating sufficient controversy and counter-pressure. Of course, other beneficial environmental outcomes may be achieved without confrontation. For instance, where local environmental groups become involved in stable partnerships with decision-makers. However, recognition of the power of opponents of environmental goals suggests, as Gould *et al.* show, that such outcomes will be marginal. If there are threats to business interests, confrontation is likely and the evidence surveyed above from a variety of countries suggests that open public opposition is the best strategy for environmental groups. Where they devote their energies to working within institutions they are unlikely to be able to balance the institutional power of their opponents with political costs.

The second contribution of local environmental campaigns is in their educative and sensitising role. As we have seen, local environmental campaigns help to build the capacity of the environmental movement as a whole. Individuals are often transformed by their experience of campaigning and raise issues in ways that were either not considered by other environmental groups, such as environmental racism, or change public perceptions about risks, encouraging further challenges to authority and scepticism about the interests of government and business.

Globalisation of the economy has led to new transnational agreements that reduce the ability of national governments to impose environmental regulations on industry or to protect national industries against competition. This apparent weakness of national governments calls into question the wisdom of relying on strategies that focus on national government. Much of the effort of EMOs is designed to secure changes in environmental policy at the national level, but these are unlikely to protect local communities facing environmental hazards in the short-term. Schlosberg (1999) argues that the network form of organisation developed by the environmental justice movement and others makes it more difficult for companies seeking to exploit the weakest local community in order to overcome opposition. This strategy can sometimes be more appropriate as a means of countering the power of transnational companies than those of the EMOs centred on national government. When local environmental groups are themselves able to keep global issues of political economy to the fore, they are better able to develop strategies capable of challenging their opponents.

Conclusion

Although there are usually cultural differences between local environmental groups and greens there are also points of connection. Few local groups view their campaign in purely local terms. As the frames that they use become broader they challenge many of the principles on which policy is made as well as the undemocratic forms of the policy process. They also demand recognition for their community, which is based upon a socially constructed idea of what that community is.

There are important differences between types of local environmental struggle. For some issues of inequality are paramount, for others lack of democracy. It is impossible to tell from the evidence available what proportion of groups undergo some form of radicalisation and whether it is only those most active who are affected.[20] Nor is it clear how long this process of radicalisation lasts. Yet, we do know that there have been processes of radicalisation in which arguments linking environmentalism and democracy have developed, for instance in the citizens' initiatives and anti-nuclear struggles in Germany, in some of the local anti-roads groups in the UK and arguments focusing on inequality which have been particularly strong in the USA.

There are reasons for optimism for greens in the linkage of local environmentalism, with egalitarianism and arguments for greater democratic rights, but the greens cannot simply appropriate these diverse forms of environmentalism as part of the green movement. There is much evidence of a cultural gulf between greens and grassroots environmentalists, best evidenced by Lichterman's studies, discussed above. Yet, the gulf between greens and grassroots environmentalists is not necessarily a weakness because it helps to reduce the isolation of greens, and increases their legitimacy.

As Castells says (1997), one of the ways in which the environmental movement challenges current forms of global capitalism that treat all localities in terms of their market value is to assert the socio-cultural importance of local space. Radical greens are too few in number to make up many of the numbers in the local communities that they want to defend against economic forces, and grassroots struggles based upon the defence of specific local environments are carried out mainly by less radical groups. Yet, when radical greens fail to find much support among local populations, as in the opposition to a new road and tunnel linking France and Spain in the Vallée de l''Aspe (Hayes 2000), there is no real chance of success. Thus, whatever their weaknesses from a radical green point of view greens need effective alliances with non-green local environmental campaigns.

8 Conclusion

The future of the green movement

In Chapter One of this book it was argued that it was possible to speak of a green movement as a transnational, though western, network of radical environmentalists linked by a collective identity. The green movement is a social movement within the more diffuse and broader environmental movement. The existence of this social movement can be discerned by the extent to which it meets four criteria of a collective identity, network ties between groups and individuals, participation in protest; and challenging existing forms of power, both cultural and structural. Green action is, like all social movement action, ideologically structured. The green ideology that was examined in Chapter Three was argued to be a broad framework based upon interdependent green commitments to ecological rationality, egalitarianism, and participatory and decentralised democracy. This ideology is indentifiable in its most elaborated form in the programmes and statements of green parties, but it can also be found in direct action networks and the more radical of the green environmental movement organisations. The sources of this ideology, as was argued in Chapter Two, seem best explained by the distinctive experiences shared by the social groups which make up the core of the activists in green movements. The importance of experience and its effect on the development of a diverse green movement, was defined through the idea of the collective learning of a political generation, embracing the new knowledge of the ecological crisis, the anti-authoritarian, global and radical democratic ideas of the New Left and the experience of greens in other social movements and in welfare-oriented professions.

Whereas the first three chapters concentrated on a model based on common ground between various types of greens, the four chapters that followed examined different types of green group. These chapters focused on the nature of green praxis and on the dilemmas specific to different types of green group. In the green parties there has been a shift towards accepting participation in government, but no resolution of the problem of how to achieve radical change from a minority position within the political system. There has been some deradicalisation in green parties but not such a major ideological change that they can be seen as having broken with green ideology. In other respects, such as ties with other parts of the movement, shared collective identity and participation in protest, most green parties remain part of the green movement, notwithstanding the specific constraints that they face as parties operating within the electoral sphere.

In the mass environmental movement organisations, institutionalisation has expanded resources but not always produced the expected deradicalisation of EMOs' values, or a decline in protest. On both these questions, however, the differences between countries were particularly marked. Some groups such as FOE in Australia clearly do fit the ideal type characteristics of the green movement, others such as FOE in Italy and Greenpeace are further from the model. Nevertheless, there remain important differences between groups such as FOE and Greenpeace and non-radical conservation groups such as WWF, which clearly lie outside the green movement.

The direct action networks in Britain, the USA and Australia examined in Chapter Six best approximate the ideal type definition of the green movement outlined earlier. Yet, even in such groups there have been conflicts over the status of social justice versus wilderness as in the USA in the 1980s and over violence, which reflect the need to negotiate a collective identity.

Local environmental groups generally do not fit within the ideal type definition of the green movement: they often do not see themselves as part of the green movement they do not usually have ties to other parts of the movement, (although greens do sometimes participate within these groups), and they do not usually share a green ideology or the culture that is part of the green identity. But, it was argued that it is still useful to analyse them using social movement concepts because this reveals features of their praxis that interest group analysis tends to miss or at least downplay. Examples of this include the effects of distrust and disaffection from decision-makers in encouraging opposition; the common patterns of radicalisation experienced by activists and the cultural questions raised by the need to engage with science and to speak for the interests of their community.

Table 8. 1 below captures two dimensions of the differences between these four types of group. Direct action groups and green parties are distinct from EMOs and most grassroots environmental groups in the breadth of their campaigning agendas. Although we saw that there are processes of radicalisation that affect many local environmental campaigns and that EMOs also retain a commitment to green ideological goals, the public campaigning of these two types of group tends to focus on specific issues. Direct action groups and grassroots local environmental groups are distinct from most green parties and most EMOs in retaining autonomy from the state. The institutionalisation of the latter has meant more participation in state institutions. For direct action groups retaining autonomy from the state is an ideological position, for local environmental groups the position varies. While most are involved in consultations with local officials, few are sufficiently long-lasting to be offered any formal status in local authorities, and as we saw of the anti-toxics and environmental justice movements, many retain their autonomy for strategic reasons.

Table 8.1 Ideology and the state in the four types of green movement

	Work partly within state	*Autonomy from state*
Ideological breadth	Green Parties	Direct Action Groups
Focus on specific issues	(most) EMOs	Local environmental groups Greenpeace and some other EMOs.

There are major differences between different types of groups defined here as green. They differ also in forms of organisation, in types of protest action undertaken, and particularly in strategy. No one can claim therefore to speak for the movement, and *no particular group or organisation 'owns' or leads the movement*. Yet, there are also commonalities of ideas, practices, network ties and experiences that mean that it is still possible to speak of a green movement. Others, viewing the evidence of what these movements are like, may prefer to still speak of green movements in the plural. Activists are likely to identify more with their specific type of green movement than with the green movement as defined here. Yet, these two identities are not incompatible and the use of green when articulated by activists as 'what we believe' refers to something more specific than the environment or the environmental movement.

A recurring argument in this book has been that the institutionalisation of part of the green movement, particularly the EMOs and green parties, does not disqualify them from being part of the green movement as a social movement, nor does it lead to wholesale de-radicalisation. While it could be argued that the EMOs and green parties have been replaced by the direct action groups as the social movement part of the environmental movement, this would mean underplaying the network ties, joint protests and common identity, and ideological framework that cut across all three types of group and which mark them as collectively distinct from the non-radical parts of the environmental movement. These boundaries are admittedly difficult to draw, however, and they vary cross-nationally and over time. Moreover, it is possible that the differentiation of the green movement will increase divisions between these groups to the extent that they no longer share common features. Thus, this movement is not predestined to remain a social movement.

The future of the green movement

The most relevant questions in assessing the likely future of the green movement seem to be the following:

- The increase in transnational activism, particularly linking green movements in the north with radical environmentalism in the south.
- The changing environmental debate and new opportunities for the greens
- Strategic divisions within the green movement

The increase in transnational activism

There are good arguments that it might be better to analyse social movements in ways that are less constrained by national boundaries (Sheller 2000; Urry 2000) because so much of social life is now about mobility of people, ideas and money across national boundaries. The green movement is a clear case in point, since greens began with concern about a global ecological crisis, global social inequalities and the power of the north in relation to the rest of the world. But to what extent can we speak of a global green movement in the sense of a social movement as defined in this book?

As defined in this book, the green movement is a phenomenon of the industrialised countries of western Europe, North America and Australasia. While there are also important environmental movements outside these countries, the green movement is a type of environmental movement specific to the more industrialised countries. Moreover, it is not simply industrialisation that leads to green movements. Although Japan has an important history of local environmental activism (Broadbent, 1998; Er Peng 1999), particularly targeted at pollution, more radical green movements, arguing for a major restructuring of society, and prepared to openly challenge the authority of political elites have only begun to emerge in the last decade and still have very little public support. For a more radical green sphere to emerge strongly in Japan would require the development of a stronger anti-authoritarian political culture.

In eastern Europe, environmental movements were one of the few forms of independent political activism tolerated by the authorities before the collapse of socialism. Environmental issues became a means for opposing the regime, as in the case of the Nagyramos dam in Hungary. But since 1989, despite continuing severe environmental problems, environmental activism has mostly declined, as the focus has been on catching up with the west economically, dealing with major structural changes as a consequence of transition to capitalism and questions of national identity.

Another cultural factor may explain the weakness of green radicalism in both eastern Europe and Japan. If the green movement emerged partly as a consequence of changes in the western Left, as argued in Chapter Two, the absence of comparable forms of left networks and traditions in Japan and eastern Europe may also explain the weakness of green movements.

The argument that the green movement is a global movement is based on two trends. First, the growth of institutions of international environmental governance in the 1990s has provided opportunities for NGOs from both north and south to become more involved in consultation on treaty negotiations. (Kellow 2000; Rootes 1999; Arts 1998). Second, there is an increasing network of ties between northern and southern environmental groups, including radical groups. There have been increasing efforts to construct alliances between western and non-western EMOs, but at the international level the distinction between non-green EMOs such as WWF and the International Union for the Conservation of Nature (now the World Conservation Union) and green EMOs such as FOE also applies. For instance, WWF, the IUCN and the 'Big

Six' US EMOs including the NRDC, the Sierra Club and the Audubon Society were increasingly co-opted within the UNCED process while FOE, Greenpeace and other EMOs avoided co-option, remained politically critical and sought to establish ties with development-oriented groups in the south. (Finger 1994).

The resources of western EMOs can make them powerful actors internationally. WWF invests millions in conservation programmes in eastern Europe and the south and is often seen as a conduit for western influence, through its close relationship with western governments. At times this leads to actions that would be questionable from the green ideological standpoint.[1] For instance, the World Wide Fund for Nature and the International Union for Nature Conservation were among the powerful western EMOs who supported the 'shoot to kill policy' used against poachers in Kenya's wildlife parks that led to the deaths of over 100 people between 1989 and 1991. The creation of Kenyan wildlife parks had involved the displacement of tribal people without compensation and a tough anti-human policy was in the interests of the political and economic elite who benefited most from Kenya's tourist trade. A former deputy director of WWF International, Henner Ehringhaus, described the fate of those who questioned this narrow environmentalism:

> The old school is interested in animals, plants and protected areas. They are interested in people and development as an instrument to protect the forests. The new school is interested in sustainable development, meeting human needs, helping people improve their agriculture, and working with local communities in a more integrated approach to nature conservation. What we have seen is those of the new school fired, and those of the old school promoted.
>
> (*The Guardian*, 18 November 1995, quoted in Purdue 2000: 88)

It is not only the non-green EMOs that are powerful actors in international governance. Kellow (2000) analyses the role of Greenpeace International, in relation to bodies such as UNEP. The latter has a budget of a mere $20m, whereas in 1997 Greenpeace had a budget of around $140m. Bodies such as UNEP benefit from the mobilising capacity and scientific expertise of groups such as Greenpeace. Moreover, Greenpeace's resources are not only financial, but also political and technical. Greenpeace can use the media effectively and it has a mass support in the most powerful western countries that can put pressure on national governments (Rootes 1999). It is probably this popular pressure that explains the coincidence of the positions advanced by Greenpeace with those of the German government in the Kyoto climate change negotiations. Yet, as Kellow (2000) suggests, when combined with the concentration of Greenpeace membership in a few countries, this qualifies the claim of Greenpeace to be global.

Links between environmental activists from the north and the south are increasing. Keck and Sikkink (1998) chart the growth of transnational networking of NGOs, particularly linking groups from the industrialised countries with those in the south. Cheaper air travel, the internet and satellite technology have made transnational networking easier. In the past such links were largely limited to NGOs with the resources to travel, but since the internet reduced the cost of communication, contact and the co-ordination of protest action between grassroots activists has increased. This has been evident in particular in multi-country protests such as those against the G8 Summits in Birmingham (1998) and Cologne (1999) and the WTO meeting in Seattle (1999). In the Summer of 1999 a transnational network of grassroots activists linked under the banner of People's Global Action, organised an Inter-Continental Caravan of activists from the south, which toured western Europe in an effort to allow activists to put their case directly to western publics and politicians. Nevertheless accounts of the caravan, also reveal some of the obstacles to the development of a global environmental movement. There were major differences of outlook and styles of organisation between some of the activists from India and their hosts in Britain and Germany and clashes over gender and hierarchical organisation. Tensions over environment and the nature of development also arose between the more radical EMOs from the north and development-oriented environmental groups from the south during the UNCED process (Finger 1994: 210).

Guha has argued that most Third World environmental groups do not share a common identity with western environmentalists:

> Where Northern environmentalism has highlighted the significance of value change (the shift to 'postmaterialism'), Southern movements seem to be more strongly rooted in material conflicts, with the claims of economic justice – that is, the rights to natural resources of poorer communities – being an integral part of green movements.
>
> (Guha 2000: 122)

But, since the ideological framework shared by western greens gives prominence to material conflicts they cannot be accurately characterised as postmaterialists. Instead their materialism is often on behalf of other groups. If alliances are to increase it is likely to be on the basis of such materialist struggles. For instance, Van der Heijden (1999) notes the common rejection of development-oriented discourses such as ecological modernisation and sustainable development by counter-cultural radicals in the West and Third World and Taylor (1995: 320) makes a similar argument in stressing that threats to human livelihood provide the most important reason for the emergence of ecological resistance movements and link grassroots activists in the Third World and local environmental groups, often from poorer communities, in richer countries (1995: 336). Direct action groups in the west have targeted businesses involved in rainforest destruction in the Third World, or dam builders, who were already being opposed by environmental groups in Third World countries, making it more difficult for western companies to get away with environmental and social injustices where they are more politically vulnerable in the west.

Third World environmentalism is often radical, and rarely single-issue environmentalism (Haynes 1999). Even movements that begin as largely environmental ones, such as that against the Narmada dam in India, broaden their agenda because arguments about the environment become arguments about appropriate development and democracy. There is no doubt, therefore, that there are radical environmental movements in the Third World and that many of these are social movements as defined in this book. The prominent role of figures such as Vandana Shiva the Indian radical environmentalist within transnational green networks supports the view that common agendas are developing, particularly in opposition to neo-liberalism.

There is as yet, however, no global green movement. Environmental conflicts in the south are bound up with the degree of democracy and repression and the opportunities for mobilisation differ considerably. In countries such as India with a strong democratic system there is more space for environmental activism, and in Latin America, community-based environmental conflicts have been an important motor of democratisation (Foweraker 1995). But, in most countries in the south where they exist at all EMOs are small, and grassroots struggles against the environmental effects of development isolated and often repressed violently. Most transnational networking still depends on the greater resources of northern EMOs and activists.

The green movement as analysed in this book is very much a product of western structures and culture. There are points of connection and common interest between western greens and radical environmentalists in the south, but also major differences of context and tradition. More certain is that the agendas of western greens and non-western environmentalists will continue to change as a result of mutual contacts and engagement with global ecological governance. Although they have long been committed to seeking global solutions, the main challenge faced by radical environmentalists is how to build an argument that combines social justice in a form that is acceptable and persuasive in both north and south. Ideas such as contraction and convergence, developed by the Global Commons Institute, in order to seek a means of furthering international agreement on climate change, have widespread support in the green movement. Contraction and convergence is based on the idea that the western countries need to reduce their emissions of greenhouse gases in order that non-western countries can expand economically, but this redistribution must occur within a framework compatible with sustainability. Through this and similar ideas such as that of 'environmental space' which, as the Danish group NOAH puts it, means 'that every person in the world has the right (but not the duty) to use the same amount of natural resources and produce the same amount of pollution' and 'ecological debt', according to which the West owes other countries for the greater ecological damage it has produced, the greens are seeking ways to develop the arguments for global ecological solutions alongside a recognition of the need for the west to reduce its consumption.

The changing environmental debate and new opportunities for the greens

According to some commentators environmentalism has achieved its primary aim in getting environmental problems onto political agendas. It now has the status of a 'masterframe', commanding universal agreement around the core principle that humanity needs a new relationship with nature to achieve a sustainable future. (Eder 1996; Blühdorn 1997; 2000).[2] This poses strategic problems for the environmental movement because it no longer dominates the terms of debate and no longer appears distinctive, particularly as environmental groups have become increasingly proficient in science and in policy discourse, oriented towards pragmatic solutions, and therefore now use the same concepts as their erstwhile opponents.

The strategy of ecological modernisation has emerged as a central feature of this development, based on the assumption that economic growth can continue but only if economies are transformed in ways that can be ecologically sustainable. The move towards more technical solutions-oriented discourse by some EMOs and the willingness of some environmentalists to explore the idea that economic growth might be compatible with ecological sustainability has allowed a new discourse coalition to emerge linking policy-makers in government, scientists and parts of the environmental movement (Hajer 1995; 1996). When all greens rejected all economic growth as ecologically damaging their arguments were simply too radical to be countenanced but if they accept the discourse of ecological modernisation it might provide them with new opportunities. And yet, the opportunity provided by ecological modernisation (as was argued in Chapter Three) is dependent upon challenging the weak interpretation of the concept favoured by government and business with a stronger interpretation in which ecological questions are linked to questions of global social justice and the radicalisation of democracy.

Dryzek (1996) links the debates about ecological modernisation with those about risk society in a form which emphasises the importance of the green movement retaining a critical capacity and avoiding full incorporation within state structures. He argues that both ecological modernisation and the debates about risk are connected to core imperatives of the state, and social movements are likely to only have significant influence upon the state when their main interest can be assimilated with its core imperatives. Ecological modernisation connects with the interest of the state in securing capital accumulation (or the successful management of the capitalist economy) and risk society connects with the role of the state in securing legitimation (the acceptance of the system by its citizens). Beck (1992; 1996) has argued that new kinds of risks, such as BSE, are likely to produce new legitimation crises. BSE undermined the legitimacy of the state because the nature of the disease was so clearly misunderstood and mishandled. However, it was not so much mistakes by the government as structural features of science and the political system that made it impossible for government to act successfully. Laboratory science could not explain the level of risk or the process of transmission. Market-driven pressures for efficiency had led to practices such as feeding offal to herbivores which seemed 'unnatural', even if at the time they appeared safe scientifically and rational economically. In BSE as in other agricultural crises such as foot and mouth, the structure of agriculture and food production is exposed as contradicting what appears natural and undermining trust in policy-makers. However, because of the dominance of scientific reasoning in determining policy when science appeared to have no clear answer the government lacked any basis for convincing the public that its policy was well founded. Beck argues that it is only by involving citizens in the management of risk that the state can avert the legitimation crisis that ecological and other risks threaten.[3] The 'sub-politics' that develop beyond formal political institutions when citizens debate and take action on such issues provide the basis for this wider involvement. It is when society is confronted with the unintended consequences of industrial society that a transition from simple modernity, to a more reflexive and critical modernity becomes possible. Struggles over risk and over ecological modernisation are contests in which the green movement can take advantage of the pressure on states and business.

But, if the argument is to be relevant to the more radical green project of social and political transformation, the politics of risk and ecological modernisation have to mean more than a process of technical adaptation by existing institutions. It would mean engaging in a political struggle against current forms of economic organisation, which would require a critical and reflexive public outside the state system. Strong ecological modernisation would have to occur in spheres outside the state, as well as within it. Where there is no strong critical green movement that has resisted incorporation the prospects for meaningful progress are poor.

This is compatible with the view advanced in this book that green movements can be both radical as a whole and varied in degrees of radicalism, both inside and outside the system. It is also compatible with the view defended in this book that the green project of social transformation requires that at least part of the environmental movement remains on the margins and maintains its capacity to expose the limits of the existing system and challenge power, that is, that the movement retains its social movement character. Third, it corresponds with the view that greens have choices about how to act. Finally, we might add that the reflexive capacity of movements also includes transnational learning. Although national structures still constrain movements, groups learn from examples overseas. We have seen this in the case of the spread of direct action networks, but it may also prompt more general strategic re-evaluations by groups that appear institutionalised. In the era of neo-liberalism when networks of power linking governments and major western transnational corporations have reduced the autonomy of the state in the west, it can be argued that the kind of critical civil society sought by Beck and Dryzek and Hunold (2001) needs to also work on a partly transnational level, at the very least by challenging the inevitability of the logic of the global market with alternative arguments based on democratic, egalitarian and ecological rationality.

Strategic divisions within the green movement

Yet, strategic divisions continue to plague the green movement and seem to militate against its ability to take advantage of new opportunities. Moreover, it has been argued in this book, that if green politics are based on a changing balance between multiple commitments, and have no permanent agent of change, nor a permanent and singular strategy, they can never be reduced to a fixed and unchanging programme of action. It is of the nature of green ideology that it needs to maintain space for debate as an end in itself. Too much of the history of radical political movements is about the closure of debate and the exclusion of rival views. The green movement creates a public sphere by engaging in the kind of debates that this book has shown are important in green discourse, such as those over ecocentrism and anthropocentrism, the relation of social justice to ecology, or the relationship between ecological concerns and democracy. This is an end in itself of green politics. Such debates should be open-ended and not oriented only to instrumental questions such as what procedures or policies will best deliver certain outcomes. But, how then, are greens to deal with their divisions over strategy?

Torgerson's analysis of the relationship between strategy and green discourse is especially helpful in throwing light on this question (1999). He takes from Arendt the idea that it is impossible to have transformative politics without debate. Torgerson argues many of the debates between greens about strategy presuppose that agreement is possible when often it is not. Many of the differences over strategy examined in this book bear this out. For instance, the question of whether green parties can achieve useful changes within the political system, whether violence can ever be justified, or whether green commitments to democracy and egalitarianism are obstacles to the effective defence of the natural world, depend on different interpretations of power, of potentials for human action and the nature of the ecological crisis. Even the most scientific of these dilemmas such as the nature of the ecological crisis is not really knowable with any certainty. Greens combine the best, but limited, knowledge of scientists with expectations about how societies will respond, in assessing what may happen and even those who are most pessimistic still disagree about whether humanity can play a positive role in averting disaster.

Torgerson is concerned to move away from conceptions of green identity which 'suggest that theoretical and practical coherence is required so that an idealised *we* can prepare for the journey ahead' (1999: 49). The three core themes of green ideology emphasised in this book, ecology, egalitarianism and grassroots democracy run the risk of being just such a closed vision. Not only this, but, as I argued in Chapter Three, failure to accept them has led to the exclusion of some from green movement groups. Yet it has also been argued that green identity and ideology is constituted and reformed by the action of greens, rather than as a result of an abstract doctrine. Thus greens have had to learn through collective debate what the boundaries are of their 'we'.

The three themes describe a broad framework found in varying degrees in different sorts of green group and which, along with emotional bonds and common experience in collective action, make it possible for greens to work together. Thus the process of creating a green 'we' creates a consensus around certain ideas and the three commitments identified here constitute the most empirically defensible version of the broad green 'we'. There is nothing in this framework to provide guidance about how clashes between commitments are to be resolved in practice, and so it is only as they confront the problems of action that greens formulate practical answers to this question. And as we have seen the choices made differ in the different types of green group.

Here, Torgerson's three categories of functional politics, constitutive politics and performative green politics are helpful in showing that greens can work in different spheres and yet still be defensibly green.[4] Functional green politics is practised by those who engage in processes of reform within the system, particularly green parties and EMOs and this involves them in professionalisation and an engagement with technocratic logics of decision-making. Constitutive green politics is found most among more radical greens, and is concerned with social transformation and is therefore most opposed to the current system. In the strategic disputes between reformists and radicals both 'tend towards exaggerated totalisations' (Torgerson 2000: 10). However, the practice of these different dimensions of green politics is not strictly divided between more reformist and more radical groups. As we have seen, many groups are themselves divided internally over such questions. Torgerson also points to the challenge posed even by reformists to 'the monological character of technological reformism' when they argue for participation in decision-making by non-expert citizens. More radical groups also play a role in affecting environmental policy, most often discursively, by influencing the terms of debate as in debates over the use of public space by cars, or the cultural codes concerning control over food in the debates over genetic modification. While their more radical ideas are usually paid only lip service, the effects of these actions as well as those of reformists are hard to predict. Torgerson says:

> adjustments that seem minor can become part of a wider pattern of outcomes that contain both unexpected and unintended elements. There is thus an ambivalent potential in efforts to enhance system adaptability. Challenging the apparent imperatives of a system will certainly bring various defence mechanisms into play. But there is no predicting whether these will prove successful. Not only are there limits to human knowledge, in this regard, but also a potential for the imperatives of the system to turn against one another.
>
> (2000: 11)

This might occur, for instance, when legitimation imperatives are used to question the economic structure of food production, which generates new forms of risk based on processes that appear unnatural and unnecessary given surplus food production. This might then open space for other green ideas such as the ethic of sufficiency and a broadening of the categories of relevant knowledge in decision-making.

Where does this leave green strategy? Some have argued that as the green movement has become more differentiated it has lost its social movement character because there is no longer a common green project (Jamison, Eyerman and Cramer 1990). But an alternative means of approaching this is to see the differentiation of the movement as a product of its collective learning. Greens have diversified as new opportunities have emerged, but they have not fragmented into a wholly unconnected movement. Acceptance of the diversity of the green movement means accepting the impossibility of a unified green movement strategy. Defined in Torgerson's terms this means that, unwilling to give up on either the possibility of change within functional politics or the more radical transformations envisaged in the constitutive sphere, greens should see their strategy as a space in which both can co-exist, as long as they are consistent with the green ideological framework.

The green movement is much more divided over strategy than over ends. Most have argued that means and ends ought to be consistent, but the nature of the relationship between the two is still open to varying interpretations. While authoritarian strategies are ruled out because they contradict green commitments to democracy and social justice and their understanding of the limits of state-imposed change, this still leaves a range of reformist and radical possibilities which are compatible with the green ideological framework. But, one radical possibility that *is* ruled out is a vanguard revolutionary position. When green ends are so open and multiple it is impossible to be certain that only one strategy can be right.

The questioning and critique of their own actions and the actions of others is also part of what constitutes the vitality of the green movement, what keeps it moving. As we have seen greens have been characterised by processes of collective learning in which they recognised mistakes of over-optimism, or unnecessary failures of trust, and capacity building, in which skills and knowledge about the consequences of particular forms of action were passed on. There is nothing inevitable about this process, there are many lacunae in understandings of the past, and losses as well as gains in the move from previous ways of acting, but overall there is evidence of progress. This can be seen in the greater understanding by green parties of the specificities of their role as party actors rather than as necessarily responsible for the whole movement; of EMOs which have been able to build up resources for more knowledgeable contests within the technocratic sphere, while retaining a critical edge, and new forms of radical critique emerging from direct action and grassroots environmental campaigners.

Conclusion

Changes in politics in the last decade have posed new challenges for the green movement, but rather than making green ideology less relevant, they provide new opportunities for the greens to play a role in a pluralist and more global form of left politics, in which transnational north/south alliances are likely to play an increasing role. Second, although ecological modernisation and the increased legitimation problems around risk provide opportunities for the greens because they are linked to core imperatives of the state, the strength of the green movement lies in its critical capacity, requiring a degree of autonomy from the state sphere and a critical civil society. Third, many of the strategic disputes in the green movement cannot be resolved. This does not mean that greens do not share an ideology, but it also means that the movement is inescapably diverse and works in different domains of politics. Green reformists and radicals complement each other and neither can offer a total strategy appropriate to achieving the full range of the movement's goals.

In explaining and assessing the development of the green movement the central concept used in this book is the idea of collective learning. The green movement, like all social movements, is reflexive. There is considerable attention to the experiences of previous radical movements and continual efforts by greens to measure their actions against their goals. Greens have built up capacities through developing repertoires of action, resources, ideological frames and by involving new groups in action and establishing green ideas as part of western public discourse. Clearly, this process is not a neat evolution. Often mistakes from the past are repeated, inaccurate misconceptions about other groups influence action, and groups argue and distance themselves from others within the same movement. However, notwithstanding this, there are processes of collective learning both within specific types of green group and across the whole movement which have shaped the evolution of green praxis. Among these, we can include a greater acceptance of the limits to personal transformation within dominant social structures and greater sensitivity in the practice of grassroots democracy, with greater awareness also of its limits. The first followed from the many experiments in the alternative milieu in new forms of lifestyle. The second followed from the cumulative experience of practice in the many movements including the greens in which the barriers to the more abstract visions of a fully participatory alternative forms of organisation became clearer. Although this occurred in many green groups it is principally explored in this book in analysis of the green parties. In terms of action repertoires, there is now greater acceptance of the complementarity of different forms of action and less confusion and division over how to combine confrontational action with non-violent principles. There has been an expansion in international networking and the greens are prepared to take part in diverse coalitions such as the networks of protesters against neo-liberal global institutions, without expecting or insisting that all participants are greens. This also suggests the kinds of pluralism which may be more appropriate to the more fluid sphere of radical politics in the new millennium.

Notes

Introduction

1 The concentration on movements defined by their types of action and organisation might seem mistakenly to exclude women's environmental movements by definition, particularly by emphasising public action over the less public forms of political action such as consciousness raising that have been an important dimension of feminist political praxis. It is not lack of awareness of this issue so much as lack of evidence of women organising separately as ecofeminists that led me to avoid treating ecofeminism as a separate type of green movement. Where women have organised autonomously from men, for instance in some green parties or in green direct action groups, I have tried to examine this as green feminist praxis. Where women have been predominant, as for instance in many local grassroots struggles, I have tried to note the gendered aspects of this. Conversely, where men have been dominant, particularly in the more professionalised EMOs, I have also examined the consequences in terms of gender. Ecofeminist activism appears to have been strongest in the USA, but as Gottlieb notes (1993: 233) it proved impossible to sustain ecofeminist organisations. The latter were most associated with peace movement and anti-nuclear struggles of the mid-1980s and did not maintain distinct organisations thereafter.

1 The green movement as a social movement

1 This definition is based on della Porta and Diani's (1999: 13–16), but differs principally in stressing the importance of the challenge to power offered by a social movement. Della Porta and Diani use 'collective action focusing on conflicts' of interest (1999: 15), but I think that this is too inclusive because it does not distinguish more limited conflicts of interest from 'struggles for or against a new societal order' (Rucht 1998: 30). Rucht's definition of a social movement based on collective protest, struggles for or against a societal order and a broad network of groups and individuals (defined both by interaction and shared identity) is much the same as the one I set out above.

2 Gerlach and Hine's (1970) classic description of the network character of social movements still works effectively to define its key elements. Social movements are segmented: with numerous groups or cells in continuous rise and decline; polycephalous, with many leaders each with only a limited following and; reticular with multiple links between autonomous groups forming a network with indistinct boundaries.

3 Alain Touraine (2000: ch. 3) is among those who argue that political parties and formal social movement organisations (SMOs) can never be part of a social movement. This is because he regards social movements as cultural innovators and believes

that they need autonomy from the state system in order to produce new social forms. But Touraine assumes that new cultural forms can only emerge outside political institutions. I think instead that it is possible to show that the innovation within the green movement occurs through debates in green networks that cut across boundaries between formal organisations and grassroots groups.

4 Individuals in such groups may be involved in social movement activity, but only by working with others outside their organisation. The organisations themselves cannot be considered part of the movement.

5 As, for instance, in this definition of a pressure group by Wyn Grant: 'A pressure group is an organisation which seeks as one of its functions to influence the formulation and implementation of public policy, public policy representing a set of authoritative decisions taken by the executive, the legislature and the judiciary, and by local government and the European Union' (Grant 2000: 14).

6 Part of this section was published previously in Doherty, Paterson and Seel 2000.

7 A distinction between environmentalists and greens is better than that sometimes made between light and dark greens. The latter distinction assumes that greens are defined by their environmental commitments, but, as will be shown in the next chapter their social and political radicalism is as important as their environmentalism and means that they often share much with other groups on the left, which distinguishes them from many 'light greens'.

2   The emergence and growth of the green movement

1 It is also important to note the mutual influences between these figures. For instance, Gandhi had been particularly influenced by the ideas of Edward Carpenter and John Ruskin (Guha 2000: 19–20).

2 Doubts over whether wilderness could be appreciated by 'the masses' led to criticism of Marshall by others in the Wilderness Society (Gottlieb 1993: 18).

3 If scientific knowledge is valued, but lay knowledge excluded from policy-making this issue can be seen as a question of power. In the sociology of science the elements of social construction involved in science, its narrative structures, sense of received authority and dogmatic commitment have all been long recognised as meaning that science is a particular structure of thought, not a transcendent and necessarily superior form of knowledge. Yet, in the models of risk society developed by Beck (1992; 1996) and Giddens (1994) the status of scientific knowledge is not fully investigated. Legitimation problems arise only when experts disagree or are unable to answer the questions. However, Wynne (1996) suggests that other, not necessarily superior, but often insightful forms of local knowledge can show the limits and particularity of scientific knowledge, and opening up decision-making to include such knowledge could make the use of knowledge in policy-making more democratic.

4 Jamison, Eyerman and Cramer also argue that critical science preceded wider public scepticism (1990: 4; 9).

5 <http://www.sunday-times.co.uk/news/pages/sti/2000/11/12/stateofnation2>.

6 This is notwithstanding the fact that their status as 'problems' depends on how issues are framed and is therefore an act of social construction, as Eder (1996) has argued.

7 Population issues have caused divisions within green movements, but advocates of population control have been regularly defeated or excluded. For instance, the chair of the Earth First! City Group criticised as racist a (failed) Sierra Club ballot to support immigration controls 'supported by Dave Foreman and various right-wing foundations and organisations' (Anon. 1998b). Foreman had been a founder of Earth First! but left in 1990 because most activists came to reject his failure to see wilderness protection in the context of social justice issues.

8 The influence of the New Left on green movements is widely acknowledged in national histories and overviews (see Finger 1992: 232–9; Gottleib 1993; Hutton and Connors 1999: 227; Prendiville 1993a and Jamison, Eyerman and Cramer 1990).

9 Those organisations that did have a claim to a leading role such as the German and US SDSs both broke up in sectarian conflicts in 1970.

10 It is still notable, however, that the New Left parties were often among the first to take up ecological themes. This is especially true of the Dutch New Left. Bram van der Lek, the Chair of the PSP, wrote an early book outlining the seriousness of the ecological crisis for the left's traditional views of politics (interview with the author 29 May 1990); the PPR made ecology one of its principal themes from its formation in 1968, defining itself as an 'action party that wants to change mentality and social structures in order to achieve freedom, self-development and solidarity with mankind and nature' (Lucardie 1980: 170).

11 For instance, regarding these ideals Piccone comments: 'Although there is still considerable disagreement about their social and political implications, it is fair to claim that, beyond flirtation with drugs, sexual experimentation and other new forms of gratification, the general political concern was with an idealism that readily translated into participatory democracy, equality, personal commitment, and a never-too-precisely spelt out "qualitatively new life". In this sense it was the rejection of an increasingly corrupt society, which seemed to be moving closer and closer to total administration and whose smooth functioning, recognised and theorised both by mainstream sociologists (such as Bell, Lipset and Aron), as well as by critics (such as Marcuse, Adorno and C.W. Mills), unexpectedly ran into a rationality crisis' (1988: 8).

12 On the USA see: Sherkat and Blocker 1997; Whalen and Flacks 1989, McAdam 1988; on the USA and Japan, Fendrich and Krauss 1978; on Italy: Lange, Tarrow and Irwin 1989; and France: Pecheron 1991.

13 The *Berufsverbot* involved a loyalty check on 3.5 million Germans and resulted in job losses or disciplinary action for 4,500 (Markovits and Gorski 1993: 98).

14 Almost every part of the left seems to have been viewed in this way. Wasmund (1986), describes the siege mentality of the alternative scene in Frankfurt and Berlin in the late 1970s and points to the importance of communal houses (*Wohngemeinschaft*) as a source of recruitment for the terrorists. Yet, while most terrorists had lived in such houses few in the alternative scene supported their actions. As Papadakis notes of the spontaneists: 'The *Spontis* claimed that, on the one hand, the terrorists had robbed them of their future hopes for socialism because they had used violence and, on the other hand, the older generation had betrayed them about their past because they had not opposed but had actually supported fascism' (1984: 47). With the exception of one very marginal and inactive group the women's movement remained strongly committed to nonviolence but the women's movement too came under suspicion. 'Women's centres which were involved in organising abortion trips to the Netherlands were regularly raided by the police for "supporting criminal organisations"' (Kaplan 1992: 121).

15 Other alternative left candidates such as Charles Piaget – who was backed by the PSU and represented those on strike in support of demands for *autogestion* at the LIP factory in Bescançon – withdrew in favour of Mitterrand and some within the ecology camp argued that Dumont should do the same (Bennahmias and Roche 1992: 37).

16 In fact, Frank Wilson argues that in general the French party system has been better placed than others to withstand 'the challenge of alternative political organisations because (1) the major parties were themselves reforming or emerging; (2) the central political issue of the 1970s was the battle between left and right in which the alternative organisations had no part; (3) class confrontation remained important in France and encouraged citizens to identify with the established vehicles for class conflict rather than to turn to alternative political organisation unrelated to the class issue; (4)

the centralised and aloof pattern of policy-making insulated the political system from the alternative groups and their issues; and (5) the tradition of political protest made the new groups of the 1970s seem little different from a long history of other groups advocating mass action' (1988: 526). The rise of the *Front National*, the decline of the *Parti Communiste Français* and the green breakthrough in the late 1980s did show that this was not itself a wholly determinant factor. The party system was fundamentally altered by these developments, but during the 1970s Wilson's analysis had more force, particularly because the left had not yet exercised power and the hopes vested in it were so much greater.

17 An English translation appears in Ely and Mayer (1997: 293–304) along with the critical response of the fundamentalist-dominated party executive.

18 This point seems to be misunderstood by Kolinsky (1989: 201). Mayer and Ely (1997: 13) in contrast see this as a dispute within feminism.

19 This trend had been evident earlier in the 1990s too: (Frankland and Schoonmaker 1992: 88; Poguntke1992: 343–4). This was less the case in France where the green vote in the 1993 legislative elections was as high among first time voters as among the 25–34 age cohort (Roche 1993: 38).

20 Britain has historically had a relatively low proportion of graduates but in 1999 the government announced that it intended that 50 per cent of eighteen year-olds should be in higher education by 2010. By May 2000 35% of eighteen year-olds were in Higher Education, a higher proportion than the USA.

21 In Germany in 1978 nearly three-quarters of university graduates who took salaried jobs went into state employment, but between 1973 and 1985 the number of unemployed graduates rose from 10,000 to over 100,000, two-thirds of whom were under 35 and 26% of whom were teachers (Betz 1993: 106).

22 This explanation can be contrasted with that put forward by Inglehart (1977; 1990a) to explain the rise in post-material values in western Europe since the 1970s. Post-materialism refers to the commitment to 'quality of life' issues such as the environment and participatory democracy as distinct from the 'old politics' issues of wealth, public order and national security. Like the explanation above, Inglehart's explanation of the rise of post-material values is based on the distinctiveness of the experience of the first post-war generation that came of age in the 1960s. He argues that the new prosperity and the lack of direct involvement in war experienced by this generation were decisive in allowing the pursuit of higher order needs such as democracy and concern for the environment to take the place of the more basic and lower order needs based on material security. These values are likely to be enduring because Inglehart believes that pre-adult socialisation is relatively permanent. This is plausible as an explanation of why the post-materialist left is young and middle class. Yet, even if we accepted that greater material security might be a possible explanation of why some groups are more likely than others to support post-material values, it does not account for the more specific character of the green social base. In particular, there is the problem of why it is not the most affluent sectors of the middle class, but those in public sector welfare employment who are most likely to be active in green movements and parties. As critics have pointed out (Rootes, 1997: 321; Martell 1994), environmental issues are often as much material as post-material. Campaigns against locally unwanted land uses (LULUs) have often focused on questions of health and safety. And arguments for ecological rationality and sustainability can be understood as about material survival. Moreover, Inglehart's post-materialist model says nothing about the specific issues identified as central in the greens' programmes. The fact that most post-materialists identify with the left is not seen by Inglehart as an indicator of continuity in values, but rather as a chance result of the fact that post-materialists are change-oriented and the left has traditionally been seen as change-oriented too. It seems more plausible instead to emphasise the continuity between the values of 'post-material' movements such as the greens and those of the left, and see in

'post-materialism' an attempt to redefine left-wing values in response to new developments such as the ecological crisis, and increased scepticism about state-centred strategies for social change.

3 Green ideology

1 See, for instance, the mission statement of Friends of the Earth UK, one of the less radical branches of FoE, which includes among its aims: 'to increase public participation and democratic decision-making. Greater democracy is both an end in itself and is vital to the protection of the environment and the sound management of natural resources.' And, FOE campaigns 'to achieve social, economic and political justice and equal access to resources for men and women on the local, national, regional and international levels' (quoted in Byrne 1997: 133).

2 In case these three parties are seen as too small a sample, it is worth pointing out that in a comparison of the programmes of green parties from the 12 Economic Community (EC) states in the mid-1980s Thomas Poguntke (1989a: 175–94) argued that common positions on ten policy issues defined all the major green parties as sharing a broadly equivalent 'new politics'. These ten issues were: rejection of economic growth; opposition to nuclear energy; defence of civil liberties; demands for the tolerance of alternative lifestyles; feminism; participatory democracy; leftism (embracing workers' control, egalitarian policies and more societal control of the economic process); redistribution of wealth from the First to the Third World; opposition to INF deployment (Cruise, Pershing and SS20s) and unilateral disarmament proposals. Of the three case study parties, Die Grünen and the Ecology Party scored 10 out of 10 and Les Verts 9 out of 10, (missing only an explicit reference to INF deployment, as distinct from unilateral disarmament in general).

3 Doubts are often raised about the benefits of analysing programmes because they so often reflect electoral exigencies or are only partial and unclear (Blondel 1978: 121–6). A lack of clarity is also characteristic of green programmes. But in their case programmes are more often shaped by internal than external electoral considerations. Party conferences rather than the party leadership are the most important decision-makers on the content of the programme and its definition, and the weight attached to particular issues is often a result of compromise between ideological positions within the party more than external events. Moreover, they are less interested than other parties in identifying specific groups who might be wooed with particular appeals. They have recognised the need to develop concrete policies that can provide short-term as well as longer-term solutions, but give most priority to long-term goals. In this sense they fit Kay Lawson's definition of a 'programmatic party' – one oriented towards an integrated long-term plan of action which embraces both present needs and future goals (Lawson 1967: 15).

4 In the preamble to the 1994 election programme Die Grünen identify the three crises of the system as ecological, social and democratic.

5 This was provoked by the Umbau (reconstruction) Program, which emphasised the role of the state. A translation of the program is included in Mayer and Ely 1997: 267–93.

6 Barry (1999: ch. 2) argues persuasively that deep ecology develops a particular account of the human self in order to justify arguments for minimising human interference with nature. This is an asocial and anti-political version of the self, which cannot then form the basis for arguments for political action. Instead Barry suggests that deep ecology could be recast in terms of ecological virtues. This position, based on the idea that virtues are developed through human action is more self-limiting and not dependent upon the claim to have solved the problem of the relationship between humanity and nature.

7   Friends of the Earth (Australia) includes in its 'Philosophy' the statement: 'FOE believes that social and environmental issues cannot be separated from each other, and therefore we operate on a collective, non-hierarchical basis' (<http://www.foe.org.au/foehome.htm>).

8   In the mid-1980s leaders of European conservation groups identified mainly with the left but saw their organisations as more centrist. In contrast, leaders of 'ecology groups' such as Friends of the Earth and Greenpeace all 'identified their personal ideology and the political orientation of their group as leftist' (Dalton 1994: 124).

9   In a similar way, Douglas Torgerson has argued that green discourse has always contained a variety of environmentalisms. The ecologism identified by Andrew Dobson as the core ideology of the green movement is one kind of environmentalism but not the only, or necessarily the most important, one. Torgerson argues that world-views based upon ecocentrism and more human-centred ecological rationality both have a place in green discourse. This latter is held together by a 'questioning of industrialism and instrumentalism' (1999: 4). He argues that attempts to resolve questions such as whether democracy and green political thought are inextricably related tend to be ahistorical because they focus on a single principle in green thought, and broad ideological principles such as ecocentrism cannot resolve the question of what is to be done and how much compromise is possible.

10 Although some environmentalist groups such as WWF and IUCN have done so.

11 On green ideology Freeden is less convincing. He argues that ecology provides the core morphology for green ideology and identifies the four components of green ideology as: the relationship between human beings and nature; the valued preservation of the integrity of nature and forms of life, including human ones; holism based on the interdependence or harmony of forms of life; and the immediate implementation of qualitative human lifestyles (1996: 527). Defining green ideology in this way might lead some greens to deny democracy or egalitarianism. They might deny both, because they might want to give power to those with the most expertise or intelligence, because these people may be more likely to favour the right ecological solution. Such 'green' authoritarians might distrust the choices of consumerist masses. But, although this framework embraces the range of possible forms of ecologically based politics, as a definition of 'green ideology' it is not rooted in the discourse of green movements. Rather than a green ideology as it exists in green movements it describes the range of logical possibilities that an ecologically-based politics entails.

12 As Klandermans (2000: 26) argues, however, while all social movement action is ideologically structured, not all ideologically structured action is social movement action. He defends the importance of protest action as distinctive of social movements. Diani goes further (2000: 21) defining movements as based upon relational ties 'between individual and organized actors mobilizing in a conflict on the basis of a specific collective identity.'

13 Another perspective is provided by Thomas Rochon (1998 22–25). He argues that the ideologies of social movements are created by 'critical communities' within movements.

14 Frame bridging can occur when new issues arise, such as the way that British peace movement activists broadened their concerns from nuclear weapons in the mid-1980s once progress in disarmament negotiations was evident to the link between the arms trade and poverty in developing countries (Maguire 1992). Alternatively, other movements might take up a single 'master frame' developed by an earlier movement. For instance, the strength of the arguments for equal rights by the civil rights movement in the 1960s meant that a 'master frame' developed around equal rights which influenced second wave feminism, gay and lesbian rights and other movements in the late 1960s and the 1970s. A second influential master frame developed around the idea of liberation, which also had roots in the support for black power strategies by those who

rejected the quiescence of the civil rights movement in the face of increasing violence in the mid-1960s.

15 See, for instance, the debate over the meaning of framing in relation to ideology between Pamela Oliver and Hank Johnston (2000a; 2000b) and David Snow and Robert Benford (2000).

4   Green parties

1 In Australia greens have only been able to gain representation in those elections where there is some degree of proportional representation, as in the Senate (Rootes 1996) and the Tasmanian state elections (Crowley 1996; 1999). The greens have not been able to develop strong state organisations outside Tasmania and Queensland, and have been particularly affected by the competition for 'new politics' votes with the Democrats.

2 It was an important part of the identity of Die Grünen, and many other green parties, to deny that they were a party. They preferred to define themselves as a movement. However, insofar as German Law required parties to adopt certain structures to be officially recognised, Die Grünen conformed to the legal status of a party.

3 These events are described in detail in 'Länderprofile' in Raschke 1993; Papadakis 1984: 157–65; Langguth 1986: 8–12; Hülsberg 1988: 77–106; Rüdig 1986: 352–412; Fogt 1989: 93–107; Markovits and Gorski 1993: 192–200.

4 For further analysis of the formation of the French green party see Prendiville 1993a and b; Sainteny 1991; Doherty 1994 and Cole and Doherty 1995.

5 People, the first name adopted by what is now the Green Party of England and Wales, was inspired by the ideas of Edward Goldsmith. *A Blueprint for Survival* published in 1972 by the *Ecologist*. The first election statement for People was criticised for its 'attacks on women who work and neglect maternal duties, its attack on an all-pervasive welfare state and its support for the family as a means of discouraging dissent' (quoted in Wall 1994a: 22). Goldsmith also supported strong immigration controls. His conservatism was challenged from 1974 by a group on the left led by Peter Allen. This led to strong left–right conflicts which led to the departure of Allen. It was not until after 1979, when party membership increased ten-fold, that the counter-cultural wing of the party became strong.

6 Position papers for Eco-socialists, Radical Ecologists, *Realos*, Eco-Libertarians and the Aufbruch group are included in English translation in Mayer and Ely 1997: 223–52.

7 In 1987 85 per cent of green voters favoured a coalition with the SPD (Poguntke 1990: 35).

8 Author's interview with Frieder Otto Wolf, ex-MEP, 17 November 1992.

9 Although it is also important to point out that these were opposed by many greens (*Financial Times*, 19 June 1999).

10 If other green lists are included the total green vote was 10.8 per cent. However, while some of these lists were green in some form, others appear to have been motivated by new regulations for party financing, in which reimbursements were provided for votes. The confusion over which lists were genuinely green seems likely to have reduced the vote for the alliance of Les Verts and GE (Holliday 1994), at the expense of some of these opportunistic lists.

11 For these debates see Libération 30 August 1990 and 2 and 5 November 1990.

12 Libération 6 March 1991

13 These elections were fought under a semi-proportional system with multi-member constituencies, requiring a threshold for success of closer to 9 per cent as opposed to the 30–40 per cent required in General Elections for parties without a strong regional base.

14 Other gains included a radical waste programme in which recycling was to be greatly increased, and a new sustainable tranport policy, including a commitment to match spending on roads with that on rail. However, the greens suffered defeat over GM foods, since a large research programme was agreed (Mittler 1999a).

15 On which, see Doherty *et al.* (2003).

16 It was not only in Germany that greens backed the NATO action, Daniel Cohn-Bendit and Dominique Voynet also supported it, although the British Green Party was opposed.

17 See <http://www.les-verts.org>.

18 Mergers resolved other cases where there were two separate green parties, as in Austria and Italy.

19 In Australia, however, the greens have suffered from competition with the Democrats who pursued a partly similar new politics agenda (Rootes 1996).

20 Mid-term rotation carried on in some green parties. In Finland (Kontinnen 2000) the cabinet minister Pekka Haavisto was replaced by Satu Hassi, though slightly earlier than anticipated due to Haavisto's pregnancy and in France Dominique Voynet was replaced as environment minister by Yves Cochet.

21 Although the Finnish greens and Agalev (Flemish) have both compromised on this principle since joining government.

22 Die Grünen made two changes in party structure after 1998 in an attempt to improve co-ordination between the different parts of the party.

23 Die Grünen held eight party conferences in the first three years of government, some of which were regular events, but others were special congresses needed in response to a crisis of policy, as occurred for instance over Kosovo.

24 Kitschelt (1989) argued that greens were divided between a logic of constituency representation and a logic of voter maximisation. Voter maximisation includes the three goals set out by Strøm. Ideological representation is similar to constituency representation, but may be more appropriate because it places emphasis on the alternative content of the green agenda, and because green parties place less emphasis now on themselves as tribunes of the social movements within the system than they did in the 1980s.

5   Mass environmental movement organisations

1 In a report entitled 'People and Planet' the UK branch of WWF noted that it now spends 50% of its budget on human poverty alleviation (*The Guardian*, 27 September 2000).

2 In Britain, Szerszynski notes that WWF and the CPRE were able to teach FOE and Greenpeace much about organisation and lobbying, while themselves adopting some of the latter's tactics (1995: 9).

3 The level of national membership of environmental groups depends on whether conservation groups such as the National Trust or Italia Nostra are included. The survey by Krohberger and Hey in 1991 (quoted in van der Heijden, Koopmans and Giugni 1992: 18) included conservation groups such as Natuurmonumenten in the Netherlands but not the National Trust in the UK. Although it could be argued that Natuurmonumenten has a stronger campaign profile, than the NT there are no clear criteria to distinguish them. In some countries older conservation groups have a more grassroots tradition. For instance, the Swedish SNF, founded at the beginning of the century experienced significant membership growth in the 1990s and although an old conservation group, it strives to retain the character of a 'people's movement' (Jamison and Ring 1999). The largest Danish EMO is Danmarks Naturefrednings-forening (DN) (Jamison, Eyerman and Cramer 1990: 118) which in 1990 had a membership of 250,000, 15% of the adult population. This rather traditional nature

conservation organisation was founded in the 1920s and had been largely overshadowed by new environmental groups such as NOAH in the 1970s. In the 1980s, however, it was revitalised by a younger and more dynamic leadership and campaigned more aggressively on pollution, though as a single issue rather than as reflecting wider structural questions (ibid.).

4 Norway is an exception. As Bortne notes (2000: 6): 'When controlling for the population sizes in the countries, we see that there are three and a half times as many environmentalists in Sweden as in Norway, and seven times as many in Denmark as in Norway.' Bortne attributes this primarily to the greater co-option of the environment by the state and political parties in Norway than in the other Scandinavian countries. The disparity between Denmark and Sweden was earlier explained by Jamison, Eyerman and Cramer (1990) as a result of the same factor: in Sweden environmentalism never became a major issue of conflict due to the early response of the state and political parties.

5 Annual Report 2000, <http://www.acfonline.org.au/index.htm>.

6 Kellow (2000) shows how the positions taken by Greenpeace in international negotiations were consistent with German government policy and German national interests. However, he does not claim that this was as a result of any direct relationship, rather it reflected the same drives of popular support for Greenpeace and popular pressure on the German government.

7 <http://www.greenpeace.org/Admin/usage>.

8 For instance, the National Trust has around 2.5 million members of whom around 7% vote in internal elections and 35,000 do voluntary work for it. (Ward, 28 October 1999, *The Guardian*).

9 <http://www.greenpeace.org.au/aboutus/>.

10 Best known because three of its activists were targeted by McDonalds in a major libel action (Vidal 1997).

11 <http://www.amisdelaterre.org>.

12 see <http://www.noah.dk/english.html> including the statements: 'NOAH says we will not support technological fixes' and 'NOAH says "No" to growth for its own sake'.

13 For instance a debate about the lack of internal democracy in 1987 led to a reorganisation in the Danish branch of Greenpeace (Jamison, Eyerman and Cramer 1990: 118).

14 One exception was Greenpeace USA. In an interview in 1990 a staff member said: 'we've got to the point where we could pay for professional people across the board and we don't do it because we make a commitment to keep people who come in to help out for a while, get interested and want to stay' (quoted in Shaiko 1993: 92).

15 The alternative in grassroots groups is not necessarily better for those with responsibilities as primary carers (still overwhelmingly women rather than men). A culture of self-exploitation characterises many groups based on low-wage and volunteer efforts which makes it difficult for those with other commitments to play a major role.

16 In many countries the environmental portfolio is combined with others, further weakening the coherence of the Ministry as a representative of environmental interests. This is the case in Britain, where the environment was combined with transport and the regions, from 1997–2001 and since then with agriculture and rural affairs.

17 Today Programme BBC Radio 4 September 1999.

18 Dalton provides extensive evidence on ideological differences between what he calls conservation and ecological orientations within the environmental movement (1994: 46–7). Dalton's category of ecologists is similar to the category of green in this book.

19 A further development is the use of EMOs to carry out tasks for government. As the Swedish government began to privatise more public services in the 1990s, the SNF, a nature conservation movement, found itself carrying more practical conservation work for the state (Jamison and Ring 1999). In Britain the RSPB, National Trust and

Wildlife Trust together manage 2.7% of the country's land and had a turnover in 1990 that was more than 50% greater than the Government's own conservation agencies. Many conservation groups, particularly in southern Europe rely heavily on government grants for their income. Government and EU grants provided the main source of income for Spanish (Jimenez 1999) and Greek EMOs (Botetzagias 2000), with the exception of Greenpeace. While EMOs in this situation argue that they are not restricted by their relationship with government, Botezagias (2000; 2001) argues that in Greece the relationship between EMOs and political elites is clientelistic. EMOs are financially dependent upon government and act in ways that are consistent with maintaining a good relationship with the principal funder. Where membership of EMOs is low, as in southern Europe, there is good reason to think that they become more dependent on government. Government funding is also important in parts of northern Europe. In West Berlin in the mid-1990s over 30% of environmental groups had some state funding (particularly from job creation programmes), compared to none in 1980 (Rucht and Roose 1999: 67).

20 'BP is the company building the first offshore oil rigs in the Arctic. ICI features prominently in the Environment Agency's [a UK Government Regulatory Body] "hall of shame" for Britain's most polluting companies. Blue Circle has been seeking permission to use tyres, chlorinated solvents and heavy metals as fuel for its cement plants. Tesco's lobbying crushed the government's planned out-of-town parking tax' (Monbiot 2001).

21 Doyle relates a similar case of 'greenwash' by WWF Australia in 1998 when 'WWF endorsed Western Mining Corporation's 'Environmental Monitoring Report'. When the company's poor practice at its copper and uranium mine (for nuclear material) and infringement of the rights of indigenous people was pointed out, WWF argued that 'it was only assessing the quality of corporate reporting, not the corporate practices' (2000: 202).

22 It also seems to be a step back from the Greenpeace position in the mid-1990s when it worked hard to create partnerships with business making statements such as 'If any progress is to be made in saving the North Sea ecosystem, it will be a result of the alliance of corporate self-interest with the vision and determination of environmental organisations' (*Greenpeace Business* June–July 1996, quoted in Doyle 2000: 83).

23 Greenpeace was also marginal in the German environmental inter-organisational network: BUND, NABU, BBU and the Öko-Institut were the most central groups. WWF was less integrated in this network although it had campaigned with all the key groups (Rucht and Roose 2000).

24 These links are difficult to measure precisely, but as evidence that such a network exists in Britain we can quote the ties between direct activists and groups such as FOE and Greenpeace. The latter have both recruited employees from Earth First! networks, Greenpeace and FOE have both provided resources to direct action campaigners and on occasion Greenpeace has used activists from Earth First! on some of its official direct action protests.

25 McCormick argued that 'certainly there have been few formal or informal links between the [Green] Party and the environmental lobby' (1991: 123) but this is partly contradicted by the a survey of network ties within British EMOs in which 'the Green Party ranked just behind the centre-left parties and ahead of the Conservatives' (Rootes and Miller 2000: 15) in assessment of groups with which EMOs have regularly exchanged information or collaborated.

26 Dalton says: 'While we identified resource and political rivalries internal to the environmental movement, the predominant impression from our study was a broad pattern of co-operation among environmental interest groups' (1994: 170). He also notes national differences in the degree of internal network density, with less co-operation in France for instance than in Denmark, the Netherlands or Britain. The

lesser co-operation in France is consistent with Rucht's (1989) stress on the absence of links between the conservation and political ecology groups.

27 Two recent instances of this in Britain can be noted as further examples. First during the fuel protests of September 2000 individuals from Earth First, the Green Party, Friends of the Earth, Greenpeace and Transport 2000, as well as others with no organisational allegiance, participated in an email list to discuss collective green responses to the fuel protests. While there was much disagreement about what should be done, the key point was that those involved began from the assumption that as 'greens' they shared enough common ground to be able to debate a collective strategy (Doherty, Paterson, Plows and Wall 2000). The second example comes from a study of local activist communities. At local level there are considerable ties between networks of activists in different groups (Doherty, Plows and Wall 2001). The weaker ties at national level are more likely a reflection of the different spheres of action than a positive rejection of radicals. In fact the then Director of FOE, Charles Secrett met with Earth First! activists on several occasions in the 1990s to discuss common strategies (Doherty 1998).

28 See ACF philosophy and aims and annual report 2000 at <http://www.acfonline. org.au/aboutacf/mission.htm>.

29 The debate about the role of selective incentives in explaining collective action is too large an issue to discuss in detail here. In the case of EMOs the principal benefits offered are of two kinds. The first material benefits are special services such as magazines, or access to nature reserves. The second non-material kind is a sense of identity and moral rectitude. For instance, direct mail campaigns by EMOs tend to focus on emotional messages which reward supporters for their commitment with thanks and evidence of results. In its concern with incentives and free riding, RMT followed the seminal work of Olson on the Logic of Collective Action. Olson's model concentrates on material benefits, but since these are so few in the case of EMOs it seems implausible to use them as the main reason to explain why people join. Jordan and Maloney (1997) argue that 'soft incentives' such as the feeling that by joining one is identifying with the cause are more relevant in explaining membership of EMOs. This is supported by their survey of members of FOE (UK) who said that they would rather see the organisation maintain its campaign activities even if this was at the cost of reducing services for members.

30 The use of 'never' may have to be qualified in the case of the National Trust given its association with campaigns to open access to common land in the late nineteenth century, but as Gould notes (1988: 100) its lack of wider radical commitments was clear even then.

31 *The Guardian*, 20 June 2001.

32 The Swedish movement had been offered political opportunities from the earliest stages by the state. This provided many career opportunities for those who elsewhere would have had to seek work in EMOs (Jamison, Eyerman and Cramer 1990: 42).

33 Hayes (2000; 2001) provides further evidence consistent with Fillieule and Ferrier's argument.

34 Sustainable development is itself a much contested concept. Radical greens are critical not only of its misuse (Sachs 1993) but also of the inegalitarian implications inherent in the emphasis on development. Development implies the enclosure of common space and its transformation into private property (Paterson 2000: 55–8). For some greens therefore it is a term to be rejected. But sustainable development can be compatible with the green ideological framework without necessarily being acceptable to all greens, but only when it is used in such a way as to be consistent with meeting the shared goals of ecological rationality, egalitarianism and the expansion of democracy. Thus the usefulness of the concept is something different kinds of greens could argue about, based on their common adherence to the three core concepts in green ideology.

6 Ecological direct action groups

1 <http://www.enviroweb.org/ef/minnehaha2.html>.

2 Sea Shepherd is another group not covered here. It was formed in 1977 by Paul Watson who broke away from Greenpeace because he considered their methods too limited. Sea Shepherd is strongly associated with the personality of Watson and its main role is to protect marine wildlife, including on occasion sinking whaling ships. Although widely admired by many environmental radicals for his confrontational style, Watson's group cannot be analysed here because it is too different in aims and style from Earth First! and similar groups and dealing with it in any detail would require more space than is available. It is more narrowly focused on marine conservation and, although supported by many EDAers is not part of the counter-cultural networks that link groups such as Earth First! For more on Sea Shepherd see <http://2.seashepherd.org> and Scarce (1990).

3 <http://www.enviroweb.org/ef/primer/Monkeywrench.html>.

4 For instance, around 3,000 took part in the Redwood Summer protests of 1990.

5 Craig Rosebraugh (2001) who writes on the ELF in the *Earth First! Journal* says that the group was inspired by UK activists.

6 <http://www.animal-liberation.net/library/facts/elf.html>.

7 The 'ELF' also claimed responsibility for bomb hoax phone calls to McDonald's outlets (van der Heijden 2000).

8 The activism at Terania Creek led to a judicial inquiry which took 18 months. Although the judge decided in favour of logging, further direct action in 1981 led to the state government to intervene to protect the rainforest (Seed 2000).

9 For instance in 1990 Judi Bari and Darryl Cherney renounced tree spiking to try to reduce the hostility of workers in the logging industry to Earth First! This move was condemned by Dave Foreman who saw it as a betrayal of the forests.

10 There were also occasional links between greens and unions in Britain, as in the actions of the National Union of Seamen in the 1970s and 1980s in opposition to dumping nuclear waste at sea, but these were much rarer than in Australia.

11 Although this was countered by the turn away from environmental themes by the ALP in the 1990s.

12 This remains the case. For instance, Earth First! Melbourne describes itself as 'an anti-nuclear group, opposing all facets of the nuclear industry', and in recent years the Jabiluka uranium mine in the Kakadu National Park in the Northern Territory has been a major focus of protest. Other important recent protests include tree sits in Northern Tasmania in 1998 in opposition to the Tasmanian Regional Forest Agreement, which facilitated woodchipping by timber companies (see <http://www.nfn.org.au/up-a-tree/>).

13 Although in the 1990s it has become less involved in protest, the Wilderness Society still requires local branches to carry out an NVDA training day every year.

14 Throughout the 1970s and 1980s *Peace News* played an important role as the main source of news and debate for direct action politics based on non-violence, ecology, feminism and a loose commitment to anarchism. The Peace News booklet *Making Non-violent Revolutions*, (Clark 1981) first published in 1977, expressed much of the culture of the movement and shows how long-standing many of the views about personal transformation and movement organisation are.

15 Welsh shows that activists from Torness were among those involved at Greenham and other peace camps (2000 ch. 6), and Wall (1999) shows how activists with long experience of non-violent direct action were important in passing on skills in the early stages of Earth First! in Britain. Research by Doherty, Plows and Wall (2001) into activists' communities at local level also shows that ties with older activists with previous experience of direct action have been important in maintaining the continuity of the local activist networks.

16 Although as Martin (1998) shows, more recent recruits to New Age Traveller networks, often join for material reasons such as homelessness.

17 This point about the poll tax was made by many interviewees in research carried out by this author with Alex Plows and Derek Wall.

18 For an analysis of the media coverage of Swampy see Paterson (2000).

19 Among those papers most hostile to GM food was the *Daily Mail*, the paper most associated with attacking 'political correctness' and the liberal and left 'establishment' from the right. A poll for the *Sunday Times* in October 2000 showed that 55% of respondents regarded the 'destruction of genetically modified (GM) crops in government-approved field trials' as justified, as opposed to 35% who said it was not justified.

20 McDonald's was a target because of low wages and poor working conditions, because it represented a commercial and cultural imperialism, reducing diversity and driving local cafes and food suppliers out of business, because it used products from non-fair trade sources and its production processes damaged the environment. But, above all for British activists, it was the lengths that McDonald's went to in legal actions to suppress such criticism, in the infamous McLibel trial, that made McDonald's a special case. On the McLibel trial see Paterson 2000b; Vidal 1997 and <http://www.Mcspotlight.org>.

21 See Christopher Zinn 'The World Economic Forum and After' on <http://www.rte.ie/news/mayday/action7a>; Jade McCullough 'Globalising Violence' *Arena*, No. 50, December 2000.

22 MOVE has built links with US Earth First! in the 1990s and MOVE activists also toured the UK in 1996 following extensive involvement by EDA activists in a campaign to overturn the death sentence on MOVE activist Mumia Abu-Jamal.

23 And this heritage is much acknowledged in activist literature.

24 Anarchist and peace movement activists had organised 'Stop the City' protests in London in 1982 and 1983, but these were on a much smaller scale and protesters lacked the skills and experience developed in EDA in the 1990s.

25 L'Association pour une Taxation des Transactions Financière pour l'Aide aux Citoyens.

26 One example of many is given in Roseneil's account of Greenham: 'On 29 October 1983, just before Cruise was due to arrive at Greenham, a mass fence-cutting action was held, organized secretly by word of mouth. Thousands of women, many dressed as witches, stood on each other's shoulders to cut down over four miles of fence, and 187 women were arrested for criminal damage. This action took the soldiers and police completely by surprise, and many women were badly beaten with sticks and truncheons, receiving broken fingers and arms, and dislocated shoulders as the men struggled to cope with this huge act of criminal damage' (Roseneil: 1995: 107).

27 While the meaning and applicability of the concept of violence is historically variable, as John Keane argues, the core of the concept lies in this idea of using physical force to do harm to a body (1996: 66). Thus the definition of violence used by EDAers has a defensible historical pedigree. However, if violence is defined in this way, this also entails activists abandoning a common misuse of the term, as for instance, when they refer to the 'violence of the system'. The system may be oppressive, but it is only violent when it does actual harm to a body. It is quite possible therefore that non-violent oppression may cause more harm than violence by its opponents, but the two must be kept separate analytically.

28 The editorial collective of the US *Earth First! Journal* commented recently … 'we should reserve the term "violence" for matters involving harm to human life … Researchers who twist the very basis of our genetic makeup for corporate profit, executives emotionlessly calculating cost-benefit analyses on controversial earth-raping projects, politicians drafting death penalty legislation for acts of ecotage, these

are the types that environmentalists should recognize and describe as violent' (Anon. 2001b).

29 A paper reflecting on 'The Overt/Covert Debate' by Rowan Tilly a Ploughshares Activist with a long record of 'accountable' non-violent direct action in the peace movement and Earth First! was widely distributed in 1997.

30 Another important reason is practical. If activists accept arrest they will spend more time in prison and find it harder to maintain their action.

31 For instance, one member of the SWP in Manchester commented that the problem with the violence in Prague was not that it occurred, but that it had not been sufficiently directed or disciplined enough.

32 One important trend in this direction has been the rise of fundamentalist politics based on Christian and Islamic identities, as well as the growth of far-right militias in the USA.

33 Gamson (1990) and Piven and Cloward (1977) have argued that groups using violence have sometimes achieved their goals. However, movements that have used violence 'successfully' have been doing so in ways that are different from current EDA. Usually they had sufficient confidence in public support that they felt that those they attacked, principally repressive employers, would be blamed. Also, they were usually focused on a specific aim – which the state or another opponent could respond to – not on bringing about a new kind of society and replacing capitalism. In fact, Gamson says, if you are aiming to displace an opponent violence doesn't work. Moreover, he drew this conclusion when researching the years 1800–1945 in the USA. As he and others note, the rise in the acceptability of non-violent protest and the increased role of the media, especially television, have now changed the context for protest.

34 For instance, when asked 'If governments don't listen, peaceful protest, blockades, and demonstrations are a legitimate way of expressing people's concerns' 81% of Britons polled agreed, and only 7% disagreed. 39% disagreed with the view that the government should not change its policies in response to such protests (as opposed to 35% who agreed); but perhaps most surprising was the support expressed by 55% for the view that the destruction of GM crops was justified and by 42% for demonstrations in Seattle and Prague (27% felt they were not justified), <http://www.sundaytimes.co.uk/news/pages/sti/2000/11/12/stateofnation2.html>. See also Dalton 1996; Curtice and Jowell 1997 for evidence of increased support for protest over time and for a contrary view see Crozat (1998). Crozat's data, however, predates the transformation noted by many in the 1990s.

7 Local environmental protest groups

1 Excluding local branches of national Environmental Movement Organisations, which can usually be assumed to have a more long-standing and less self-interested environmental interest.

2 John Stewart of co-ordinator of ALARM UK (a Federation of UK local anti-roads groups; ALARM stands for 'All London Against the Roads Menace'). commented that after campaigns end, particularly when they end successfully, 'a lot of road groups turn to something else especially individuals from groups' (author's interview, 5 October 1995).

3 As Pulido (1996) points out, struggles against environmental racism or for environmental justice did not begin only when these terms were articulated in the 1980s. Earlier struggles such as those of Chicano farm workers against pesticide use, or by rural Hispanic communities such as the community development group Granados del Valle in New Mexico, for sustainable use of natural resources are part of a diverse history. Such cases were viewed by mainstream environmentalists as economic or civil

rights issues. However 'subaltern environmentalism' demands that environmental issues be defined as occurring within relations of inequality and domination. The environmentalism of the poor, including that of people of colour, cannot then be dismissed as concerned with different issues from that of other environmentalists, since all environmental struggles occur within relations of domination.

4 Although the group began in London it later became a national network.

5 On Greece (Kousis 1999); on Spain (Jimenez 2001) and on Australia (Doyle 2000: 34).

6 In the only quantitative study of large numbers of local environmental protests Kousis (1999: 184) is able to show a significant rise in local protest in Greece, Spain and Portugal since the end of dictatorships in all three countries. Carmin's analysis of 1975–90 records a surge in levels of local protest in the USA between 1976 and 1980, a decline and then peaks in 1985, 1987 and 1989, pointing to a cyclical pattern. Others (Rootes 2000; Rucht and Roose 2001) have commented on how the numbers of local protests appear to have risen, even as the number of people involved has not, but most are cautious about how far these events can be surveyed accurately.

7 One interesting exception is the survey of Alarm UK activists by Wallace McNeish (2000).

8 Most of the case studies mentioned here are from Britain, the USA and Germany. Where evidence from other countries is relevant, cases from these countries are also mentioned. Cross-national differences are therefore not examined in any detail.

9 John Stewart of ALARM UK, said of the local anti-roads groups: 'they're not all started up by people with experience, but in many cases you will find people who've got a radicalism coming from somewhere else, but not always' (interview with the author, 5 October 1995). For evidence from Italy see Donati (1994), for Greece, Spain and Portugal see Kousis (1999)

10 The quantitative examination of local environmental protest in Greece, Spain and Portugal contradicted the emphasis on opposition to new sites to some extent. Only one third of campaigns were against new sites, although opposition to construction was still important. Over half of both short and more sustained campaigns involved claims related to exposure to pollution (Kousis 1999: 189)

11 On the effects of 'suddenly imposed grievances', see Walsh, E. (1983) on local opposition to Three Mile Island nuclear plant in Pennsylvania.

12 This account is based on a meeting with the initial campaigners and emails from activists.

13 Another example in a campaign against the construction of a series of dams along the Loire arose when the developers proposed a leisure park for summer holiday-makers to use the river. Locals pointed out that the river would be mainly mud in summer (Hayes 2001).

14 See for instance, the Donegal Alternative to Pylons case above. Also, Keith Buchan of the environmental consultancy the Metropolitan Transport Research Unit played an important role in proposing alternative traffic management systems in public inquiries that led to proposed roads being cancelled in Norwich, Hereford and Crosby in Britain in the mid-1990s (Stewart, Bray and Must 1995: 10–11).

15 For instance, activists campaigning against a new motorway in Berglund's (1998) study of a northern German city were moved to question at root the kind of society that accepted unrestricted growth in motor traffic as unavoidable.

16 See also Szasz (1994: 153) on the transformation from ordinary lives to political leaders, new worldviews, new skills and new social networks of activists in anti-toxics movements.

17 For instance, Penny Newman, the focus of a study by Gottleib (1993, ch. 6) organised opposition to a toxic waste dump once she realised that her family and community were under threat.

18 Some anti-nuclear demonstrations in the 1970s were exceptions. In the USA the Clamshell Alliance was one, but its base was concentrated among radicals living in New England rather than a broad cross-section of the community (Epstein 1991). In Italy protests against the nuclear site at Montalto di Castro had initial local support but this diminished later (Diani 1995). In France, the attack by police on demonstrators at Malville in 1976 made local farmers more sympathetic – but the escalation of confrontation the following year undermined this (Rüdig 1990). In the USA environmental justice groups also carried out sit-ins blockaded roads and disrupted meetings on many occasions, combining this with more conventional tactics such as rallies, marches and public meetings (Schlosberg 1999: 129).

19 Nevertheless, this was not simply a legal battle, it was combined with many mass demonstrations, petitions and procedural complaints so that a local issue became a question of national policy (Rüdig 1990: 156–65).

20 Halfacre and Matheny (1999) examined the effects of environmental discourses on locals not involved in activism in two communities in the USA. This showed that most did not view the problem in NIMBY terms, retained faith in science, and were willing to play a positive role in helping to achieve practical policy solutions. They were not for the most part disillusioned with the political process.

8 Conclusion: the future of the green movement

1 A detailed account of how WWF programmes have led to the displacement of many tribal peoples in the south is given in the British Earth First! Journal *Do or Die* (see Anon. 1998b) 'World Wide Fraud' *Do or Die*, 7: 76–8.

2 Jamison (1996) makes a similar argument when he sees the social movement phase of the environmental movement as having passed since the movement became more institutionalised. Environmental groups now produce information rather than new knowledge.

3 Beck is more convincing in outlining the novelty of new kinds of risk than he is in showing that this is likely to lead to an expansion of the reflexive and participatory involvement of citizens in managing risks. His optimism that risk society can be the basis for the expansion of the reflexive capacity of modernity, shared with Giddens, is really a hope that would require the kind of action and discourse engaged in by the green movement and others, and not a likely consequence of structural changes alone. Thus, while it is an argument worth making, it is not something that can be demonstrated to be occurring.

4 Functional politics, parallels Arendt's conception of labour, based on work in the formal economy, and is concerned with the maintenance and operation of the political system. Constitutive politics is concerned with constructing the cultural artefacts of a civilisation, paralleling Arendt's category of work, which means constructing artefacts of enduring value. Performative politics are concerned only with the intrinsic value of politics as a form of action, similar to Arendt's category of action in which 'people are with others and neither for nor against them – that is, in sheer human togetherness' (Torgerson 1999: 17).

Bibliography

Abbey, E. (1975) *The Monkey Wrench Gang*, London: Robin Clark

Affigne, A.D. (1990) 'Environmental Crisis, Green Party Power: Chernobyl and the Swedish Greens', in Rüdig, W. (ed.) *Green Politics One*, Edinburgh: Edinburgh University Press.

Aguitton, C. (2001) 'French Connections', *Red Pepper*, July: 28–9.

Alber, J. (1989) 'Modernization, Cleavage Structures and the Rise of Green Parties and Lists in Europe', in Muller-Rommel, F. (ed.) *New Politics in Western Europe: The Rise and Success of Green Parties and Alternative Lists*, Boulder CO: Westview Press.

Amis de la Terre (2000) 'Rapport moral et d'activités 2000', online at<http://www.amis delaterre.org/pages/presentation/moral_2000.html>.

Andersen, A. (1997) *Media, Culture, and the Environment*, London: Routledge.

Anderson, B. (1983) *Imagined Communities: Reflections on the Origin and Spread of Nationalism*, London: Verso.

Anon. (1998a) 'No Escape from Patriarchy: Male Dominance on Direct Action Camps', *Do or Die*, 7: 10–13.

—— (1998b) 'How to Work on Environmental and Population Issues Without Scapegoating Immigrants', *Earth First! Journal*, 18, 3.

—— (1999) 'Friday June 18th Confronting Capital and Smashing the State', *Do or Die!*: 1–12.

—— (2001a) 'Confonting Oppression at the Fall Creek Tree Village', *Earth First! Journal*, 21, 7: 10–11.

—— (2001b) 'A Look into Activism, Tactics and Strategy', *Earth First! Journal*, 21, 7: 2.

Arendt, H. (1970) *On Violence*, London: Allen Lane.

Arts, B. (1998) *The Political Influence of Global NGOs: Case Studies on the Climate and Biodiversity Conventions*, Utrecht: International Books.

Aufheben (1998) 'The politics of anti-road struggle and the struggles of anti-road politics: the Case of the no M11 Link Road campaign', in McKay, G. (ed.) *DiY Culture: Party and Protest in Nineties Britain*, London: Verso.

Australian Conservation Foundation (2001) *Annual Report 2000*, online at <http://www.acfonline.org.au>.

Baggott, R, (1998) 'Nuclear Power at Druridge Bay', *Parliamentary Affairs*, 51: 3: 384–96.

Bagguley, P. (1992) 'Social Change, the Middle Class and the Emergence of New Social Movements: a Critical Analysis', *Sociological Review*, 40, 1: 26–48.

Bahro, R. (1979) *The Alternative in Eastern Europe*, London: New Left Books.

—— (1982) *Socialism and Survival*, London: Heretic Books.

—— (1984) *From Red to Green*, London: Verso.

—— (1986) *Building the Green Movement*, London: Heretic Press.

—— (1986) 'Relationships to Life', *Green Line*, 44: 12–14.

—— (1994) *Avoiding Social and Ecological Disaster: the Politics of World Transformation*, Bath: Gateway Books.

Barnes, S. and Kaase, M. (eds) (1979) *Political Action: Mass Participation in Five Western Democracies*, London: Sage Publications.

Barry, J. (1999) *Rethinking Green Politics: Nature, Virtue and Progress*, London: Sage.

Barry, J. and Doherty, B. (2001) 'The Greens and Social Policy', *Social Policy and Adminstration*, 35: 587–607.

Beck, U. (1992) *Risk Society: Towards a New Society*, London: Sage.

—— (1996) 'Risk Society and the Provident State', in Lash, S. Szerszynski, B. and Wynne, B. (eds) *Risk, Environment and Modernity: Towards a New Ecology*, London: Sage.

Beetham, D. (1987) *Bureaucracy*, Milton Keynes: Open University Press.

Bennahmias, J.L. and Roche, A. (1992) *Des verts de toutes les couleurs*, Paris: Albin Michel.

Bennie, L.G., Rüdig, W. and Franklin, M.N. (1995) 'Green Dimensions: The Ideology of the British Greens', in Rüdig, W. (ed.) *Green Politics Three*, Edinburgh: Edinburgh University Press.

Bennulf, M. (1995) 'Sweden: The Rise and Fall of Miljöpartiet de Gröna', in Rootes, C. and Richardson, D. (eds) *The Green Challenge: the development of green parties in Europe*, London: Routledge

Berglund, E. (1998) *Knowing Nature, Knowing Science: An Ethnography of Local Environmental Activism*, Cambridge: the White Horse Press.

Betz, H.G. (1992) 'Postmodernism and the New Middle Class', *Theory, Culture and Society*, 9, 2: 93–114.

Bevir, M. (2000) 'New Labour: A Study in Ideology', *British Journal of Politics and International Relations*, 2, 3: 277–301.

Bey (1996) *TAZ: The Temporary Autonomous Zone, Ontological Anarchy, Poetic Terrorism*, Camberley: Green Anarchist Books.

Biehl, J. (1988) 'Deep Ignorance', *Green Line*, 59: 11–14.

—— (1991) 'Farewell to the German Greens', *Green Line*, 88: 12–14.

—— (1993) 'La mythologie de la déesse dans l'écologie politique', *Ecologie et politique*, 9: 145–58.

Biorcio, R. (2000) 'Greens in Italy: Changes in Organization and Activism', paper presented at the European Consortium for Political Research, Copenhagen, 14–19 April 2000.

Birnbaum, N. (1983) 'All is Calm', *Telos*, 55: 185–8.

Blondel, J. (1978) *Political Parties*, London: Wildwood House.

Blühdorn, I. (1995) 'Campaigning for Nature: Environmental Pressure Groups in Germany and Generational Change in the Ecology Movement', in Blühdorn, I. Krause, F. and Scharf, T. (eds) *The Green Agenda: Environmental Politics and Policy in Germany*, Keele: Keele University Press.

—— (1997) 'A Theory of Post-Ecologist Politics', *Environmental Politics*, 6, 2: 125–47.

—— (2000) *Post-Ecologist Politics: Social Theory and the Abdication of the Ecologist Paradigm*, London: Routledge.

—— (2001) 'Further to Fall? Adverse Conditions for the Survival of the Greens', paper presented at the European Consortium for Political Research, Grenoble 6–11 April.

Boardman, B. with Bullock, S. and McLaren, D. (1999) *Equity and the Environment: Guidelines for Green and Socially Just Government*, London: Catalyst and Friends of the Earth.

Böltken, F. and Jagodzinski, W. (1985) 'Postmaterialism in the European Community, 1970–1980: Insecure Value Orientations in an Environment of Insecurity', *Comparative Political Studies*, 17, 4: 453–84.

Bomberg, E. (1992) 'The German-Greens and the European Community: Dilemmas of a Movement-Party', *Environmental Politics*, 1, 4: 160–85.

—— (1998) *Green Parties and Politics in the European Union*, London: Routledge.

Bomberg, E. and Burns, C. (1999) 'The Environment Committee of the European Parliament: New Powers, Old Problems', *Environmental Politics*, 8, 1: 174–9.

Bookchin, M. (1974) *Post-Scarcity Anarchism*, London: Wildwood House.

—— (1982) *An Ecology of Freedom*, Palo Alto, CA: Cheshire Books.

—— (1990) *Remaking Society*, Montreal: Black Rose Books.

—— (1991) 'Redefining Politics', *Green Line*, 93: 10–11.

—— (1992) 'Ecology as a Dismal Science', *Green Line*, 96: 11–12.

Bortne (2000) 'The Local Community Perspective in Norwegian Environmentalism', paper presented at the European Consortium for Political Research, Copenhagen, 14–19 April 2000.

Botezagias, I. (2000) 'Patterns of Networking and Interaction for Greek ENGOs', paper presented at the European Consortium for Political Research, Copenhagen, 14–19 April 2000.

—— (2001) '"Nobody does it better": Intra-movement conflict concerning species conservation in Greece', paper presented at the European Consortium for Political Research, Grenoble, 6–11 April 2001.

Bourdieu, P. (1984) *Distinction: A Social Critique of the Judgement of Taste*, London: Routledge.

Bowler, S. and Farrell, D. (1992) 'The Greens at the European Level', *Environmental Politics*, 1, 1: 132–7.

Boy, D. (1981) 'Le Vote Ecologiste en 1978', *Revue Française de Science Politique*, 31: 394–416.

—— (1989) 'L'Ecologisme en France: évolutions et structures', paper presented at the Joint Sessions of the ECPR, Paris.

—— (2002) 'French Greens in Government', *Environmental Politics*, 11, 1, (forthcoming).

Brand, K.W. (1990) 'Cyclical Aspects of New Social Movements: Waves of Cultural Criticism and Mobilization Cycles of New-Middle Class Radicalism', in Dalton, R.J. and Kruechler, M. (eds) *Challenging the Political Order: New Social and Political Movements in Western Democracies*, Cambridge: Polity Press.

—— (1999) 'The Dialectics of Institutionalisation: The Transformation of the Environmental Movement in Germany', *Environmental Politics*, 8, 1: 35–58.

Broadbent, J. (1998) *Environmental Politics in Japan: Networks of Power and Protest*, Cambridge: Cambridge University Press.

Bryner, G.C. (2001) *Gaia's Wager: Environmental Movements and the Challenge of Sustainability*, Oxford: Rowman and Littlefield.

Bull, M.J. (1993) 'Review Article: The Crisis of European Socialism: Searching for a (Really) Big Idea', *West European Politics*, 16, 3: 413–23.

Bündnis 90/Die Grünen (1998) *New Majorities Only with Us*, (Election Manifesto) Bonn. Online at <http://www.gruene.de/aktuell/english>.

Burgmann, M and Burgmann, V. (1998) *Green Bans, Red Union: Environmental Activism and the New South Wales Builders Labourers' Federation*, Sydney: University of New South Wales Press.

Bürklin, W.P. (1987) 'Governing Left Parties Frustrating the Radical Non-Established Left: The Rise and Inevitable Decline of the Greens', *European Sociological Review*, 3, 2: 109–26.

—— (1988) 'A Politico-Economic Model instead of a Sour Grapes Logic: A Reply to Herbert Kitschelt's Critique', *European Sociological Review*, 4, 2: 161–166.

Burns, R and van der Will, W. (1988) *Protest and Democray in West Germany: Extra-Parliamentary Opposition and the Democratic Agenda*, London: Macmillan Press.

Byrne, P. (1989) 'The Green Party', in Müller-Rommel, F. (ed.) *New Politics in Western Europe: The Rise and Success of Green Parties and Alternative Lists*, London: Westview.

—— (1997) *Social Movements in Britain*, London: Routledge.

Calhoun, C. (1995) 'New Social Movements of the Early Nineteenth Century', in Traugott, M. (ed.) *Repertoires and Cycles of Collective Action*, Durham, NC: Duke University Press.

Camacho, D.E. (1998) (ed.) *Environmental Injustice: Political Struggles*, Durham, NC: Duke University Press.

Capra, F. (1985) *The Turning Point: Science, Society and the Rising Culture*, London: Fontana.

Capra, F. and Spretnak, C. (1984) *Green Politics: The Global Promise*, London: Hutchinson.

Carey, J. (1998) 'Fresh Flavour in the Media Soup: the Story of SQUALL magazine', in McKay, G., (ed.) *DiY Culture: Party and Protest in Nineties Britain*, London: Verso.

Carmin, J. (1999) 'Voluntary Associations, Professional Organisations and the Environmental Movement in the United States', *Environmental Politics*, 8, 1: 101–121.

Carroll, M.K. and Ratner, R.S. (1996) 'Master Framing and Cross-Movement Networking in Contemporary Social Movements', *Sociological Quarterly* 37: 601–25.

Carson, R. (1962) *Silent Spring*, Boston: Houghton Mifflin.

Carter, N. (1992a) 'The Greening of Labour', in Smith, M.J. and Spear, J (eds) *The Changing Labour Party*, London: Routledge: 118–32.

—— (1992b) 'Whatever Happened to the Environment? The British General Election of 1992' *Environmental Politics*, 1, 3: 442–8.

—— (1999) 'The Greens in the 1999 European Parliamentary Elections', *Environmental Politics*, 8, 4: 160–7.

Castells, M. (1997) *The Power of Identity* Oxford: Blackwell.

Cathles, G. (2000) 'Friends and Allies: the role of local campaign groups', in Seel, B., Paterson, M. and Doherty, B. (eds) *Direct Action in British Environmentalism*, London: Routledge.

Caute, D. (1988) *68, The Year of the Barricades*, London: Paladin.

Chafer, T. (1982) 'The Anti-Nuclear Movement and the Rise of Political Ecology', in Cerny, P. (ed.) *Social Movements and Protest in France*, London: Frances Pinter: 202–20.

—— (1985) 'Politics and the Perception of Risk: A Study of the Anti-Nuclear Movements in Britain and France', *West European Politics*, 8, 1: 5–23.

Chandler, W.M. and Siaroff, A. (1986) 'Postindustrial Politics in Germany and the Origins of the Greens', *Comparative Politics*, 18, 3: 303–25.

Chase, S. (1991) 'Whither the Radical Ecology Movement?', in Bookchin, M., Forman, D., Chase, S. and Levine, D. (eds) *Defending the Earth: A Dialogue Between Murray Bookchin and Dave Foreman*, Boston, MA: South End Press: 7–24.

Christie, I. and Warburton, D. (2001) *From Here to Sustainability*, London: Earthscan.

Christoff, P. (1996) 'Ecological Modernisation: Ecological Modernities', *Environmental Politics*, 5, 3: 476–500.

Clark, H. (1981) *Making Nonviolent Revolution*, Nottingham: Spokesman.

Clark, H, Crown, S., McKee, A. and McPherson, H. (1984) *Preparing for Nonviolent Direct Action*, Nottingham: *Peace News*.

Cole, A. and Doherty, B. (1995) 'Pas comme les autres: The French Greens at the Cross-roads', in Richardson, D. and Rootes, C. (eds) *The Green Challenge: the development of green parties in Europe*, London: Routledge.

Commission on Racial Justice (1987) *Toxic Wastes and Race in the United States*, Cleveland, OH: United Church of Christ.

Cotgrove, S. (1982) *Catastrophe or Cornucopia: the Environment, Politics and the Future*, Chichester: John Wiley.

Cotgrove, S. and Duff, A. (1980) 'Environmentalism, Middle-Class Radicalism and Politics', *Sociological Review*, 28: 333–51.

—— (1981) 'Environmentalism, Values and Social Change', *British Journal of Sociology*, 32, 1: 92–110.

Crowley, K. (1996) 'The Tasmanian State Election 1996: Green Power and Hung Parliaments', *Environmental Politics*, 5, 3: 530–5.

—— (1999) 'A Failed Greening? The Electoral Routing of the Tasmanian Greens', *Environmental Politics*, 8, 4: 186–93.

Crozat, M. (1998) 'Are the Times a-Changin? The Acceptance of Protest in Western Democracies', in Meyer, D. and Tarrow, S. (eds) *The Social Movement Society*, Oxford: Rowman and Littlefield.

Curtice and Jowell, (1997) 'Trust in the Political System' in Jowell, R., Curtice, J., Park, A., Brook, L. and Thomson, K. (eds.) *British Social Attitudes; the 14th Report*, Aldershot: Ashgate.

Dalton, R.J. (1994) *The Green Rainbow*, New Haven, CT: Yale University Press.

—— (1996) *Citizen Politics in Western Democracies*, Chatham: Chatham House.

della Porta, D. (1995) *Social Movements, Political Violence and the State*, Cambridge: Cambridge University Press.

della Porta, D. and Andretta, M. (2000) 'National Environmental Organisations in the Italian Political System: Lobbying, Concentration and Political Exchange', paper presented at the European Consortium For Political Research, Copenhagen, 14–19 April.

—— (2001) 'Environment, Territory and the City', paper presented at the European Consortium For Political Research, Grenoble, 6–11 April.

della Porta, D. and Diani, M (1999) *Social Movements: An Introduction*, Oxford: Blackwell

della Porta, D. and Reiter, H. (eds) (1998) *Policing Protest: The Control of Mass Demonstrations in Western Democracies*, Minneapolis: University of Minnesota Press.

Demirovic, A. (1997) 'Grassroots Democracy', in Mayer, M. and Ely, J. *The German Greens*, Philadelphia: Temple University Press.

Devall, B. and Sessions, G. (eds) (1985) *Deep Ecology: Living as if Nature Mattered*, Layton, UT: Peregrine and Smith.

Diani, M. (1990) 'The Italian Ecology Movement Between Moderation and Radicalism', in Rüdig, W. (ed.) *Green Politics One*, Edinburgh: Edinburgh University Press.

—— (1992a) 'The Concept of Social Movement', *Sociological Review*, 40: 1–25.

—— (1992b) 'Analysing Social Movement Networks', in Diani, M. and Eyerman, R. (eds) *Studying Collective Action*, London: Sage.

—— (1995) *Green Networks: A Structural Analysis of the Italian Environmental Movement*, Edinburgh: Edinburgh University Press.

—— (2000) 'Organizational Diversity in Italian Environmental Protest, 1988–97', paper presented at the European Consortium For Political Research, Copenhagen, 14–19 April.

Diani, M. and Donati, P. (1999) 'Organisational Change in Western European Environmental Groups: A Framework for Analysis', *Environmental Politics*, 8, 1: 13–34.

Die Grünen, (1985) *Programme of the German Green Party*, London: Heretic Books.

Die Grünen (undated) 'For a Multicultural Society: Against Right-Wing Extremism and Xenophobia', Bonn.

Ditfurth, J. (1988) 'Die Grünen – Future Prospects', *Green Line*, 67: 10–12.

Dobson, A. (2000) *Green Political Thought*, London: Unwin Hyman.

Dobson, A. (ed.) (1991) *The Green Reader*, London: André Deutsch.

Doherty, B. (1992a) 'The *Fundi-Realo* Controversy: An Analysis of Four European Green Parties', *Environmental Politics*, 1, 1: 95–120.

—— (1992b) 'The Autumn 1991 Conference of the UK Green Party', *Environmental Politics*, 1, 2: 292–8.

—— (1994) 'Ideology and the Green Parties of Western Europe: A Thematic Analysis of the Green Parties of Britain, France and Germany 1973–93', unpublished Ph.D., Manchester University.

—— (1998) 'Opposition to Road-Building', *Parliamentary Affairs*, 51, 3: 370–83.

—— (1999a) 'Paving the Way: the rise of direct action against road-building and the changing character of British Environmentalism', *Political Studies*, 47, 2: 275–91.

—— (1999b) 'Manufactured Vulnerability: Eco-Activist Tactics in Britain', *Mobilization*, 4, 1: 75–89.

—— (1999c) 'Change the World via Email', *New Statesman*, 1 November 1999.

Doherty, B., Paterson, M., Plows, A., and Wall, D. (2000) 'Constructing the Fuel Protests: New Populist Movements in British Politics', paper presented at the BSA/ISA Conference on Social Movements, Manchester University, November 3–5.

—— (2003, forthcoming) 'Constructing the Fuel Protests', *British Journal of Politics and International Relation*, 5, 1: n.p.

Doherty, B., Paterson, M. and Seel, B. (2000) 'Direct Action in British Environmentalism', in Seel, B. Paterson, M. and Doherty, B. (eds) *Direct Action in British Environmentalism*, London: Routledge.

Doherty, B. Plows, A. and Wall, D. (2001) 'Comparing Radical Environmental Activism in Manchester, Oxford and North West Wales'. paper presented at the European Consortium for Political Research, Grenoble 6–11 April.

Doherty, B. and Rawcliffe, P. (1995) 'British Exceptionalism: Comparing the Environmental Movement in Britain and Germany', in Blühdorn, I. Krause, F. and Scharf, T. (eds) *The Green Agenda: Environmental Politics and Policy in Germany*, Keele: Keele University Press.

Dominick, R. III (1992) *The Environmental Movement in Germany: Prophets and Pioneers, 1871–1971*, Bloomington and Indianapolis: Indiana University Press.

Donati, P. (1994) 'Media Strength and Infrastructural Weakness: Recent Trends in the Italian Environmental Movement', EUI Working Papers SPS 94/14, European University Institute, Florence.

Dowie, M. (1995) *Losing Ground: American Environmentalism at the End of the Twentieth Century*, Cambridge, MA: MIT Press.

Doyle, T. (1994) 'Direct Action in Environmental Conflict in Australia: A Re-Examination of Non-Violent Action', *Regional Journal of Social Issues*, 28: 1–13.

—— (2000) *Green Power: the Environmental Movement in Australia*, Sydney: University of New South Wales Press.

Dryzek, J.S. (1996) 'Political Inclusion and the Dynamics of Democratization', *American Political Science Review*, 90, 1: 475–87.

Dryzek, J. and Hunold, C. (2001) 'Greening the State? Ecological Modernization Between State and Movement in the USA, UK, Germany and Norway', paper presented at the European Consortium For Political Research, Grenoble, 6–11 April.

—— (1997) *The Politics of the Earth: Environmental Discourses*, Oxford: Oxford University Press.

Eckersley, R. (1989) 'Green Politics and the New Class: Selfishness or Virtue?', *Political Studies*, 37, 2: 205–23.

—— (1992) *Environmentalism and Political Theory: Towards an Ecocentric Approach*, London: University College London Press.

The Ecologist (1972) *A Blueprint for Survival*, Harmondsworth: Penguin.

Ecology Party (1983) *Politics for Life*, London.

Eder, K. (1985) 'The New Social Movements: Moral Crusades, Political Pressure Groups, or Social Movements?', *Social Research*, 52, 4: 869–90.

—— (1993) *The New Politics of Class*, London: Sage Publications.

—— (1996) 'The Institutionalisation of Environmentalism: Ecological Discourse and the Second Transformation of the Public Sphere', in Lash, S. *et al.* (eds) *Risk, Environment and Modernity: Towards a New Ecology*, London: Sage.

Eder, K. and Kousis, M. (2001) (eds) *Environmental Politics in Southern Europe*, Dordrecht: Kluwer.

Ehenreich, B. and Ehenreich, J. (1979) 'The Professional-Managerial Class', in Walker, P. (ed.) *Between Capital and Labour*, Brighton: Harvester: 5–45.

Ehrlich, P. (1968) *The Population Bomb*, New York: Ballantine

Elander, I. (2000) 'Towards a Green Welfare Economy? The Green Party in Sweden since the 1998 Parliamentary Election', *Environmental Politics*, 9, 3: 137–44.

Ely, J. (1997) 'Green Politics in Europe and the United States' in Mayer, M. and Ely, J. (eds) *The German Greens*, Philadelphia: Temple University Press.

Epstein, B. (1991) *Political Protest and Cultural Revolution: Nonviolent Direct Action in the 1970s and 1980s*, Berkeley, CA: University of California Press.

—— (1997) 'The Environmental Justice/Toxics Movement: Politics of Race and Gender', *Capitalism, Nature, Socialism*, 8, 3: 63–87.

Evans, G. (1993) 'Hard Times for the British Green Party', *Environmental Politics*, 2, 2: 327–33.

Eyerman, J. and Jamison, A. (1991) *Social Movements: A Cognitive Approach*, Cambridge: Polity Press.

Faucher, F. (1992) 'Le Militantisme Vert à Aix-en-Provence', *DEA de Science Politique Comparative*, Université d'Aix-Marseilles, 3.

—— (1999) *Les Habits Verts de la Polititique*, Paris: Presses de Sciences Po.

Fendrich, J.M. and Krauss, C. (1978) 'Student Activism and Adult Left-Wing Politics: A Causal Model of Political Socialization for Black, White and Japanese Students of the 1960s Generation', in Kriesberg, L. (ed.) *Research in Social Movements, Conflicts and Change*, 1, Greenwich, CT: JAI Press: 231–55.

Fichter, T. (1991) 'Political Generations in the Federal Republic', *New Left Review*, 186: 78–88.

Fillieule, O. (2000) 'Between the Market and the State: French Environmental Organisations', paper presented at the European Consortium For Political Research, Copenhagen, 14–19 April.

Fillieule, O. and Ferrier, F. (1999) 'Ten Years of Environmental Activism in France', in Rootes, C. (ed.) Environmental Protest in Seven European Union States, Transformation of

Environmental Activism Project, unpublished Supplementary Report, University of Kent.

—— (2001) 'Local Environmental Politics in France. The case of the Louron valley, 1984–1996', paper presented at the European Consortium for Political Research, Grenoble 6–11 April.

Finger, M. (1992) (ed.) *The Green Movement Worldwide, Research in Social Movements: Conflict and Change*, Greenwich, CT: JAI Press.

Fischer, F. (2000) *Citizens, Experts and the Environment: the Politics of Local Knowledge*, Durham, NC: Duke University Press.

Flood, M. and Grove-White, R. (1976) *Nuclear Prospects: A Comment on the Individual, the State and Nuclear Power*, London: Friends of the Earth.

Foweraker, J. (1995) *Theorizing Social Movements*, London: Pluto.

Fogt, H. (1989) 'The Greens and the New Left: Influences of Left-Extremism on Green Party Organisation and Policies', in Kolinsky, E. (ed.) *The Greens in West Germany: Organisation and Policy-Making*, Oxford: Berg.

Foreman (1985) *Ecodefense: a field guide to monkeywrenching*, Tucson, AZ: Ned Ludd Books.

Foucault, M. (1977) *Discipline and Punish*, London: Allen Lane.

Foucault, M. and Rabinow, P. (eds) (1986) *The Foucault Reader*, Harmondsworth: Penguin.

Frankland, G.E. (1988) 'The Role of the Greens in West German Parliamentary Politics, 1980–87', *The Review of Politics*: 99–121.

—— (1989) 'Federal Republic of Germany: "Die Grünen"' in Müller-Rommel, F. (ed.) *New Politics in Western Europe: The Rise and Success of Green Parties and Alternative Lists*, Boulder, CO: Westview Press.

—— (1990) 'Does Green Politics have a Future in Britain? An American Perspective', in Rüdig, W. (ed.) *Green Politics One*, Edinburgh: Edinburgh University Press.

Frankland, G.E. and Schoonmaker, D. (1992) *Between Protest and Power: The Green Party in Germany*, Boulder, CO: Westview Press.

Franklin, M.N. and Rüdig, W. (1992) 'The Green Voter in the 1989 European Elections', *Environmental Politics*, 1, 4: 129–59.

Freeden, M. (1996) *Ideologies and Political Theory: A Conceptual Approach*, Oxford: Oxford University Press.

—— (2000) 'Practising Ideology and Ideological Practices', *Political Studies*, 48, 2: 302–322.

Fuchs, D. and Rucht, D. (1990) 'Support for New Social Movements in Five Western European Countries', paper presented at ESF/ESRC Conference on Political Participation, University of Manchester.

Gallet, G. (1999) 'The Transformations of Environmental Activism: Greenpeace France Case', paper presented at the European Consortium for Political Research, Mannheim.

Gamson, W. (1990) *The Strategy of Social Protest*, Belmont, CA: Wadsworth, 2nd edn.

Gerlach, L. and Hine, V. (1970) *People, Power and Change*, Indianapolis: Bobbs-Merrill.

Gerth, H.H. and Mills, C.W. (1991) *From Max Weber: Essays in Sociology*, London: Routledge.

Gibbs, L. (1982) *Love Canal: My Story*, Albany: State University of New York Press.

Giddens, A. (1984) *The Constitution of Society*, Cambridge: Polity.

—— (1990) *The Consequences of Modernity*, Cambridge: Polity.

—— (1994) *Beyond Left and Right: The Future of Radical Politics*, Cambridge: Polity.

Goodin, R. (1992) *Green Political Theory*, Cambridge: Polity.

Goodwin, J., Jasper, J. and Polleta, F. (2000) 'Return of the Repressed: the Rise and Fall of Emotions in Social Movement Theory', *Mobilization*, 5, 1: 65–84.

Gorz, A. (1989) *A Critique of Economic Reason*, London: Verso.

Gottlieb, R. (1993) *Forcing the Spring: The Transformation of the American Environmental Movement*, Washington, DC: Island Press.

Gould, K.A., Schnaiberg, A. and Weinberg, A. (1996) *Local Environmental Struggles: Citizen Activism in the Treadmill of Production*, Cambridge; Cambridge University Press.

Gould, P.C. (1988) *Early Green Politics: Back to Nature, Back to the Land and Socialism in Britain 1889–1900*, Brighton: Harvester.

Gouldner, A. (1979) *The Failure of Intellectuals and the Rise of the New Class*, London: Macmillan.

Grant, W. (2000) *Pressure Groups and the British Political System*, London: Palgrave.

Grau, Mireia and Egea, A. (2001) 'Responsiveness or Environmental Accountability? The Analysis of Citizens' Protests around Local Environmental Issues in Murcia', paper presented at the European Consortium for Political Research, Grenoble 6–11 April.

Gray, J. (1993) *Beyond the New Right*, London: Routledge.

Green Alternative European Link (GRAEL) (1988) *Rainbow Politics: Green Alternative Politics in the European Parliament*, Brussels.

Green Party (2001) *Manifesto for a Sustainable Society* (updated annually) London: Green Party, online at <http://www.greenparty.org.uk/policy.mfss>.

Green, S. (2000) 'Beyond Ethnoculturalism? German Citizenship in the new millennium, *German Politics*, 9,3: 105–24.

—— (2001) 'The Greens and the Reform of German Citizenship Law', paper presented at the European Consortium for Political Research, Grenoble 6–11 April.

Greenpeace (1999) *Greenpeace Global Annual Report*, online at <http://www.greenpeace.org/report99/>.

Griggs, S., Howarth, D. and Jacobs, B. (1998) 'Second Runway at Manchester', *Parliamentary Affairs*, 51, 3: 358–69.

Guha, R. (2000) *Environmentalism: A Global History*, New York: Longman.

Guha, R. and Martinez-Alier, J. (1997) *Varieties of Environmentalism: Essays North and South*, London: Earthscan.

Gurin, D. (1979) 'France: Making Ecology Political and Politics Ecological', *Contemporary Crises*, 3, 2: 149–66.

Halfacre, A. and Matheny, A.R. (1999) in Holder, J. and McGilvray (eds) *Locality and Identity: Environmental Issues in Law and Society*, Aldershot: Ashgate

Hajer, M.A. (1995) *The Politics of Environmental Discourse*, Oxford: Oxford University Press.

—— (1996) 'Ecological Modernisation as Cultural Politics', in Lash, S., Szerszynski, B. and Wynne, B. (eds) *Risk, Environment and Modernity*, London: Sage.

Hanf, K. and Jansen, A.I. (1998) (eds) *Governance and Environment in Western Europe*, Harlow: Longman.

Hardin, G. (1968) 'Tragedy of the Commons', *Science*, 162: 1243–8.

Harvey, D. (1999) 'The Environment of Justice', in Fischer, F. and Hajer, M. (eds) *Living with Nature*, Oxford: Open University Press

Hayes, G. (1994) 'In Splendid Isolation: the French Greens', *Environmental Politics*, 3, 1: 170–3.

—— (2000) 'Exeunt Chased by Bear: Structure, Action and the Environmental Opposition to the Somport Tunnel', *Environmental Politics*, 9, 2: 126–48.

—— (2001) 'The Opposition to Serre de al Fare: understanding local environmental protest in a multi-level policy process', paper presented at the European Consortium for Political Research, Grenoble 6–11 April.

Haynes, J. (1999) 'Power, Politics and Environmental Movements in the Third World', *Environmental Politics*, 8, 1: 222–42.

Hayward, T. (1995) *Ecological Thought: An Introduction*, Oxford: Polity.

Heath, A.F. and Evans, G. (1989) 'Working-Class Conservatives and Middle-Class Socialists', in Jowell, R., Witherspoon, S. and Brook, L. (eds) *British Social Attitudes: the Fifth Report*, Aldershot: Gower.

Heilbronner, R.L. (1974) *An Enquiry into the Human Prospect*, New York: Norton.

Hetherington, K. (1998) *Expressions of Identity: Space, Performance, Politics*, London: Sage.

Hoad, D. (1999) 'Local Environmental Campaigns: the Failure of the Anti-Ijburg Campaign', unpublished M.Phil. thesis, Keele University.

Hoffman, J. (1999) 'From a Party of Young Voters to an Ageing Generation Party? Alliance '90/The Greens after the 1998 Federal Elections', Environmental Politics, 8, 3: 140–6.

Hofrichter, J. and Reif, K. (1990) 'Evolution and Environmental Attitudes in the European Community', *Scandinavian Political Studies*, 13, 2: 119–46.

Holliday, I (1993) 'Une Présidente Verte in Nord-Pas de Calais: The First Year's Experience', *Environmental Politics*, 2, 3: 486–94.

—— (1994) 'Dealing in Green Votes: France, 1993', *Government and Opposition*, 29, 1: 64–79.

Hooghe, M. and Rihoux, B. (2000) 'The Green Breakthrough in the Belgian General Election of June 1999', *Environmental Politics*, 9, 3: 129–36.

Hülsberg, W. (1988) *The German Greens: A Social and Political Profile*, London: Verso.

Hunold, C (2001) 'Nuclear Waste in Germany: Environmentalists Between State and Society', *Environmental Politics*, 10, 3: 127–33.

Hutton, D. and Connors, L. (1999) *A History of the Australian Environmental Movement*, Cambridge: Cambridge University Press.

Inglehart, R. (1977) *The Silent Revolution: Changing Values and Political Styles Among Western Publics*, Princeton, NJ: Princeton University Press.

—— (1990a) *Culture Shift in Advanced Industrial Society*, Princeton, NJ: Princeton University Press.

—— (1990b) 'Values, Ideology and Cognitive Mobilization in New Social Movements', in Dalton, R.J. and Krechler, M. (eds) *Challenging the Political Order: New Social and Political Movements in Western Democracies*, Cambridge: Polity Press: 43–66.

Irvine, S. (1993) *Towards a Politics of Ecology*, Newcastle: Real World.

Irvine, S. and Ponton, A. (1988) *A Green Manifesto: Policies for a Green Future*, London: Optima.

—— (1992) 'Crisis in the English Green Party', unpublished paper.

Jacobs, M. (1996) *The Politics of the Real World*, London: Earthscan.

Jahn, D. (1993) 'The Rise and Decline of New Politics and the Greens in Sweden and Germany', *European Journal of Political Research*, 24, 2: 177–94.

Jakubal, M. (1998) 'Nonviolence Forever', *Earth First! Journal*, Eostar.

Jamison, A. and Ring, M. (1999) 'Ten Years of Environmental Activism in Sweden', in Rootes, C. (ed.) Environmental Protest in Seven European Union States, Transformation of Environmental Activism Project, unpublished Supplementary Report, University of Kent.

Jamison, A. Eyerman, R. and Cramer, J. (1990) *The Making of the New Environmental Consciousness*, Edinburgh: Edinburgh University Press.

Jasper, J. M. (1997) *The Art of Moral Protest: Culture, Creativity and Biography in Social Movements*, Chicago: University of Chicago Press.

Jasper, J.M. and Nelkin, D. (1992) *The Animal Rights Crusade: the Growth of a Moral Protest*, New York: Free Press.

Jensen, D. (1998) 'Action Speaks Louder than Words', *Earth First! Journal*, Beltane.

Jimenez, M. (1999a) 'Consolidation Through Institutionalisation? Dilemmas of the Spanish Environmental Movement in the 1990s', *Environmental Politics*, 8, 1: 149–71.

—— (1999b) 'Ten Years of Environmental Protest in Spain: Issues, Actors and Arenas', in Rootes, C. (ed.) Environmental Protest in Seven European Union States, Transformation of Environmental Activism Project, unpublished Supplementary Report, University of Kent.

—— (2001) 'National Policies and Local Struggle: Environmental Politics over Industrial Waste in Spain', paper presented at the European Consortium for Political Research, Grenoble 6–11 April.

Jones, P., Pestorius, M. and Law, B. (1995) 'Networking for Nonviolence in Australia', *Peace News*, February.

Jopke, C. (1991) 'Nuclear Power Struggles After Chernobyl: The Case of West Germany', *West European Politics*, 13, 2: 178–91.

—— (1992) 'Explaining Cross-National Variations of Two Anti-Nuclear Movements: A Political Process Perspective', *Sociology*, 26, 2: 311–31.

Jordan, G. (1998) 'Politics Without Parties', *Parliamentary Affairs*, 51, 3: 314–28.

Jordan, G. and Maloney, W. (1997) *The Protest Business? Mobilizing Campaign Groups*, Manchester: Manchester University Press.

Jowers, P. Dürrschmidt, J. O'Doherty, R and Purdue, D. (1999) 'Affective and Aesthetic Dimensions of Contemporary Social Movements in South West England', *Innovation*, 12, 1: 99–118.

Kaplan, G. (1992) *Contemporary Western European Feminism*, London: VCL Press.

Keane, J. (1996) *Reflections on Violence*, London: Verso.

—— (1988) *Democracy and Civil Society*, London: Verso.

Keck, M.E. and Sikkink, (1998) *Activists Beyond Borders: Advocacy Networks in International Politics*, Ithaca, NY: Cornell University Press.

Kellow, A. (2000) 'Norms, Interests and Environmental NGOs: the Limits of Cosmopolitanism', *Environmental Politics*, 9, 3: 1–22.

Kelly, P. (1984) *Fighting for Hope*, London: Chatto and Windus.

Kemp, P. and Wall, D. (1990) *A Green Manifesto for the 1990s*, Harmondsworth: Penguin.

Kemp, P., Wolf, F.O., Juquin, P., Antunes, C., Stengers, I. and Telkamper, W. (1992) *Europe's Green Alternative: A Manifesto for a New World*, London: Green Print.

Kenny, M. (1995) 'The First New Left in Britain', London: Lawrence and Wishart.

Kitschelt, H. (1986) 'Political Opportunity Structures and Political Protest: Anti-Nuclear Movements in Four Democracies', *British Journal of Political Science*, 16, 2: 58–98.

—— (1989) *The Logic of Party Formation*, Cornell: Cornell University Press.

—— (1990) 'The Medium is the Message: Democracy and Oligarchy in Belgian Ecology Parties', in Rüdig, W. (ed.) *Green Politics One*, Edinburgh: Edinburgh University Press: 82–114.

Kitschelt, H. and Hellemans, S. (1990) *Beyond the European Left: Ideology and Political Action in the Belgian Ecology Parties*, Durham, NC: Duke University Press.

Klandermans, B. (2000) 'Mobilization Forum: Must we Redefine Social Movements as Ideologically Structured Action?', *Mobilization*, 5, 1: 25–30.

Kolinsky, E. (1989) 'Women in the Green Party', in Kolinsky, E. (ed.) *The Greens in West Germany: Organisation and Policy-Making*, Oxford: Berg: 189–221.

Kousis, M. (1993) 'Collective Resistance and Sustainable Development in Rural Greece: The Case of Geothermal Energy on the Island of Milos', *Sociologia Ruralis*, 33, 1: 3–24.

——— (1999) 'Sustaining Local Environmental Mobilizations: Groups, Actions and Claims in Southern Europe', *Environmental Politics*, 8, 1: 172–98.

Konttinnen, A. (2000) 'From Grassroots to the Cabinet: the Green League of Finland', *Environmental Politics*, 9, 4: 129–34.

Krauss, Celine (1993) 'Blue Collar Women and toxic waste protests: the process of politicisation', in Richard Hofrichter, (ed.) *Toxic Struggles: the theory and Practice of Environmental Justice*, Philadelphia: PA, New Society.

Kriesi, H.P. (1989) 'New Social Movements and the "New Class"', *American Journal of Sociology*, 94, 1: 1078–1116.

Kriesi, H.P., Koopmans, R., Dynendak, J.W. and Giugni, M.G. (1995) *New Social Movements in Western Europe: A Comparative Analysis*, London: UCL Press.

Laclau, E. and Mouffe, C. (1985) *Hegemony and Socialist Strategy: Towards a Radical Democratic Politics*, London: Verso.

Ladrech, R. (1991) 'Social Movements and Party Systems: The French Socialist Party and New Social Movements', *West European Politics*, 12, 3: 262–79.

Lamb, R. (1996) *Promising the Earth*, London: Routledge.

Lange, P., Tarrow, S. and Irwin, C. (1989) 'Mobilization: Social Movements and Political Party Recruitment: the Italian Communist Party Since the 1960s', *British Journal of Political Science*, 20, 1: 15–42.

Langguth, G. (1986) *The Green Factor in German Politics*, Boulder and London: Westview Press.

Lawson, K (1967) *The Comparative Study of Political Parties*, New York: St Martin's Press.

——— (1988) 'When Linkage Fails', in Lawson, K. and Merkl, P. (eds) *When Parties Fail: Emerging Alternative Organizations*, Princeton: Princeton University Press.

Lean, G. (2001) 'Blame Ego Politics, Not Eco-Politics', *New Statesman* 16 July 2001.

Lee, M.F. (1995) *Earth First! Environmental Apocalypse*, Syracuse, NY: Syracuse University Press.

Lees, C. (1999) 'The Red-Green Coalition', *German Politics*, 8, 2: 174–97.

Leopold, A. (1948) *A Sand County Almanac*, Oxford: Oxford University Press, 1968.

Les Verts (undated) *Historique de l'écologie*, available online <http://www.les-verts.org>.

Lichtermann, P. (1996) *The Search for Political Community: American Activists Reinventing Tradition*, Cambridge: Cambridge University Press.

Lipietz, A. (1993) *Vert espérance: L'avenir de l'écologie politique*, Paris: Éditions la Découverte.

——— (2001) 'Statement for internal Primaries in Support of Presidential Candidature', online at <http://www.les-verts.org>.

Littletree, A. (2000) 'EF! Survives Agents of Repression' *Earth First! Journal*, 21, 1, November–December 2000.

Logue, J. (1982) *Socialism and Abundance: Radical Socialism in the Danish Welfare State*, Minneapolis: University of Minnesota Press.

Lucardie, P. (1980) 'The New Left in the Netherlands (1960–77): A Critical Study of New Political Ideas and Groups in the Netherlands with Comparative References to France and Germany', Ph.D. Thesis, Queen's University, Kingston, Ontario.

—— (1989) 'The Liturgy of the Logocrats: New Left Politics and New Middle Class Interests', paper presented at the Joint Sessions of the ECPR, Paris.

Lucardie, P. Voerman, G. and van Schuur, W. (1993) 'Different Shades of Green: A Comparison Between Members of *Green Links* and *De Groenen*', *Environmental Politics*, 2, 1: 40–62.

Lukes, S. (1990) 'Socialism and Capitalism, Left and Right', *Social Research*, 57, 3: 571–8.

McAdam, D. (1982) *Political Process and the Development of Black Insurgency, 1930–1970*, Chicago: University of Chicago Press.

—— (1988) *Feedom Summer*, New York: Oxford University Press.

McAdam, D. McCarthy, J.D, and Zald, M.N. (eds) (1996) *Comparative Perspectives on Social Movements: Political Opportunities, Mobilizing Structures, and Cultural Framings*, Cambridge: Cambridge University Press.

McAdam, D. and Rucht, D. (1993) 'The Cross-National Diffusion of Movement Ideas', in Dalton, R.J. (ed.) 'Citizens, Protest and Democracy', *The Annals of the American Academy of Political and Social Science*, 528: 56–74.

McAdam, D., Tarrow, S. and Tilly, C. (1996) 'To Map Contentious Politics', *Mobilization* 1: 17–34.

McCarthy, J.D. and Zald, M.N. (1977) 'Resource Mobilization and Social Movements: A Partial Theory', *American Journal of Sociology*, 82: 1212–41.

McCormick, J. (1991) *British Politics and the Environment*, London: Earthscan.

McCullough, J. (2000) 'Globalising Violence', *Arena*, No.50.

MacIntyre, A. (1985) *After Virtue: a study in moral theory*, London: Duckworth, 2nd edn.

McLellan, D. (1986a) *Ideology*, Milton Keynes: Open University Press.

McNeish, W. (2000) 'The Vitality of Local Protest: Alarm UK and the British Anti-Roads Protest Movement', in B. Seel, Paterson, M. and Doherty, B. (eds) *Direct Action in British Environmentalism*, London: Routledge.

Maenz, K. 2000 'The Life and Times of Our Beloved Journal – A Not-So-Brief History', *Earth First! Journal* November-December, 21.

Maguire, D. (1992) 'When the Streets Begin to Empty: The Demobilisation of the British Peace Movement after 1983', *West European Politics*, 15, 4: 75–94.

Maier, J. (1990) *The Green Parties in Western Europe – A Brief History, Their Successes and Their Current Problems*, Bonn: Die Grünen.

Mannheim (1952) 'The Problem of Generations', in P. Kecskemetis (ed.) *Essays on the Sociology of Knowledge*, London: Routledge.

Marcuse, H. (1964) *One Dimensional Man*, London; Routledge.

Markovits, A.S. and Gorski, P.S. (1993) *The German Left: Green and Beyond*, Cambridge: Polity Press.

Marshall, P. (1992) *Nature's Web: Rethinking Our Place on Earth*, London: Cassell.

Martell, L. (1994) *Ecology and Society: An Introduction*, Cambridge: Polity.

Martin, G. (1998) 'Generational Differences Among New Age Travellers', *Sociological Review*, 46: 735–56.

Mattausch, J. (1991) 'The Political Implications of CND's Social Basis of Support', paper presented at the PSA Conference, University of Lancaster.

Mayer, M. and Ely, J. (eds) *The German Greens*, Philadelphia: Temple University Press.

Meadows, D., Meadows, D. Randers, J and Behrens III, W. (1974) *The Limits to Growth*, London: Pan.

Mejer, J. (1993) 'The Events of May-June 1968', *Theory, Culture and Society*, 10, 1: 53–74.

Mellor, M. (1992a) *Breaking the Boundaries: Towards a Green Feminist Socialism*, London: Virago.

—— (1992b) 'Green Politics: Ecofeminist, Ecofeminine or Ecomasculine?', *Environmental Politics*, 1, 2: 229–51.

Melucci, A. (1989) *Nomads of the Present*, London: Radius.

—— (1996) *Challenging Codes: Collective Action in the Information Age*, Cambridge: Cambridge University Press.

Mewes, H. (1997) 'A Brief History of the German Green Party', in Mayer, M. and Ely, J. (eds) *The German Greens: Paradox Between Movement and Party*, Philadelphia: Temple University Press.

Meyer, D. and Tarrow, S. (1998) 'A Movement Society: Contentious Politics for a New Century', in Mayer, D. and Tarrow, S. (1998) (eds) *The Social Movement Society: Contentious Politics for a New Century*, Oxford: Rowman and Littlefield.

Michels, R. (1911) *Political Parties: A Sociological Study of the Oligarchical Tendencies of Modern Democracy*, New York: The Free Press, 1962.

Mittler, D. (1999a) 'German Eco-Tax Success? Hardly…', *Link*, April/June.

—— (1999b) 'Eclipse of the German Greens', online source <http://www.the ecologist.org/GermanGreens.html>, accessed on 21 November 2000.

Mol, A.P.J. and Spaargaren, G. (2000) 'Ecological Modernisation Theory in Debate: A Review', *Environmental Politics*, 9, 1: 17–49.

Monbiot, G. (2000) *Captive State: the Corporate Takeover of Britain*, London: Macmillan.

—— (2001) 'Sleeping with the Enemy' *The Guardian*, 4 September.

Müller-Rommel, F. (1990) 'New Political Movements and New Political Parties in Western Europe', in Dalton, R. and Küchler (eds) *Challenging the Political Order*, Cambridge: Polity Press.

Müller-Rommel, F. (1989) 'Green Parties and Alternative Lists Under Cross-National Perspective', in Müller-Rommel, F. (ed.) *New Politics in Western Europe: The Rise and Success of Green Parties and Alternative Lists*, London: Westview Press.

Müller-Rommel, F. and Poguntke, T. (1989) 'The Unharmonious Family: Green Parties in Western Europe', in Kolinsky, E. (ed.) *The Greens in West Germany: Organisation and Policy-Making*, Oxford: Berg Press.

Mulley, Marie (1999) *The Guardian*, 22 June.

Mushaben, J. (1985) 'Cycles of Peace Protest in West Germany', *West European Politics*, 8, 1: 24–40.

Naess, A. (1973) 'The Shallow and the Deep, Long-Range Ecology Movement: A Summary', *Inquiry*, 16: 95–100.

—— (1988) 'The Basics of Deep Ecology', *Resurgence*, 126: 14–17.

Nanterre Students (1969) 'Why Sociologists?', in Cockburn, A. and Blackburn, R. (eds) *Student Power*, Harmondsworth: Penguin.

Nelkin, D. and Pollak, M. (1981) *The Atom Beseiged: Extraparliamentary Dissent in France and Germany*, Cambridge, MA: MIT Press.

Newman, R. (2001) 'Making Environmental Politics: Women and Love Canal Activism', *Women's Studies Quarterly*, 24, 1–2: 65–84.

Oberschall, A. (1973) *Social Conflict and Social Movement*, Englewood Cliffs: Prentice Hall.

Offe, C. (1985) 'New Social Movements: Changing Boundaries of the Political', *Social Research*, 52, 4: 816–68.

—— (1990) 'Reflections on the Institutional Self-Transformation of Movement Politics: A Tentative Stage Model', in Dalton, R.J. and Kuechler, M. (eds) *Challenging the Political Order: New Social and Political Movements in Western Democracies*, Cambridge: Polity Press.

Oliver, P and Johnston, H. (2000a) 'What a Good Idea! Ideology and Frames in Social Movement Research', *Mobilization*, 5, 1: 37–54.

—— (2000b) 'Mobilization Forum: Breaking the Frame', *Mobilization*, 5, 1: 61–64.

Oliver, P. and Marwell, G. (1992) 'Mobilizing Technologies for Collective Action', in Morris, A.D. and Mueller, C.M. (eds) *Frontiers in Social Movement Theory*, New Haven, CT: Yale University Press.

Olson, M. (1965) *The Logic of Collective Action*, Cambridge: MA: Harvard University Press.

Ophuls, W. (1977) 'Frugality and Freedom', in Dobson, A. (ed.) *The Green Reader*, London: Andre Deutsch: 107–11.

Padgett, S. and Paterson, W. (1991a) *A History of Social Democracy in Post-War Europe*, Harlow: Longman.

—— (1991b) 'The Rise and Fall of the West German Left', *New Left Review*, 186: 46–77.

Paehlke, R. (1989) *Environmentalism and the Future of Progressive Politics*, New Haven, CT: Yale University Press.

Panebianco, A. (1988a) *Political Parties: Organization and Power*, Cambridge: Cambridge University Press.

Papadakis, E. (1984) *The Green Movement in West Germany*, London: Croom Helm.

—— (1988) 'Social Movements, Self-Limiting Radicalism and the Green Party in West Germany', *Sociology*, 22, 3: 433–54.

Parkin, F. (1968) *Middle Class Radicalism*, Manchester: Manchester University Press.

—— (1988) 'Green Strategy', in Dodds, F. (ed.) *Into the Twenty-First Century: An Agenda for Political Realignment*, Basingstoke: Green Print: 163–80.

—— (1989) *Green Parties: An International Guide*, London: Heretic Books.

—— (ed.) (1991) *Green Light on Europe*, London: Heretic Books.

Parry, G., Moyser, G. and Day, N. (1992) *Political Participation and Democracy in Britain*, Cambridge: Cambridge University Press.

Paterson, M. (2000) 'Swampy Fever: Media Constructions and Direct Action Politics', in Seel, B. Paterson, M. and Doherty, B. (eds) *Direct Action in British Environmentalism*, London: Routledge.

Peng-Er, L. (1999) *Green Politics in Japan*, London: Routledge.

Pepper, D. (1984) *The Roots of Modern Environmentalism*, Beckenham: Croom Helm.

—— (1991) *Communes and the Green Vision: Counter-Culture, Lifestyles and the New Age*, Basingstoke: Green Print.

—— (1993a) *Eco-Socialism: From Deep Ecology to Social Injustice*, London: Routledge.

—— (1993b) 'Anthropocentrism, Humanism and Eco-Socialism: A Blueprint for the Survival of Ecological Politics', *Environmental Politics*, 2, 3: 428–52.

Percheron, A. (1991) 'La mémoire des generations: La guerre d'Algerie – Mai 68', in Duhamel, O. and Jaffré, J. (eds) *SOFRES: L'état de l'opinion 1991*, Paris: Seuil.

Phillips, A. (1987b) *Divided Loyalties: Dilemmas of Sex and Class*, London: Virago.

—— (1991) *Engendering Democracy*, Cambridge: Polity Press.

Piccone, P. (1988) 'Reinterpreting 1968: Mythology on the Make', *Telos*, 77: 7–43.

Pickerill, J. (2000) Environmental Activists' Use of Computer Mediated Communication', unpublished Ph.D., University of Newcastle-upon-Tyne.

Pilat, J.F. (1982) 'Democracy or Discontent? Ecologists in the European Electoral Arena', *Government and Opposition*, 17, 2: 222–34.

Piven, F.F. and Cloward, R. (1977) *Poor People's Movements: Why they Succeed, How they Fail*, New York: Pantheon Press.

Plows, A. (1998) 'Earth First! Defending Mother Earth, direct-style', in McKay, G., (ed.) *DiY Culture: Party and Protest in Nineties Britain*, London: Verso.

Plows, A., Wall, D. and Doherty, B. (2001) 'From the Earth Liberation Front to Universal Dark Matter: the challenge of covert repertoires to social movement research', paper presented at the European Consortium for Political Research, Grenoble 6–11 April.

Poguntke, T. (1987) 'The Organisation of a Participatory Party: The German Greens', *European Journal of Political Research*, 15: 609–33.

—— (1989a) 'The "New Politics Dimension" in European Green Parties', in Müller-Rommel, F. (ed.) *New Politics and New Politics Parties in Western Europe: The Rise and Success of Green Parties and Alternative Lists*, Boulder, CO: Westview Press.

—— (1989b) '*Basisdemokratie* and Political Realities: The German Green Party and the Limits to Grassroots Democracy', paper presented at the Conference of the American Political Science Association, Atlanta.

—— (1990) 'Party Activists Versus Voters: Are the German Greens Losing Touch with the Electorate', in Rüdig, W. (ed.) *Green Politics One*, Edinburgh: Edinburgh University Press.

—— (1992) 'Between Ideology and Empirical Research: The Literature on the German Green Party', *European Journal of Political Research*, 21, 4: 337–56.

—— (1993a) *Alternative Politics: The German Green Party*, Edinburgh: Edinburgh University Press.

—— (1993b) 'Goodbye to Movement Politics? Organisational Adaptation of the German Green Party', *Environmental Politics*, 2, 3: 379–404.

Porritt, J. (1984) *Seeing Green*, Oxford: Basil Blackwell.

Prendiville, B. (1989) 'France: "Les Verts"', in Müller-Rommel, F. (ed.) *New Politics in Western Europe: The Rise and Success of Green Parties and Alternative Lists*, Boulder, CO: Westview: 87–99.

—— (1992) 'The French Greens, Inside Out', *Environmental Politics*, 1, 2: 283–7.

—— (1993a) *Écologie, la politique autrement? Culture, sociologie et histoire des écologistes*, Paris: Editions Harmattan.

—— (1993b) 'The "Entente Ecologiste" and the French Legislative Elections of March 1993', *Environmental Politics*, 2, 3: 479–86.

Prendiville, B. and Chafer, T. (1990) 'Activists and Ideas in the Green Movement in France', in Rüdig, W. (ed.) *Green Politics One*, Edinburgh: Edinburgh University Press: 177–209.

Pronier, R. and le Seigneur, V.J. (1992) *Génération verte: les écologistes en politique*, Paris: Presses de la Renaissance.

Pulido, L. (1996) *Environmentalism and Economic Justice: Two Chicano Struggles in the Southwest*, Tucson, AZ: the University of Arizona Press.

Purdue, D.A. (2000) *Anti-GenetiX: the emergence of the anti-GM Movement*, Aldershot: Ashgate.

Purkis, J. (2000) 'Modern Millenarians?: anti-consumerism and the new urban environmentalism', in Seel, B. Paterson, M. and Doherty, B. (eds) *Direct Action in British Environmentalism*, London: Routledge.

Raschke, J. (1993) *Die Grünen: wie sie wurden, was sie sind*, Köln: Bund Verlag.

Rawcliffe, P. (1998) *Environmental Pressure Groups in Transition*, Manchester; Manchester University Press.

Restless, R. (2000) 'First American Tripod is Constructed in Cove/Mallard, 1992' *Earth First! Journal* 21, 1 November–December.

Rihoux, B. (2000) 'Governmental Participation and the Organisational Transformation of Green Parties: a Comparative Enquiry', paper presented at the Political Studies Association Conference, 10–13 April, London.

Roberts, G.K. (2001) 'How Greens Govern: the Experience of the Green Party in Länder Government', paper presented at the European Consortium for Political Research, Grenoble 6–11 April.

Robinson, N. (2000) 'The politics of the car: the limits of actor-centred models of agenda setting', in Seel, B. Paterson, M. and Doherty, B. (eds) *Direct Action in British Environmentalism*, London: Routledge.

Robinson, M. and Boyle, M.J. (1987) 'Nuclear Energy in France', in Blowers, A. and Pepper, D. (eds) *Nuclear Power in Crisis*, Beckenham: Croom Helm: 55–84.

Roche, A. (1993) 'Les écologistes à la veille des législatives: l'heure des choix', *Revue Politique et Parlémentaire*, 963: 29–34.

—— (1993) 'Mars 1993: un révélateur des faiblesses des écologistes', *Revue Politique et Parlémentaire*, 964: 34–41.

Rochon, T. (1988) *Mobilizing for Peace: Antinuclear Movements in Western Europe*, Princeton, NJ: Princeton University Press.

—— (1990) 'The West European Peace Movement and the Theory of New Social Movements' in Dalton, R.J. and Kuechler, M. (eds) *Challenging the Political Order: New Social and Political Movements in Western Democracies*, Cambridge: Polity Press.

—— (1998) *Culture Moves: Ideas, Activism and Changing Values*, Princeton, NJ: Princeton University Press.

Rohrschneider, R. (1988) 'Citizen Attitudes Toward Environmental Issues: Selfish or Selfless?' *Comparative Political Studies*, 21, 3: 347–67.

—— (1990) 'The Roots of Public Opinion Towards New Social Movements: An Empirical Test of Competing Explanations', *American Journal of Political Science*, 34, 1: 1–30.

—— (1993) 'Environmental Belief Systems in Western Europe', *Comparative Political Studies*, 26, 1: 3–29.

Rootes, C. (1992) 'The New Politics and the New Social Movements: Accounting for British Exceptionalism', *European Journal of Political Research*, 22, 2: 171–92.

—— (1994) 'A New Class? The Higher Educated and the New Politics', in Maheu, L. (ed.) *New Actors, New Agendas*, London: Sage.

—— (1995a) 'Britain: Greens in a Cold Climate', in Richardson, D. and Rootes, C. (eds) *The Green Challenge: the development of green parties in Europe*, London: Routledge.

—— (1995b) 'Environmental Consciousness, Institutional Structures and Political Competition in the Formation and Development of Green Parties', in Richardson, D. and Rootes, C. (eds) *The Green Challenge: the development of green parties in Europe*, London: Routledge .

—— (1997a) 'Environmental Movements and Green Parties in western and eastern Europe', in Redclift, M. and Woodgate, G. (eds) *The International Handbook of Environmental Sociology*, Cheltenham: Edward Elgar.

—— (1997b) 'From Resistance to Empowerment: the Struggle Over Waste Management and its implications for Environmental Education', paper presented at IRNES Conference, Imperial College.

—— (1999) 'Acting Globally, Thinking Locally? Prospects for a Global Environmental Movement', *Environmental Politics*, 8, 1: 290–310.

—— (2000) 'Environmental Protest in Britain 1988–97', in Seel, B. Paterson, M. and Doherty, B. (eds) *Direct Action in British Environmentalism*, London: Routledge.

—— (2001) 'Discourse, Opportunity or Structure? Determining outcomers of local mobilisations against waste incinerators in England', paper presented at the European Consortium for Political Research, Grenoble 6–11 April.

Rootes, C. and Miller, A. (2000) 'The British Environmental Movement: Organisational Field and Network of Organisations', paper presented at the European Consortium for Political Research, Copenhagen, 14–19 April 2000.

Rootes, C. and Adams, D. and Saunders, C. (2001) 'Local Environmental Politics in England: Environmental Activism in South East London and East Kent Compared', paper presented at the European Consortium for Political Research, Grenoble 6–11 April.

Rootes, C. Seel, B. and Adams, D. (2000) 'The old, the new and the old new: British Environmental Organisations from Conservationism to Political Ecologism', paper presented at the European Consortium for Political Research, Copenhagen, 14–19 April 2000.

Roselle, M. (1998) 'Building Movement Basics', *Earth First! Journal*, Eostar.

Roseneil, S. (1995) *Disarming Patriarchy: Feminism and Political Action at Greenham*, Buckingham: Open University Press.

Roth, R. and Murphy, D. (1997) 'From Competing Factions to the Rise of the Realos', in Mayer, M. and Ely, J. (eds) *The German Greens*, Philadelphia: Temple University Press.

Rowbotham, S., Segal, L. and Wainwright, H. (1981) *Beyond the Fragments*, Spokesman, Nottingham.

Rowell, A. (1996) *Green Backlash*, London: Routledge.

Rucht, D. (1989) 'Environmental Movement Organizations in West Germany and France: Structure and Interorganizational Relations', in Klandermans, B. (ed.) *International Social Movement Research*, 2: 61–94.

—— (1990) 'The Strategies and Action Repertoires of New Movements', in Dalton, R.J. and Kuechler, M. (eds) *Challenging the Political Order: New Social and Political Movements in Western Democracies*, Cambridge: Polity Press.

—— (1995) 'Ecological Protest as Calculated Law Breaking: Greenpeace and Earth First! in Comparative Perspective', in Rüdig, W. (ed.) *Green Politics Three*, Edinburgh: Edinburgh University Press.

—— (1998) 'The Structure and Culture of Collective Protest in Germany since 1950', in Meyer, D. and Tarrow, S. (eds) *The Social Movement Society*, Oxford: Rowman and Littlefield.

Rucht, D. and Roose, J. (1999) 'The German Environmental Movement at A Crossroads?', *Environmental Politics*, 8, 1: 59–80.

—— (2000) 'Neither Decline nor Sclerosis: the Organisational Structure of the German Environmental Movement', paper presented at the European Consortium for Political Research, Copenhagen, 14–19 April 2000.

—— (2001) 'The Transformation of Environmental Activism in Berlin', ECPR paper presented at the European Consortium for Political Research, Grenoble 6–11 April.

Rüdig, W. (1986) 'Energy, Public Protest and Green Parties: A Comparative Analysis', Ph.D. Thesis, University of Manchester.

—— (1989) 'Explaining Green Party Development: Reflections on a Theoretical Framework', paper presented at the UK Political Studies Association Conference, Warwick.

—— (1990) *Anti-Nuclear Movements: A World Survey of Opposition to Nuclear Energy*, Harlow: Longman.

—— (2000) 'Phasing Out Nuclear Energy in Germany', *German Politics*, 9, 3: 43–80.

—— (2002) 'German Greens in Government, *Environmental Politics*, 11, 1 (forthcoming).

Rüdig, W., Bennie, L.G. and Franklin, M.N. (1991) *Green Party Members: A Profile*, Glasgow: Delta Publications.

—— (1992) 'Local and National Activism in the British Green Party', paper presented at the PSA Conference, Belfast.

—— (1993) *Green Blues: The Rise and Decline of the British Green Party*, Strathclyde Papers on Government and Politics 95.

Rüdig, W. and Franklin, M.N. (1992) 'Green Prospects: The Future of Green Parties in Britain, France and Germany; in Rüdig, W. (ed.) *Green Politics Two*, Edinburgh: Edinburgh University Press.

Rüdig, W. and Lowe, P. (1986) 'The Withered "Greening" of British Politics', *Political Studies*, 34, 2: 262–84.

Ryle, M. (1988) *Ecology and Socialism*, London: Radius.

Sachs, W. (1993) 'Global Ecology and the Shadow of "Development" ', in Sachs, W. (ed.) *Global Ecology*, London: Zed.

Sainteny, G. (1991) *Que sais-je? Les Verts*, Paris: Presses Universitaires de France.

Sale, K. (1985) *Human Scale*, London: Secker and Warburg.

—— (1996) *Rebels Against the Future: The Luddites and their war on the Industrial Revolution, Lessons for a Computer Age*, London. Quartet Books.

Sandweiss, S. (1998) 'The Social Construction of Environmental Justice', in Camacho, D.E. (ed.) *Environmental Injustice: Political Struggles*, Durham, NC: Duke University Press.

Sargisson, L. (2000) *Utopian Bodies and the Politics of Transgression*, London: Routledge.

Scarce, R. (1990) *Eco-Warriors: Understanding the Radical Ecology Movement*, Chicago, IL: Noble Press.

Scharf, T. (1991) 'The German Greens and the New Local Politics', Ph.D. Thesis, University of Aston.

—— (1994) *The German Greens: Challenging the Consensus*, Oxford: Berg.

Schlosberg, D. (1999) *Environmental Justice and the New Pluralism*, Oxford: Oxford University Press.

Schmitt-Beck, R. (1992) 'A Myth Institutionalised: Theory and Research on New Social Movements in Germany', *European Journal of Political Research*, 21, 4: 357–84.

Schoonmaker, D. (1988) 'The Challenge of the Greens to the West German Party System', in Lawson, K. and Merkl, P. (eds) *When Parties Fail: Emerging Alternative Organisations*, Princeton: Princeton University Press: 41–75.

Schumacher, E.F. (1974) *Small is Beautiful*, London: Abacus.

Seed, J. (2000) 'EF! Gets Down (Under)', *Earth First! Journal* 21 (1): 20–1.

Seel, B. (1999) 'Strategic Identities: Strategy, Culture and Consciousness in the New Age and Road Protest Movements', unpublished Ph.D. thesis, Keele University.

Shaiko (1993) 'Greenpeace USA, Something Old, Something New', in Dalton, R.J. (ed.) 'Citizens, Protest and Democracy', *The Annals of the American Academy of Political and Social Science*, 528.

Sheller, M. (2000) 'Social Networks and Social Flows', paper presented at the ESRC Conference on Social Movement Analysis: the Network Perspective, Loch Lomond.

Sherkat, D.E. and Blocker, T.J. (1997) 'Explaining the Political and Personal Consequences of Protest', *Social Forces*, 75: 1049–76.

Shull, T. (1992) 'The Ecologists in the Regional Elections: Strategies Behind the Split', *French Politics and Society*, 10, 2: 13–29.

Smelser, N. (1962) *Theories of Collective Behaviour*, London: Routledge and Kegan Paul.

Snap Dragon (1998) 'Beyond Civil Disobedience', *Earth First! Journal, Eostar*.

Snow, D. and Benford, R. (1992) 'Master Frames and Cycles of Protest', Morris, A.D. and Mueller, C.M. (eds) *Frontiers in Social Movement Theory*, New Haven, CT: Yale University Press.

Snow, R. and Benford, R. (2000) 'Mobilization Forum: Clarifying the Relationship Between Framing and Ideology', *Mobilization*, 5, 1: 56–60

Snow, D. Rocheford, D.E, Worden, S. and Benford, R. (1986) 'Frame Alignment Processes, Micromobilization and Movement Participation', *American Sociological Review*, 51: 464–81.

Stansill, P. and Mairowitz, D.Z. (eds) (1971) *BAMN: By any Means Necessary, Outlaw Manifestos and Ephemera, 1965–70*, Harmondsworth: Penguin.

Steinberg, M. (1998) 'Tilting the frame: Considerations on collective action framing from a discursive turn', *Theory and Society*, 27, 6: 845–72.

Stewart, J. Bray, J. and Must, E. (1995) *Roadblock: How People Power is Wrecking the Road Programme*, London: ALARM UK.

Stock, M.J. (1991) 'The Portuguese Greens: Growing Pains or Compromised Future', paper presented at the Joint Sessions of the ECPR, Colchester.

Strange, P. (1983) *It'll Make a Man of You: A Feminist View of the Arm's Race*, Nottingham: Mushroom Books.

Strøm, K. (1990) 'A behavioural theory of competitive political parties', *American Journal of Political Science*, 34: 365–98.

Students Against Nuclear Energy (1980) *Anti-Nuclear Now…Or Never*, London.

Sutton, P.W. (2000) *Explaining Environmentalism: In Search of a New Social Movement*, Aldershot: Ashgate.

Szarka, J. (1999) 'The Parties of the French "Plural Left"', *West European Politics*, 22, 4: 20–37.

Szasz, A. (1994) *Eco-Populism: Toxic Waste and the Movement for Environmental Justice*, London: UCL Press.

Szerszynski, B. (1995) 'Environmental NGOs in Britain: Communication Activities and Institutional Change', unpublished MS, CSEC, Lancaster University.

Tarrow, S. (1990) 'The Phantom at the Opera: Political Parties and Social Movements of the 1960s and 1970s in Italy', in Dalton, R.J. and Kuechler, M. (eds) *Challenging the Political Order: New Social and Political Movements in Western Democracies*, Cambridge: Polity Press.

—— (1998) *Power in Movement*, Cambridge: Cambridge University Press, revised, 2nd edn.

Taylor, B. (ed.) (1995) *Ecological Resistance Movements*, New York: State University of New York Press.

Taylor, R. and Pritchard, C. (1980) *The Protest Makers: The British Nuclear Disarmament Movement 1958–65 Twenty Years On*, Oxford: Pergamon Press.

Tesh, S. (2000) *Uncertain Hazards: Environmental Activists and Scientific Proof*, Durham, NC: Duke University Press.

Tilly, C. (1995) Contentious Repertoires in Great Britain, 1758–1834, in Traugott, M. (ed.) *Repertoires and Cycles of Collective Action*, Durham, NC: Duke University Press.

Torgerson, D. (1999) *The Promise of Green Politics*, Durham, NC: Duke University Press.

—— (2000) 'Farewell to the Green Movement? Political Action and the Green Public Sphere', *Environmental Politics*, 9, 4: 1–19.

Touraine, A. (1981) *The Voice and the Eye: An Analysis of Social Movements*, Cambridge: Cambridge University Press.

—— (1985) 'An Introduction to the Study of Social Movements', *Social Research*, 52, 4: 749–87.

—— (2000) *Can We Live Together: Equality and Difference*, Cambridge: Cambridge University Press.

Touraine, A., Hegedus, Z. and Wieviorka, M. (1983) *Anti-Nuclear Protest: The Opposition to Nuclear Energy in France*, Cambridge: Cambridge University Press.

Urry, J. (2000) *Sociology Beyond Societies: mobilities for the twenty-first century*, London: Routledge.

Van der Heijden, H.A. (1997) 'Political Opportunity Structures and the Institutionalisation of the Environmental Movement', *Environmental Politics*, 6, 4: 25–50.

—— (1999) 'Environmental Movements, Ecological Modernisation and Political Opportunity Structures', *Environmental Politics*, 8, 1: 199–221.

—— (2000) 'Dutch Environmentalism in the 1990s', paper presented at the European Consortium for Political Research, Copenhagen, 14–19 April 2000.

Van der Heijden, Koopmans, R. and Giugni, M. (1992) 'The West European Environmental Movement', in Finger, M. (ed.) *The Green Movement Worldwide*, Greenwich, CT, JAI Press.

Veen, H.J. (1989) 'The Greens as a Milieu Party', in Kolinsky, E. (ed.) *The Greens in West Germany: Organisation and Policy-Making*, Oxford: Berg: 31–60.

Veillet, G. (2001) 'The Seige Mentality: Radical Local Activism and the Animal Rights Crusade in London', paper presented at the European Consortium for Political Research, Grenoble 6–11 April.

Vidal, J. (1997) *McLibel: Burger Culture on Trial*, London: Macmillan.

Vincent, A. (1992) *Modern Political Ideologies*, Oxford: Blackwell.

Voerman, G. (1995) 'The Netherlands: Losing Colours, Turning Green', in Richardson, D. and Rootes, C. *The Green Challenge: the development of green parties in Europe*, London: Routledge.

Vuillamy, E. (1998) 'Eco-Activists Turn Up the Heat', *The Observer*, 25 October.

Wainwright, H. (1987) *Labour: A Tale of Two Parties*, London: Hogarth Press.

—— (1994) *Arguments for a New Left: Answering the Free Market Right*, Oxford: Blackwell.

Wall, D. (1989) 'Kinnockites for Life on Earth?', *New Ground*, Autumn.

—— (1990) *Getting There: Steps to a Greener Society*, Basingstoke: Green Print.

—— (1991) 'Goodbye to the Green Party?' *Green Line*, 88: 114–5.

—— (1994a) *Weaving a Bower Against Endless Night…An Illustrated History of the UK Green Party*, Hereford: Green Party.

—— (1994b) *Green History: A Reader in Environmental Literature, Philosophy and Politics*, London: Routledge.

—— (1999) *Earth First! and the Anti-Roads Movement: Radical Environmentalism and Comparative Social Movements*, London: Routledge.

—— (2000) 'Snowballs, Elves and Skimmingtons: Genealogies of Direct Action', in Seel, B., Paterson, M. and Doherty, B. (eds) *Direct Action in British Environmentalism*, London: Routledge.

Waller, M. and Millard, F. (1992) 'Environmental Politics in Eastern Europe', *Environmental Politics*, 1, 2: 159–85.

Walsh, E. and Warland, R. (1983) 'Social Movement Involvement in the Wake of a Nuclear Accident: activists and Free Riders in the Three Mile Island Area', *American Sociological Review*, 48: 764–81.

Walsh, E. Warland, R. and Clayton Smith, D. (1993) 'Backyards, NIMBYs and Incinerator Sitings', *Social Problems*, 40, 1: 25–38.

Wapner, P. (1996) *Environmental Activism and World Civic Politics*, Albany: State University of New York.

Warcry (2001) 'The Criminalisation of Ecology', *Earth First! Journal*, 21 7: 5–7.

Ware, A. (1987) *Citizens, Parties and the State*, Cambridge: Polity Press.

Wasmund, K. (1986) 'The Political Socialization of West German Terrorists', in Merkl, P. (ed.) *Political Violence and Terror: Motifs and Motivations*, Berkeley: University of California Press.

Weale, A. (1992) *The New Politics of Pollution*, Manchester: Manchester University Press.

Welsh, I. (2000) *Mobilizing Modernity: The Nuclear Moment*, London: Routledge.

Welsh, I. and Chesters, G. (2001) 'Re-Framing Social Movements: Margins, Meanings and Governance', Working Paper 19, School of Social Sciences, Cardiff University.

Weston, J. (1986) *Red and Green*, London: Pluto.

Whalen, J. and Flacks, R. (1989) 'Echoes of Rebellion: the Liberated Generation Grows Up', *Journal of Political and Military Sociology*, 12: 61–78.

Whiteside, K. (1997) 'French Ecosocialism: From Utopia to Contract', *Environmental Politics*, 6, 3: 99–124.

Whittier, N. (1995) *Feminist Generations*, Philadelphia: Temple University Press.

Wiener, M. (1981) *English Culture and the Decline of the Industrial Spirit, 1850–1950*, Cambridge: Cambridge University Press.

Wiesenthal, H. and Ferris, J. (eds) (1993) *Realism in Green Politics: Social Movements and Ecological Reform in Germany*, Manchester: Manchester University Press.

Williams, R. (1989) *Resources of Hope*, London: Verso.

Wilson, F.L. (1988) 'When Parties Refuse to Fail: The Case of France', in Lawson, K. and Merkl, P. (eds) *When Parties Fail: Emerging Alternative Organizations*, Princeton: Princeton University Press.

Wissenburg, M. (1998) *Green Liberalism: the Free and the Green Society*, London: UCL Press.

Wynne, B. (1996) 'May the Sheep Safely Graze? A Reflexive View of the Expert-Lay Knowledge Divide', in Lash, S., Szerszynski, B. and Wynne, B. (eds) *Risk, Environment and Modernity: Towards a New Ecology*, London: Sage.

York Earth First! (2000) 'Who Owns the Movement', unpublished leaflet, adapted from a paper by Michael Albert, available at <http://www.lbbs.org/who_owns.htm>.

Young, S. (1992) 'The Different Dimensions of Green Politics', *Environmental Politics*, 1, 1: 9–44.

—— (1993) *The Politics of the Environment*, Manchester: Baseline Books.

Zakin, S. (1993) *Coyotes and Town Dogs: Earth First! and the Environmental Movement*, New York: Penguin.

Zald, M.N. (2000) 'Ideologically Structured Action: An Enlarged Agenda for Social Movement Research', *Mobilization*, 5, 1: 1–16.

Zinn, C. (n.d) 'The World Economic Forum and After' at <http://www.rte.ie/news/mayday/action7a>.

Green Movement Sources

The most up to date sources are now groups' websites. A selection are listed below:

Green Parties

European Federation of Green Parties: <http://www.europeangreens.org> (includes links to member parties).

England and Wales: <http://www.greenparty.org.uk>

France: <http://www.les-verts.org>

Germany: <http://www.gruene.de>

USA: <http://www.greenparties.org>
Australia: <http://www.greens.org.au>
New Zealand: <www.greens.org.nz>

International EMOs

Friends of the Earth International: <http://www.foei.org> (includes links to national branches).
Greenpeace: <http://www.greenpeace.org> (links for national branches)
World Wide Fund for Nature: <www.wwwf.org>

National Environmental Associations

Australian Conservation Foundation: <www.acfonline.org.au>
The Wilderness Society (Aus): <http://www.wilderness.org.au>
Sierra Club: <http://www.sierraclub.org>

Direct Action Groups:

Earth First! (UK): <http://www.eco-action.org/efau> (with links to groups in other countries).
Earth First! Journal (USA): <http://www.earthfirstjournal.org>
People's Global Action: <http://www.apg.org>

Alternative news media

Indymedia (with links to national sites): <http://www.Indymedia.org>
Schnews, weekly newsheet from UK direct action networks: <www.schnews.co.uk>

Index

Secrett, Charles 151
Seed, John 163
Seel, Ben 10, 129
Sellafield (Windscale) 32, 164
Shell Oil 22, 137
Shiva, Vandana 16, 216
Sierra Club 29, 121, 124, 131, 214, 224
Snow, David 88
social characteristics of green activists 57–63
social movement(s): compared to interest groups 19–20; definition of 7; green movement as a 17–20; and local environmental protest groups 183–6
socialism and green ideology 79–82; Blatchford, Morris and Carpenter 29–30; early socialist strategy and greens' compared 118; limits of 62–3 70; in New Left 35; *see also* French eco-socialism; left-wing: parties and green movements
Spain: EMOs 123, 125, 132; green party 92; local environmentalism 192, 237; protest in 149
SPD (Social Democratic Party, Germany) 37–8, 41–2, 94–5, 99, 105–7, 111
spirituality 76–7, 167
Steinberg, Marc 88–9
Stewart, John 189, 199, 236
strategy: direct action groups 169–172, 181; divisions over within green movement 219–21; EMOs 150–2; green parties 97–8, 102–4, 117–20; local protests 204–5
sustainable development 150–1, 215, 233
Sutton, Phillip 30, 152
Swampy 167
Sweden: EMOs 123, 145, 149; green party 83, 91; growth of green movement 50
Sydney 163–4
Szerszynski, Bron 128

tactics: Australian direct action innovations 163; transnational diffusion of 53, 164–5, 166, 168–9
Tarrow, Sidney 10, 13, 23, 24–5, 35, 202
Tasmanian Wilderness Society (TWS) 162–3
Terania Creek 161
Tesh, Sylvia 195
Third World, green parties and 71; radical environmentalism in 213–16; *see also*

India; People's Global Action; Shiva, Vandana
Tilly, Charles 10, 23, 171
Torgerson, Doug 219–21, 228
Torness 164, 234
Torrance, Jason 163
Touraine, Alain 17, 20, 223–4
trade unions: Australia 162; Britain 175, 234; Germany 41
Transformation of Environmental Activism Project 149–50
tree spiking 161; *see also* monkey-wrenching
Trittin, Jürgen 107, 111, 135
Twyford Down, Britain 133, 167, 192–3, 202

Unabomber 77, 179
United Nations Conference on Environment and Development (UNCED) 137, 214
USA: EMOs and Clinton Administration 135–7; green movement in 49, 51–2, 146; green party 91; *see also* Earth First!; Environmental Defense; environmental justice campaigns; Greenpeace; New Left; Vietnam War

van der Heijden, Hein-Anton 122, 144–5, 146–7, 150, 215
vegetarianism 8
Verts, Les 47–8, 69; formation 95–7; in government 108–11, 116; growth 101–3
Vietnam War 34
violence and non-violence 53–4, 72, 95–6, 107–8, 156, 163, 167, 175–80, 192, 235–6
Voynet, Dominique 102–3, 108–9, 111–12, 135

Wackersdorf, Germany 43, 97, 191, 203
Waechter, Antoine 101–2
Wall, Derek 30, 54, 58, 166
Weber, Max 17, 141
welfare professionals and the green movement 61–3
welfare state green views of 71
Welsh, Ian 25, 90, 164–5, 199
Whittier, Nancy 27
Whyl, Germany 53, 191, 199–201, 205
Wiesenthal, Helmut 119
wilderness 29, 39, 49, 76, 79, 131, 147, 156–9, 162–3, 166, 179, 224
Wilson, Frank 225–6